Smart Innovation, Systems and Technologies

Volume 335

The Smart Innovation, Systems and Technologies book series encompasses the topics of knowledge, intelligence, innovation and sustainability. The aim of the series is to make available a platform for the publication of books on all aspects of single and multi-disciplinary research on these themes in order to make the latest results available in a readily-accessible form. Volumes on interdisciplinary research combining two or more of these areas is particularly sought.

The series covers systems and paradigms that employ knowledge and intelligence in a broad sense. Its scope is systems having embedded knowledge and intelligence, which may be applied to the solution of world problems in industry, the environment and the community. It also focusses on the knowledge-transfer methodologies and innovation strategies employed to make this happen effectively. The combination of intelligent systems tools and a broad range of applications introduces a need for a synergy of disciplines from science, technology, business and the humanities. The series will include conference proceedings, edited collections, monographs, handbooks, reference books, and other relevant types of book in areas of science and technology where smart systems and technologies can offer innovative solutions.

High quality content is an essential feature for all book proposals accepted for the series. It is expected that editors of all accepted volumes will ensure that contributions are subjected to an appropriate level of reviewing process and adhere to KES quality principles.

Indexed by SCOPUS, EI Compendex, INSPEC, WTI Frankfurt eG, zbMATH, Japanese Science and Technology Agency (JST), SCImago, DBLP.

All books published in the series are submitted for consideration in Web of Science.

Siddhartha Bhattacharyya · Jyoti Sekhar Banerjee · Debashis De
Editors

Confluence of Artificial Intelligence and Robotic Process Automation

Springer

Editors
Siddhartha Bhattacharyya
Algebra University College
Zagreb, Croatia

Rajnagar Mahavidyalaya
Rajnagar, Birbhum
West Bengal, India

Jyoti Sekhar Banerjee
Department of Computer Science
and Engineering (AI & ML)
Bengal Institute of Technology
Kolkata, India

Debashis De
Department of Computer Science
and Engineering
Maulana Abul Kalam Azad University
of Technology
Kolkata, West Bengal, India

ISSN 2190-3018 ISSN 2190-3026 (electronic)
Smart Innovation, Systems and Technologies
ISBN 978-981-19-8295-8 ISBN 978-981-19-8296-5 (eBook)
https://doi.org/10.1007/978-981-19-8296-5

This Springer imprint is published by the registered company Springer Nature Singapore Pte Ltd.
The registered company address is: 152 Beach Road, #21-01/04 Gateway East, Singapore 189721, Singapore

Siddhartha Bhattacharyya would like to dedicate this volume to his loving wife Rashni.

Jyoti Sekhar Banerjee would like to dedicate this volume to his dear parents, wife (Arpita), and loving son (Rishik).

Debashis De would like to dedicate this volume to his father Late Dilip Kumar De.

Preface

Nowadays, the term "Robotic Process Automation" (RPA) has become a buzzword in discussions about disruptive technology. Robotic Process Automation (RPA) refers to the practice of using computer programs as a "virtualized workforce" to carry out tasks formerly performed by humans at a computer terminal. This rule-based reasoning allows robots to replace humans in doing mundane process jobs. The convergence of AI and RPA will allow software robots to automate an increasing amount of human activity, which will have societal and economic repercussions. Quick and predictable cost reduction and the possibility for scalable near-real-time services are only two of RPA's notable advantages, along with the usual drawbacks that come with any automation strategy.

Compared to other technological prototypes and models, RPA is unique in a number of ways. Robotic Process Automation has potential applications in a variety of fields, including finance, education, health care, intellectual property rights, supply chain management, privacy, security, and detecting fraud and scams. Personal financial management is another potential use of RPA, by processing information on the user's preferred stocks. RPA may be used in a number of academic contexts, as has been discovered. RPA systems have also shown to be effective in several contexts, including cybersecurity, data maintenance, and data privacy. Furthermore, it can track very sophisticated crimes like falsified insurance claims, etc.

Many large and medium-sized businesses make the mistake of assuming that the IT department has a monopoly on new technologies like Robotic Process Automation, Cognitive Automation (CA), and Artificial Intelligence (AI). However, it is important to remember that RPA initiatives that succeed provide value to the business and operational teams via the use of a digital workforce. Therefore, the owners of business functions are in the greatest position to lead the way and identify the areas of concern that may be addressed via technological enablement with RPA. Teams in charge of company operations are given complete leeway to assess the effects of potential changes to staffing levels and make decisive moves in response. The information presented in this book will serve as a link between academic studies of RPA and practical applications of the technology.

This volume presents the most recent research and innovative developments regarding AI-enabled RPA. This volume also focuses on the different automation aspects in various fields, the need for automation utilizing RPA techniques, and the introduction of revolutionary solutions to automation-related issues. The volume comprises 14 contributory chapters apart from the introductory and concluding chapters to report the latest developments in this direction.

Robotic Process Automation (RPA), a fast-evolving automation technology that bridges the gap between Artificial Intelligence (AI) and Business Process Management, enables business organizations to now automate high-volume activities (BPM). The actions of a human user on a computer's interface may be recorded using robotic process automation (RPA) tools, enabling a software robot to perform those activities in the user's place. By using additional AI techniques and algorithms for process optimization, forecasting, pattern recognition, and data extraction, RPA operations may be made more accurate and effective. RPA is still a new area, thus there isn't a lot of published research at the moment. So, the goal of Chap. 1 is to look at how the research fraternity defines RPA and how fully the literature has looked at its implementation, state, and tendencies. The authors also conduct a cutting-edge literature review of recent RPA investigations. The authors have also contributed to assessing the advantages and disadvantages of this formidable tool and speculating on its possible future applications. Numerous viewpoints on cutting-edge technologies like RPA are presented in this chapter.

In the era of Industry 4.0, this opens up new opportunities for integrating cutting-edge technologies into IT Audit tasks. A journey into the realm of automation has already been undertaken by a number of organizations, both large and small. Departments that have long struggled with tedious activities and manual procedures may now assign the majority of the "grunt labor" to digital workers who don't mind putting in long hours and repeating tasks. IT auditing is the process of examining and rating an organization's IT policies, procedures, and infrastructure. To provide findings or artifacts as proof of an organization's compliance with a standard, auditors, particularly those working in the area of information security, must do several tedious tasks including gathering information on different workstations and establishing checklists. With an emphasis on the need for Intelligent Auditing, Chap. 2 introduces the idea of robotic process automation (RPA) and how it affects the fields of accounting and auditing. The chapter also outlines the role of automation and AI in finance and auditing with an explanation of how intelligent audits are conducted in light of multiple data and security concerns. Finally, the chapter discusses how RPA may be used to extend Intelligent Auditing in the future.

The effects of RPA and AI on the banking industry are discussed in Chap. 3. In this chapter, the author analyzes the present issues in the banking sector, and the main disruptors, and sets the framework to explain how Intelligent Automation using AI and RPA might aid in resolving these disruptions. Banking was one of the early adopters of RPA and is now adopting AI. The next section of the chapter focuses on how these technologies will change how banks run. The chapter then delves further into real-world use examples of banking fraud, contact center operations, and lending operations. The chapter for each of these areas outlines the business issue, provides a brief overview of the general process and common issues connected to it, identifies the automation levers that can be used to address the issue, and outlines the solution with high-level components specified and the benefits realized as a result of the solution. To attain the advantages that AI and RPA may bring to the workplace, Chap. 3 concludes by outlining the steps that businesses must take and the practices they must develop. It also suggests ideal operational models for delivery, support, and governance.

Businesses are very cautious about their supply chain. Any delay or disruption may lead to unimaginable loss. The current COVID-19 scenario has taken this concern to an epic level. Businesses could not meet their customers' needs due to the lack of human resources during the prolonged lockdown. Companies had to either downsize or shut down. Businesses have started searching for means to operate their processes with as few resources as possible. Thus, many have started adopting Robotic Process Automation, though it is still evolving. There is literature that discusses the implementation of RPA in various streams like accounting, finance, HR, and others. Yet there is a very limited study that has focused on RPA implementation in supply chain processes. In Chap. 4, the authors discuss the importance of the adoption of RPA in the various stages of the supply chain. A complete real-time automation use case on invoice processing is also discussed. Chapter 4 aims to help researchers, RPA enthusiasts, and entrepreneurs to realize the time, effort, and economic benefits of implementing RPA. As a result, they are encouraged to use RPA in their supply chain processes.

Automating organizational processes typically involves document processing techniques for a large document set. For that purpose, the Intelligent Document Processing (IDP) paradigm has been studied for decades. With the fast emergence of Robotic Process Automation (RPA) in the process automation landscape, the industrial solution of IDP with RPA integration has assumed significant importance in the last few years. However, there is no up-to-date overview of the available knowledge in this area. Therefore, in Chap. 5, the authors summarize the current scientific knowledge about IDP and its integration into RPA through a systematic literature review that analyzed 77 primary studies. In addition, an industry review was performed, analyzing and characterizing 37 industrial tools. Although the results confirm the growth in the research interest in IDP in different dimensions, they also identify a lack of proposals that integrate IDP and RPA paradigms in confrontation with the industrial solutions that have increasingly led to its integration.

Robotic Process Automation (RPA) is a technology used by banks and other financial institutions to automate manual business operations so that they can compete in today's market. Onboarding clients, ongoing services, credit card processing, mortgage processing, loan processing, and information retrieval are just a few of the banking activities that necessitate a huge workforce and take time. The necessity for a large workforce for these procedures is rapidly shrinking with the development of Robotic Process Automation and other cognitive technologies. Through automation, the bank's analysts may devote more time to advanced-value tasks, such as evaluating automated outcomes and assessing complex loans that were previously too complex to automate, which can improve process precision, decrease operating time per loan, and provide the bank with additional analyst capacity for client support. Chapter 6 addresses the current challenges faced in the banking industry and provides solutions in the form of use cases, thereby highlighting the importance of RPA and AI in the banking industry.

RPA is a modern innovation that automates repetitive, regular, rule-based human tasks for the advantage of enterprises that choose to use such software. RPA is a commercially available technology, although there is little scientific research on the issue. As a result, the goal of Chap. 7 is to examine how the academic community defines RPA and how much of its status, trends, and applications have been studied in the literature. The differences between RPA and business process management are also covered. The findings of a systematic literature review (SLR) on RPA are presented in this chapter along with an overview of the concepts and applications of RPA in real-world contexts and the benefits of adopting RPA in various economic sectors.

Health care that was previously practiced in the traditional mode has been replaced in the digital mode with various scientific innovations like robotics, revealing new ways of thinking and perception among the general public about how the healthcare system actually operates at a time when the entire world is under stress due to an impending variant of the Novel Corona Virus. RPA, or robotic process automation, is increasingly important in the modern medical system. The authors' goal in Chap. 8 is to provide a thorough discussion of the development of Robotic Process Automation and its connection to Intellectual Property Management. Chapter 8 begins with the authors tracing recent developments of robotics in health care through automation during COVID-19 as it has significant inception. Secondly, the authors also elaborate on how the management of IP assets created out of such a digital revolution due to accelerated and ramped-up innovation comes into the picture quite significantly. It should enable readers to gradually grasp the connection between the two fields.

The rate of advancement in robotic process automation (RPA) during the last several years has been remarkable. Automation of applications has increased efficiency, slashed operating costs, and sped up research, development, and manufacturing. The fourth industrial revolution is being referred to as an intelligent automated industry with little human interaction that has already included RPA in its workflow. The healthcare sector is well ahead of many other businesses in this drive toward change. When COVID-19 was quickly expanding, it withstood the test of time and was also tenacious despite all difficulties. The system did go through an unparalleled crisis that showed its frailty, vulnerability, and lack of readiness. A new paradigm compelled the healthcare system to change. Despite the fact that there were casualties and economic losses, we were able to grow as a society as a result of this catastrophe. In Chap. 9, the authors describe the function of RPA and discuss how the technology might help healthcare professionals with their daily tasks. They also discuss what the fourth industrial revolution will entail for the healthcare industry. The automated, intelligent system would provide a smooth procedure for obtaining data through multiple channels, processing it, and supporting medical professionals in providing high-quality care.

In Chap. 10, the authors introduce how AI and RPA work in real-time scenarios such as financial fraud and money laundering. It discusses how AI builds the knowledge graph and recommends products and services for each customer. This knowledge is implemented and delivered using RPA. The AI application gained prominence in every banking business segment, such as equity, personal, investment, and loans. The application of RPA is present in all business segments, although the percentage is increasing yearly. AI and RPA can help banks to convert challenges into opportunities. There have been various challenges, and the application of AI and RPA combinations is the key to solving the inefficiencies. Advanced analytical techniques on open-source data have been used in this chapter.

The transition beginning from early Industry 1.0 to the current Industry 4.0 can be seen as a movement from traditional methods to the present digital processes. These digital processes claim to be the foundation of Industry 4.0 and have led to the creation of vast volumes of data, collectively referred to as Big Data. While this was a boon to developing technology, storing and effectively utilizing this data, which required unimaginable human efforts, was a crisis in and of itself. Here's where Robotic Process Automation (RPA) and Artificial Intelligence (AI) play their respective roles. While AI techniques mimic human thinking, RPA processes automate repetitive tasks. The integration of RPA and AI leads to Intelligent Automation, thus streamlining human efforts along with increasing speed, productivity, and quality. Apart from these benefits, combining AI algorithms, strategies, and techniques with RPA tools improves RPA processes' effectiveness and precision in recognition, data extraction, forecasting, classification, and process optimization. In this regard, the main goal of Chap. 11 is to provide a brief history of the evolutions leading up to Industry 4.0 and to discuss the importance of integrating AI and RPA as well as the benefits to humanity. It continues to explain the challenges faced in the effective integration of the two booming technologies, elucidate a few misconceptions related to it, and also illustrate the use cases of the same. In addition, Chap. 11 throws light upon different RPA tools in great detail.

The reason behind the formation of the smart city involves the application of intelligence in various sectors. The application of intelligence having the phenomenon of automation will increase smartness and gives rise to the formation of a new smart environment which includes vehicles, homes, government sectors, market complexes, hospitals, institutions, etc. The smartness of the device specifies the usage of the Internet in various devices which generally denotes the Internet of Things (IoT). IoT involves the communication between humans and devices. In Chap. 12, the authors describe that in near future, the usage of Artificial Intelligence (AI) and Robotic Process Automation (RPA) will give scope for realization. Around 2030, there will be the completion of fulfillment for various targets and hence the conversion from IoT to the Internet of Everything (IoE) will be done. This period specifies the development from 5G to 6G. In 6G, the device will able to communicate with another device. As a result, huge facilities will be available through the application of the Internet.

Chapter 13 provides an overall view of Robotic Process Automation (RPA) evolution and its use cases in the IT industrial sector. In recent times, RPA has evolved and democratized all sectors of the economy for better and more efficient production and usage of products, processes, and services. Chapter 13 shows the reasons for the flourishment of technology. The evolution of industrial sectors from steam engines to unattended automatic robots taking business intelligence decisions is driven by one major quotient, i.e., automation. Chapter 13 comprises the entire journey that RPA has taken to be now one of the most affordable and efficient solutions out of all. The ultimate aim of every corporation is to generate revenue and to achieve it through the different processes that happen around it. Each of the processes involves different user personas and applications and thus humans act as the binding chains and deliver the end product or service to the consumer. As humans cannot handle complex computations and tedious monotonous tasks, automation minimizes these tasks and makes their work easier. This chapter also covers the sectors where automation is a critical savior for industries. It also talks about the pros and cons of automation over a period and how it could be a disruptive technology in the upcoming years.

In Chap. 14, the authors describe the advent of emerging Technologies in the healthcare industry and the growth in revenue generation. Patients, doctors, insurance companies, and other entities are the key players in Healthcare Industry. It is critically necessary to establish new, correct back-office procedures in order to preserve equilibrium among the increasing number of patients, the documentation needed for development, and insurance prerogatives, among other things. Hence, Robotic Process Automation (RPA) can assist healthcare organizations in increasing effective proficiency, lowering expenses, and reducing the risk of human error when handling data such as doctor credentialing, staffing, and patient fitness, as well as medical record maintenance, payables and denial recovery, and patient scheduling.

The emerging trends of cognitive Internet of Things (CIoT) are disrupting industrial process automation by infusing intelligence within the pervasive interactions and process automation of enterprise assets. Robotic Process Automation (RPA) is another fascinating technology trend playing a pivotal role in accelerating operational excellence across industries. RPA solutions are designed to orchestrate service workflows that automate repetitive and rule-driven voluminous tasks. While the CIoT facilitates intelligent cyber-physical integration to enhance ubiquitous operational intelligence, RPA introduces automated workflows within the connected enterprise to maximize agility and resilience. As industrial computing is inclining toward maximizing situational awareness and autonomous operations, the integration of AI-powered IoT and intelligent RPA is paving the path to disrupting innovations in the Industry 4.0 era. Chapter 15 delves into key technology components and architectural patterns that introduce a new breed of Cognitive enterprise systems, enabling intuitive operations and need-based control functions beyond complex decision support and pervasive interlocking of Industrial IoT. The authors present unique architectural semantics that introduces RPA capabilities within CIoT to transform the actionable insights into context-aware process flows, promote interoperability, and execute prescriptive actions. The objective of Chap. 15 is to present the design rationale of next-generation industrial automation, compelling industrial IoT use cases, and the research directions on autonomous systems achieved through such convergence of CIoT and RPA.

Finally, Chap. 16 presents the concluding remarks on various RPA applications. Increasing software intelligence is a hot issue in the research field today. When seen in this light, the fascinating challenge posed by the integration of RPA with Artificial Intelligence (AI) in many domains of application, such as the execution of unstructured versus organized activities, becomes clear. RPA might benefit from using artificial intelligence (AI) or ideas like Machine Learning and Data Mining so that it does not have to depend only on rule-based approaches. There is a perception that RPA was created to help businesses save money on labor expenditures associated with routine work. To ensure that the expense of implementing and maintaining RPA is offset by the savings it yields, it would be useful to analyze how RPA use impacts other aspects of a company's R&D, development, competency, etc.

This volume explains that RPA tools refer to a collection of approaches aimed at streamlining labor by automating repetitive activities. Along with the usage of RPA, the integration of Artificial Intelligence (AI)—algorithms and methodologies—enables the execution of automated processes to be more precise. Industry 4.0 refers to a collection of technology and sensors that enable further advancements in processes and applications of automation of AI applications for organizational operations, therefore improving performance and providing new prospects. The primary audience of this volume includes researchers, professors, graduate students, scientists, policymakers, professionals, and developers working in IT and ITeS, i.e., people working on emerging technologies. Additionally, the book will be very useful to many professors, who may wish to adopt this book as a text and supplement independent study projects.

Finally, we, the editors, want to express our gratitude to all the reviewers for submitting their valuable comments and suggestions in due time. Finally, we would like to express our gratitude to Mr. Aninda Bose, the Senior Editor at Springer, for his kind support in publishing this book.

Birbhum, India Siddhartha Bhattacharyya
Kolkata, India Jyoti Sekhar Banerjee
Kolkata, India Debashis De

Contents

Editors and Contributors

About the Editors

Dr. Siddhartha Bhattacharyya [FRSA, FIET (UK), FIEI, FIETE, LFOSI, SMIEEE, SMACM, SMAAIA, SMIETI, LMCSI, LMISTE] is currently the Principal of Rajnagar Mahavidyalaya, Birbhum, India. He is also serving as the Scientific Advisor of Algebra University College, Zagreb, Croatia. Prior to this, he was a Professor at CHRIST (Deemed to be University), Bangalore, India. He also served as the Principal of RCC Institute of Information Technology, Kolkata, India. He has served VSB Technical University of Ostrava, Czech Republic as a Senior Research Scientist. He is the recipient of several coveted national and international awards. He received the Honorary Doctorate Award (D.Litt.) from The University of South America and the SEARCC International Digital Award ICT Educator of the Year in 2017. He was appointed as the ACM Distinguished Speaker for the tenure 2018–2020. He has been appointed as the IEEE Computer Society Distinguished Visitor for the tenure 2021–2023. He is a co-author of 6 books and the co-editor of 94 books and has more than 400 research publications in international journals and conference proceedings to his credit.

Dr. Jyoti Sekhar Banerjee B.Tech., M.E, Ph.D. (Engg.), is currently serving as the Head of the Department in the Computer Science and Engineering (AI & ML) Department at the Bengal Institute of Technology, Kolkata, India and visiting researcher (Post Doc) at Nottingham Trent University, UK. Additionally, He is also the Professor-in-Charge, R & D and Consultancy Cell of BIT. He has teaching and research experience spanning 18 years and completed one IEI funded project. He is a life member of the CSI, IEEE, ISTE, IEI, ISOC, IAENG and fellow of IETE. He is the present honorary Secretary-cum-Treasurer, of the ISTE WB Section. He is the current honorary Secretary of the Computer Society of India, Kolkata Chapter. He is also the Executive Committee Member of the IETE, Kolkata Centre. He has published over fifty papers in various international journals, conference proceedings, and book chapters. He is the lead author of "A Text Book on Mastering Digital

Electronics: Principle, Devices, and Applications". He has also co-authored another book and is currently processing six edited books in reputed international publishers like Springer, CRC Press, De Gruyter, etc. Presently he is also processing two more textbooks; those are now in press. His areas of research interests include Computational Intelligence, Cognitive Radio, Sensor Networks, AI/ML, Network Security, Different Computing Techniques, IoT, WBAN (e-healthcare), Expert Systems.

Debashis De is a professor in the Department of Computer Science and Engineering and the director of School of Computational Science of MAKAUT, WB, India. He received M.Tech. from the University of Calcutta, 2002, and a Ph.D. from Jadavpur University in 2005. He is a senior member-IEEE, fellow IETE, and life member CSI. He was awarded the prestigious Boyscast Fellowship by the Department of Science and Technology, Government of India, to work at the Herriot-Watt University, Scotland, UK. He received the Endeavour Fellowship Award from 2008–2009 by DEST Australia to work at the University of Western Australia. He received the Young Scientist Award both in 2005 at New Delhi and in 2011 in Istanbul, Turkey, from the International Union of Radio Science, Belgium. In 2016, he received the JC Bose research award by IETE, New Delhi. In 2019, he received Siksha Ranta Award by the Govt. of West Bengal. He established the "Centre of Mobile cloud computing" (CMCC) for IoT applications. He published 310 journals and 100 conference papers, fifteen books, and filed eight patents. His h index is 31, citation 5000. Listed in Top 2% Scientist List of the world, Stanford University, USA.

Contributors

Quazi Mohmmad Alfred Aliah University, Kolkata, India

Michael Sebastian Almeida Oracle India Pvt. Ltd. Bannnergahatta Road, Bangalore, India

Shaeril Michael Almeida CHRIST (Deemed to be University), Bangalore, India

S. Anjana Department of Computer Science and Business Systems, Rajalakshmi Engineering College, Chennai, India

Anvith Department of Computer Science and Engineering, M.S. Ramaiah Institute of Technology (Affiliated to VTU), Bangalore, Karnataka, India;
Department of Electronics and Communication Engineering, M.S. Ramaiah Institute of Technology (Affiliated to VTU), Bangalore, Karnataka, India

R. Babu SRM Institute of Science and Technology, Chennai, India

Kesevan Sekar Balaji Department of Computer Science and Engineering, M. S. Ramaiah Institute of Technology, (Affiliated to VTU), Bangalore, Karnataka, India

Jyoti Sekhar Banerjee Department of Computer Science and Engineering (AI & ML), Bengal Institute of Technology, Kolkata, India

Nilesh Harshit Barla PerceptronAI, HSR Layout, Bangalore, India

Ahmed Bawazir Department of Computer Science and Engineering, M. S. Ramaiah Institute of Technology, (Affiliated to VTU), Bangalore, Karnataka, India

Rabindranath Bera Sikkim Manipal Institute of Technology, Sikkim Manipal University, Sikkim, India

Prasenjit Bhadra Ranial Systems Inc., New York, NY, USA

Siddhartha Bhattacharyya Algebra University College, Zagreb, Croatia; Rajnagar Mahavidyalaya, Rajnagar, Birbhum, West Bengal, India

Arpita Chakraborty Department of ECE, Bengal Institute of Technology, Kolkata, India

Shilpi Chakraborty Ranial Systems Inc., New York, NY, USA

Debanjana Dasgupta Open Group Certified Distinguished Architect, New Delhi, India

Debashis De Department of CSE, Maulana Abul Kalam Azad University of Technology, Kolkata, West Bengal, India

K. Devaki Department of Computer Science and Engineering, Rajalakshmi Engineering College, Chennai, India

Jagjit Singh Dhatterwal Department of Artificial Intelligence and Data Science, K L Educational Foundation, Vaddeswaram, Andhra Pradesh, India

Nabanita Ghosh School of Commerce, Finance and Accountancy, Christ (Deemed to be University), Bangalore, India

J. González-Enríquez Departamento de Lenguajes y Sistemas Informáticos, Escuela Técnica Superior de Ingeniería Informática. Avenida Reina Mercedes, s/n., Sevilla, Spain

Kushagra Gupta Department of Computer Science and Engineering, M.S. Ramaiah Institute of Technology, (Affiliated to VTU), Bangalore, Karnataka, India

K. Jayashree Panimalar Engineering College, Chennai, India

A. Jiménez-Ramírez Departamento de Lenguajes y Sistemas Informáticos, Escuela Técnica Superior de Ingeniería Informática. Avenida Reina Mercedes, s/n., Sevilla, Spain

Neha A. Kantikar Department of Computer Science and Engineering, M.S. Ramaiah Institute of Technology (Affiliated to VTU), Bangalore, Karnataka, India; Department of Electronics and Communication Engineering, M.S. Ramaiah Institute of Technology (Affiliated to VTU), Bangalore, Karnataka, India

Karthik K. N. Department of Computer Science and Engineering, M.S. Ramaiah Institute of Technology (Affiliated to VTU), Bangalore, Karnataka, India; Department of Electronics and Communication Engineering, M.S. Ramaiah Institute of Technology (Affiliated to VTU), Bangalore, Karnataka, India

Kuldeep Singh Kaswan School of Computing Science and Engineering, Galgotias University, Greater Noida, India

Lakshya Khandelwal Department of Computer Science and Engineering, M.S. Ramaiah Institute of Technology, (Affiliated to VTU), Bangalore, Karnataka, India

Shivi Khanna School of Business and Management, Christ (Deemed to be University), Bangalore, India

Naresh Kumar School of Computing Science and Engineering, Galgotias University, Greater Noida, India;
G. L. Bajaj Institute of Technology and Management, Greater Noida, UP, India

Shiv Onkar Deepak Kumar Head of Data Science, Straive, Chennai, India

Sunita Kumar School of Business and Management, Christ (Deemed to be University), Bangalore, India

J. M. López-Carnicer Departamento de Lenguajes y Sistemas Informáticos, Escuela Técnica Superior de Ingeniería Informática. Avenida Reina Mercedes, s/n., Sevilla, Spain

Mufti Mahmud Department of CS, Nottingham Trent University, Nottingham, UK

A. Martínez-Rojas Departamento de Lenguajes y Sistemas Informáticos, Escuela Técnica Superior de Ingeniería Informática. Avenida Reina Mercedes, s/n., Sevilla, Spain

V. Murali Bhaskaran Department of Computer Science and Engineering, Rajalakshmi Engineering College, Chennai, India

Aranya Nath Phd Scholar, Damodaram Sanjivayya National Law University, Visakhapatnam, India

Gowri Rajagopal Rapid Acceleration partners Pvt. Ltd, Chennai, Tamil Nadu, India

Raghuraman Ramamoorthy Rapid Acceleration partners Pvt. Ltd, Chennai, Tamil Nadu, India

Lokesh Ramdas Department of Computer Science and Engineering, M. S. Ramaiah Institute of Technology, (Affiliated to VTU), Bangalore, Karnataka, India

Jayanta Kumar Ray Sikkim Manipal Institute of Technology, Sikkim Manipal University, Sikkim, India

Subhajit Saha Ranial Systems Inc., New York, NY, USA

Usha Saha Phd Scholar, Damodaram Sanjivayya National Law University, Visakhapatnam, India;
LLM IPR & Cyber Law, GITAM School of Law GITAM University, Visakhapatnam, India

Panagiotis Sarigiannidis Department of Informatics and Telecommunication Engineering, University of Western Macedonia, Kozani, Greece

A. Sathya Rajalakshmi Engineering College, Chennai, India

Ganeshayya Shidaganti Department of Computer Science and Engineering, M.S. Ramaiah Institute of Technology (Affiliated to VTU), Bangalore, Karnataka, India; Department of Electronics and Communication Engineering, M.S. Ramaiah Institute of Technology (Affiliated to VTU), Bangalore, Karnataka, India

Sanjib Sil A.K. Choudhury School of Information Technology, University of Calcutta, Kolkata, India

S. P. Srinivasan Rajalakshmi Engineering College, Chennai, India

Rogina Sultana Aliah University, Kolkata, India

J. M. Sánchez-Oliva Servinform, S.A. Parque Industrial PISA, Calle Manufactura, 5, 41927 Mairena del Aljarafe, Sevilla, Spain

Himanshu Vaswani Department of Computer Science and Engineering, M.S. Ramaiah Institute of Technology, (Affiliated to VTU), Bangalore, Karnataka, India

Chapter 1
Intelligent Automation Framework Using AI and RPA: An Introduction

Arpita Chakraborty, Siddhartha Bhattacharyya, Debashis De, Mufti Mahmud, and Jyoti Sekhar Banerjee

Abstract Business organizations may now automate high-volume processes with the help of Robotic Process Automation (RPA), a rapidly developing automation technology that bridges the gap between Artificial Intelligence (AI) and Business Process Management (BPM). Tools for robotic process automation (RPA) may record the actions of a human user on a computer's interface, allowing a software robot to do those actions in the user's place. RPA procedures may be made more precise and efficient by the supplementary use of AI methods and algorithms for process optimization, forecasting, pattern recognition and data extraction. Since RPA is still a newer discipline, there is not a lot of research published on the subject just now. Hence, the purpose of this chapter is to examine how the research fraternity defines RPA and how thoroughly the implementation of RPA, state and trends have been explored in the literature. The authors also do a state-of-the-art literature evaluation of current studies about RPA. The authors have also made contributions to evaluating the benefits and drawbacks of this powerful instrument and making predictions about its potential future uses. This chapter offers many perspectives on cutting-edge technologies like RPA.

A. Chakraborty
Department of ECE, Bengal Institute of Technology, Kolkata, India

S. Bhattacharyya
Algebra University College, Zagreb, Croatia

Rajnagar Mahavidyalaya, Rajnagar, Birbhum, West Bengal, India

D. De
Department of CSE, Maulana Abul Kalam Azad University of Technology, Kolkata, West Bengal, India

M. Mahmud
Department of CS, Nottingham Trent University, Nottingham NG11 8NS, UK

J. S. Banerjee (✉)
Department of Computer Science and Engineering (AI & ML), Bengal Institute of Technology, Kolkata, India
e-mail: tojyoti2001@yahoo.co.in

S. Bhattacharyya et al. (eds.), *Confluence of Artificial Intelligence and Robotic Process Automation*, Smart Innovation, Systems and Technologies 335,
https://doi.org/10.1007/978-981-19-8296-5_1

1.1 Introduction

One of the most innovative and practical technologies in the Business Process Automation technology [1] trend is RPA [2], or robotic process automation. Robotic Process Automation always comes in help when we wish to automate a repetitive, tedious and assembly-based work as business leaders, tech builders or even business analysts. We may rely on these software-based systems to do repetitive and manual tasks as they don't call for any form of innovation or variety. As a result, the company leaders are better able to utilize the whole organization by investing both their people and financial resources into improved business operations. Business Process Management (BPM) is a key area of use for Robotic Process Automation (RPA). BPM comprises ideas, methodologies and techniques to assist the design, administration, configuration, enactment and analysis of business processes, as defined by Weske [3]. Relationships between BPM management tasks are laid forth in the BPM life-cycle model. The lifecycle consists of five stages that occur inside a single process: observe, development, redesign, investigation, discovery (see Fig. 1.1) [4].

There are distinguishing features of RPA that set it apart from other technical prototypes and models. The financial sector, the educational industry, privacy and security, and the detection of fraud and scams are all areas where robotic process automation may be put to use. RPA may be used to manage personal finances; the RPA systems can operate on chosen stocks and provide the user with an ideal portfolio [5]. It has been identified that RPA may be employed in a variety of academic contexts [6, 7]. RPA employs Intelligent algorithms that may be utilized in classrooms to get to know pupils much better. The educator or the school administration might utilize this information to improve the kids' educational standard. Cybersecurity, data upkeep and data privacy are just a few examples of areas where RPA systems have shown to be very useful. RPA software can keep an eye on a website that collects

Fig. 1.1 Lifecycle of BPM

a user's private information and alert the user if the website is the subject of many complaints. RPA technologies make it simple to determine if a business is legitimate or fraudulent. Additionally, it can trace complex crimes like false insurance claims, etc.

While RPA is a relatively recent development in the computer industry, it is already seeing widespread use in a variety of MNCs and other institutions that need meticulous record-keeping [8]. Many facets of Robotic Process Automation are covered in this chapter, which the authors have provided to shed light on the topic. The authors have made an attempt to envision the possibility of RPA's use in Business Process Management, Finance and Auditing, etc. The authors also addressed some of the best places where RPA could be used more widely to get the best results.

This chapter is constructed as follows. Section 1.2 deals with the Literature Review. In Sect. 1.3, the chapter presents What is RPA? Sect. 1.4 presents AI and Industry 4.0. Section 1.5 contains the comparison between standard IT implementation and RPA. Section 1.6 also provides the comparison between Conventional Automation and RPA. Section 1.7displays the advantages of RPA. Section 1.8 represents the Future of RPA. Finally, the Conclusion is discussed in Sect. 1.9

1.2 Literature Review

In recent years, Robotic Process Automation has been extensively researched and utilized to manage Business Process Management (BPM). The potential for RPA in the IT industry is enormous. As business leaders recognize that wasting manpower and money on the same repetitive tasks over and over can be detrimental to the business, they are more open to RPA. A state-of-the-art literature review about RPA is presented here.

Wright et al. [9] detailed their efforts to ascertain the state of RPA and potential application domains. In order to learn more about the issue and identify the steps that may be automated, surveys are conducted. They draw the conclusion that RPA has a place in some contexts, such as the billing procedure and the upkeep of supplier data. However, it draws the conclusion that poor database quality is now preventing RPA from being widely used.

RPA is theoretically reviewed by Frank [10]. RPA adds a level of automation by virtue of the following: (1). Cognitive; (2). Autonomic; (3). Orchestration; (4). Scripting; (5). Manual Execution. Despite the review's claims, no actual cases are shown, nor is there any indication of how to implement the technique systematically.

Although Lacity [11] also discusses the state of RPA adoption in business, they approach it from a somewhat different angle. The authors draw the conclusion that RPA adoption has been sluggish, at least up until 2014 and 2015. It is recommended that academics determine which business issues may be addressed using RPA. Finally, they suggest that studies be conducted to learn what tasks are best suited for humans, how processes may be improved via redesign, and where and how RPA can be most effectively deployed. Here, Ansari et al. [12] conduct a comparison analysis of the technical features of popular RPA tools on the market today. In addition, the authors detail how RPA is being used by businesses across a variety of industries, including but not limited to education, healthcare and banking.

Moreover, [13] examines the literature on four emerging innovations (advanced manufacturing, networking, robotics and artificial intelligence) and their possible effects on the corporate world and the industry from three perspectives (government publications, professional experiences, academic journals). While it's unclear who exactly will benefit and who will suffer from this next technological revolution, one thing is certain: its influence will be far-reaching, even surpassing that of the Industrial Revolution. Robotic Desktop Automation, i.e. RDA is a new idea introduced in this paper. Compared to RPA, RDA automates a broader range of front-office tasks.

The purpose of the study presented by Gami et al. [14] is to shed light on the significance of RPA and offer material on potential avenues for further exploration of this technology. Accurately, they concentrate on two ideas, SPA, i.e. Smart Process Automation and AI, both of which are often seen as natural extensions of RPA. The present status and difficulties of AI and RPA in the domain of auditing and accounting are described by Gotthardt et al. [15]. The authors provide two examples to demonstrate the variety of currently available services. At the end of the session, a table is shown and evaluated to highlight certain automation and business risks brought on by the usage of AI and RPA, such as governance, automation strategy, cyber threats, or privacy and data leakage. However, the related studies in this research do not demonstrate any kind of logical procedure.

Robotic process automation (RPA) is distinguished from Business process management (BPM) and similar technologies by the authors of [16]. In this way, RPA might be seen as a convergent technology for the automobile industry. Unlike BPM Technologies, RPA may be implemented using a simple 'Drag, Drop, Link' algorithm, making it ideal for automating routine tasks. According to [17], RPA software is used to eliminate the need for humans to do routine, standardized tasks when a computer is available. By simulating human interaction with software, bots may do tasks such as reading email attachments, completing online forms, re-entering data and more. Robots may be seen as a shared resource for use in data centres positioned in the middle and rear of an organization. According to Jovanovi et al. [18], the RPA implementation process is broken down into stages, the first of which involves idea evaluation and approval, and the second, which involves fine-tuning the robot and the process. The last step is user adoption of RPA.

1.3 What is Robotic Process Automation?

Enterprise Resource Planning (ERP) systems in the 1970s marked the beginning of the development of automation capacity. In the middle of the '80s, firms began using Business Process Management (BPM) technologies. Finally, beginning in the middle

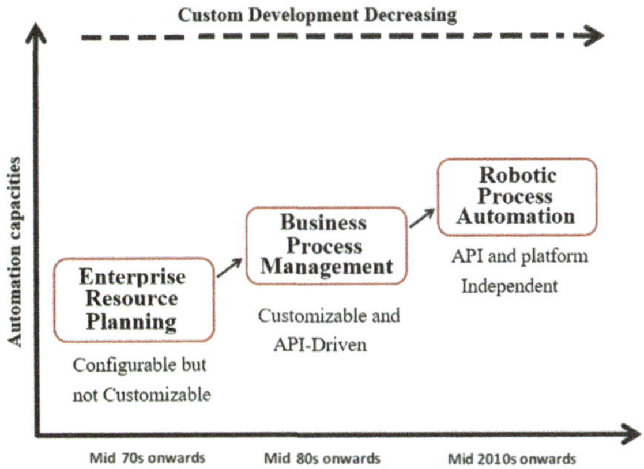

Fig. 1.2 Developments in process automation techniques [19]

of the 2010s, the introduction of Robotic Process Automation (RPA) has begun (see Fig. 1.2).

Robotic process automation (RPA) is a cutting-edge technology for automating mundane, repetitive work, often found in the administrative support departments of modern businesses. Several jobs in a firm are boring, repetitive, and, in the end, a complete waste of resources. These jobs are often carried out in accordance with a predetermined algorithm. Using RPA, we can automate these processes using fixed-algorithms in a manner that is both quick and efficient, without wasting any human resources. Since RPA is an IT-based tool, it is only usable in software systems, yet the word 'Robotic' in its language could lead some people to believe that it is a concept with hardware systems like Robots inside its systems. The three main categories of RPA are hybrid RPA, unattended automation and attended automation.

1.3.1 Attended Automation

This particular form of bot runs on the user's computer and is often activated by the user. The tasks that are initiated at hard-to-detect programming moments are best suited for attended automation.

1.3.2 Unattended Automation

Data processing operations are carried out in the background using these bots, much as cloud-based batch processes. Using unattended automation is a great way to lessen the workload of support staff. A number of methods exist for kicking off automation without human intervention, including data input at a designated point, timed intervals, orchestrator activation, and bot activation.

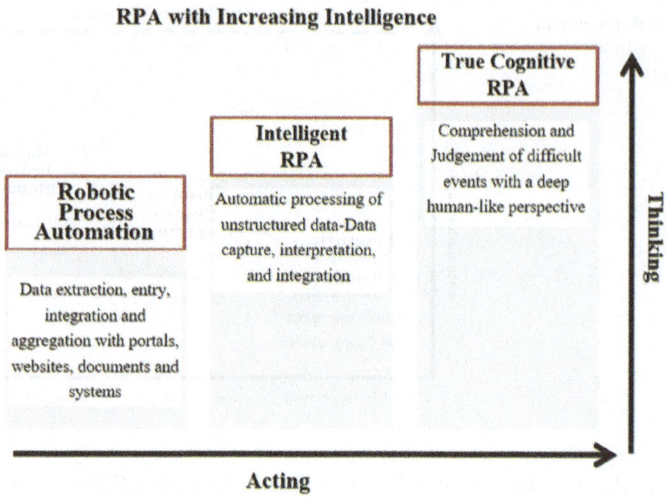

Fig. 1.3 Present and future of RPA [23]

1.3.3 Hybrid RPA

This kind of RPA combines unattended and attended RPA bots to enable automation for both front office and back office tasks. This enables process automation [20, 21] from beginning to finish.

Russ Gould, vice president of KOFAX, predicts that True Cognitive RPA and Intelligent RPA will bring automation to the highest level of cognition and behaviour in the future, while also highlighting the advantages of existing RPA system, such as cost of production, velocity, and operational quality in organizations [22] (see Fig. 1.3).

1.4 AI and Industry 4.0

There was a period when the field of Artificial Intelligence (AI) [24–33] was divided into many distinct specializations. Some of such disciplines were intelligent data retrieval, automated theorem proving, computer vision, robotics, automatic programming, natural language processing, etc. These days, every one of these uses is so broad that it merits its own discipline. The term 'artificial intelligence' (AI) is today most accurately used to refer to a set of fundamental concepts that provide the basis for a wide variety of these applications [34]. Industry 4.0 [35, 36] and smart factories are based on the premise that machines may utilize AI to increase productivity, efficiency and quality in the production of products and services [37]. With the support of cyber-physical systems [38], artificial intelligence (AI) technologies are spreading across the industrial sector, blurring the lines between the digital and the real. With the use of AI, the manufacturing sector can adapt to new conditions and meet difficulties such as customer preferences, faster product cycles and more sensors in machinery [39]. Manufacturing a wide variety of goods is simplified by using adaptable robots and AI. Large amounts of real-time data acquired by different sensors may be analyzed using AI approaches (like data mining) [40–43].

1.5 Comparison Between Standard IT Implementation and RPA

As shown in Table 1.1, a number of characteristics set the RPA deployment strategy apart from common methods for automating new business unit needs.

1.6 Conventional Automation versus RPA

Traditional automation and RPA aren't similar. We might programme a machine to do any function and carry out any phase of an operation using conventional automation. RPA, on the other hand, is a kind of automation that stays in our system's front end and completes tasks without ever needing to access the back end (see Table 1.2).

Table 1.1 Comparison between standard IT implementation and RPA

Challenges	Standard IT implementation	RPA
To meet the needs of emerging business units, required technical skillset	Extremely complex; calls for software engineers and architects with deep expertise in the required technologies, enterprise suites, and methods.	Moderate; analysts and process modellers can accomplish it with only a short-term training on RPA software.
Procedure for progress	Massive; integrating at either the data layer or the application layer is complicated and, at times, fragile.	Minimal; leverages the security and logic of preexisting programs while using just the presentation layer
Reusing portions	Strong, yet difficult and costly to create	Strong; existing components may be utilized to build whole newer robots.

Table 1.2 Comparison between conventional automation and RPA

Conventional automation	RPA
Some adjustments to the current IT infrastructure are needed.	There is no need to change anything about the current setup.
The capacity to behave like a human is not part of this. It does nothing except carry out the specified programmatic commands.	It is useful for automating routine, rule-based processes. To accomplish its goals, it imitates human behaviour.
To automate features using conventional automation, users need to have coding knowledge. The language prerequisite for using an automation tool is based on its intended use. User memorization of language syntax and scripting is required.	Users don't need any prior programming experience to get started using RPA. With RPA, automation tasks may be followed a simple flowchart. Users won't need to study syntax or scripts for any languages. They need just devote attention to the automated features provided.
Installing conventional automation might take time. Researching the viability of a test and planning for it might take a while. Automation using more conventional means is less expensive to implement initially. It's cheaper in the short term, but the long-term expenses are far higher.	When using RPA, deployment is rapid and painless. RPA software is process-driven; therefore it reduces the time needed to complete tasks. To begin with, RPA might be a little pricey. However, in the long term, you will save a significant amount of energy, money and time.
When compared to RPA, conventional automation is seen as a vital and difficult technology due to the high degree of flexibility it offers in terms of customization. Due to API constraints, conventional automation's ability to integrate disparate systems is hindered.	Depending on the user's preferences, RPA may be set up to do a variety of tasks. Integration with other programmes (e.g. CRM, ERP, email, calendar, etc.) allows for data synchronization and the generation of pre-programmed responses.
For conventional automation to be effective, additional manpower, effort and time are all necessities. In contrast, users may be required to make a number of script modifications while working with Conventional Automation. Therefore, keeping up with the latest versions of this technology might be challenging.	In comparison, RPA is the superior choice since it can implement changes immediately. Because RPA is so intuitive, it can be used to modernize almost any business process.

1.7 Popular RPA Advantages

The benefits of RPA are obvious and straightforward, as mentioned below

1. Gains in efficiency may be seen in a matter of weeks or months, leading to virtually immediate savings.
2. The licensing fees and initial investment are manageable, and both can be reasonably estimated (RoI).
3. If the cost of a more permanent technological development is even too expensive, this tactical measure might be used in the meantime.
4. Robots may be used around the clock, seven days a week.
5. While the use of RPA may result in incremental process improvements, no or little modifications to the underlying business process are required.

6. There will be no or very few adjustments to the application.
7. It can be scaled up easily and realizes cost advantages as a result.
8. Reduced risk and failure as compared to human employees; higher quality production.
9. Maintaining consistent and continuous records of compliance.
10. Possible process enhancements during deployment as a result of uncovered deficiencies, shortfalls, etc.
11. Since the process repeatability generates a lot of data—necessary for six sigma—and eliminates people as potential causes of mistakes, RPA may be useful for Lean Six Sigma programmes when used for highly standardized activities.

1.8 Future of RPA

Market experts indicate that RPA will continue to expand. The future of RPA has a wide range of potential outcomes that are being considered. Adoption of RPA is facilitated by the present economic climate. Robotic process automation is predicted to become a key component of businesses as artificial intelligence and machine learning develop at a fast rate. Consequently, it makes the adoption of RPA by businesses much simpler. As a result of RPA, which is a data-driven automation approach that helps an enterprise perform repetitive tasks more easily and efficiently by reducing time consumption in these types of activities as well as minimizing error, paperwork is currently being reduced quickly across all large enterprises. According to several studies, the automation sector will choose RPA models that are simple to adopt and have built-in features for process automation, which will significantly cut labour force and paperwork [11]. By examining the advancements RPA has made, it is clear that this technology is developing and evolving. RPA tools will be used more often for data rekeying, data input, data analysis tasks. Automation of business operations will need fewer employees, and RPA technologies will be used in its place. So here is another development that may occur in the future. A number of future developments are anticipated for RPA, including the ones listed below.

1. **The effects of RPA will be seen more strongly inside businesses**: As more organizations use RPA, the industry at large will feel its effects. RPA is going to be used in more contexts and across more processes inside enterprises than it now is. Intelligent robots, for instance, may greatly increase efficiency in areas such as the sorting of incoming email. This indicates that RPA's effect on a company's bottom line eventually became more significant. When RPA projects go beyond prototypes and into full-scale manufacturing, this may be a factor. Although many or maybe most companies have dabbled with RPA, its utility has been shown in a wide variety of contexts, prompting many to consider taking it to the next level.
2. **RPA will eventually include a plethora of other technologies**: The integration of RPA with other technologies will become more frequent in the future. This

is often due to the fact that businesses are coming to terms with the fact that automation technologies are not best used in isolation and must be connected with other tools in order to get optimal results. It is the fact that RPA is a powerful, massive and adaptable tool in the toolkit. Though the user will be able to do much with this one tool, the actual benefit will go to the company, which will be able to achieve its goals by using the combined form of tools at the time of developing anything.

3. **Combination of AI and RPA**: In the coming decade, RPA and AI will permeate many aspects of our lives, with far-reaching consequences for many sectors. Although experts may generally agree on where technology is headed, they differ widely on the effect that developments in RPA and AI will have on the economy and the job market over the next decade. It has been predicted by some that automation would eliminate many middle-class and high-income occupations in the not-too-distant future. Increases in wealth disparity, disruption and unemployment of the established social order are among the potential outcomes many experts fear may result from more computerization of the workplace. Many, however, believe that, in the next decade, technology will generate more employment than it destroys. Since the beginning of industrialization, software has been both a job killer and job creator, and it has provided several benefits to society, a number of which are now considered standard across many sectors.
4. **Low-cost RPA:** The commercialization of RPA has intensified since Microsoft entered the RPA industry, according to executives of RPA businesses. By 2020, approximately a hundred organizations will be providing a feature that was only provided by a few suppliers only a few years ago. The cost of developing a competitive RPA system is decreasing. It is anticipated that need for accessible RPA solutions would increase. When all else is equal, businesses prefer open-source solutions because they offer more visibility and are often more affordable since they simply need to purchase services. These changes promote RPA that is open-source. A business might capture a significant share of the RPA industry if it can implement a comparable RPA solution.
5. **Paperwork is replaced by robot:** RPA is a data-driven automation method that helps businesses manage the dull, repetitive, time-consuming operations that are now handled by people. The software robots have the ability to reskill and equip the human labour in an effective and efficient manner because automation is quickly growing. According to some forecasts, the automation industry will be swamped with pre-built RPA models that are simple to implement and can automate a variety of administrative processes in an organization by 2025. The use of industrial robots has been the most overt indicator of labour automation, and supply has increased globally in recent times [13].

1.9 Conclusion

In addition to being one of the most quickly expanding technologies, RPA is also one of the most dynamically developing technologies. Businesses have experienced it simpler to develop and adapt as a result, according to market trends. As the economy now stands, RPA adoption is more feasible. Industries will rely heavily on robotic process automation as the rise of machine learning and artificial intelligence makes it possible to automate routine tasks with little effort using RPA solutions. As a result, it facilitates the incorporation of RPA into organizations. But RPA is not without its problems and difficulties. When there are several automated programmes operating together, the technology might be difficult to control. Further, there may be issues with scalability and security throughout the business. Nevertheless, RPA looks to be a fundamental tool that will be there for the long term despite all of these concerns.

References

1. Hildebrandt, T., van Dongen, B. F., Röglinger, M., Mendling, J.: Business process management. Lect. Notes Comput. Sci, 11675 (2019)
2. González Enríquez, J., Jiménez Ramírez, A., Domínguez Mayo, F.J., García García, J.A.: Robotic process automation: a scientific and industrial systematic mapping study. IEEE Access **8**, 39113–39129 (2020)
3. Weske, M.: Business process management architectures, pp. 305–343. Springer, Berlin Heidelberg (2007)
4. Mendling, J., Decker, G., Hull, R., Reijers, H.A., Weber, I.: How do machine learning, robotic process automation, and blockchains affect the human factor in business process management? Commun. Assoc. Inf. Syst. **43**(1), 19 (2018)
5. Reddy, K.N., Harichandana, U., Alekhya, T., Rajesh, S.M.: A study of robotic process automation among artificial intelligence. Int. J. Sci. Res. Publ. **9**(2), 392–397 (2019)
6. Saha, O., Chakraborty, A., & Banerjee, J. S.: A fuzzy AHP approach to IT-based stream selection for admission in technical institutions in India. In Emerging technologies in data mining and information security (pp. 847–858). Springer, Singapore, (2019)
7. Saha, O., Chakraborty, A., & Banerjee, J. S.: A decision framework of IT-based stream selection using analytical hierarchy process (AHP) for admission in technical institutions. In 2017 4th International Conference on Opto-Electronics and Applied Optics (Optronix) (pp. 1–6). IEEE, (2017)
8. König, M., Bein, L., Nikaj, A., Weske, M. (, September). Integrating robotic process automation into business process 2020 management. In: International conference on business process management, pp. 132–146. Springer, Cham (2020)
9. Wright, D., Witherick, D., Gordeeva, M.: The robots are ready. Are you? Untapped advantage in your digital workforce. Deloitte, 28. (2017)
10. Frank, C.: Introduction to robotic process automation. In: Proceeding of. Institute Robot Process Automation, p. 35 (2015)
11. Lacity, L., Willcocks, M.: Robotic process automation: The next transformation lever for shared services. Technical Report No. 16/01, School Economy of Political Science, Outsourcing Unit Working Research Paper Series., London, U.K (2012)
12. Ansari, W.A., Diya, P., Patil, S., Patil, S.: A review on robotic process automation-the future of business organizations. In: 2nd International conference on advances in science & technology (ICAST), (2019)

13. Llewellyn Evans, G.: Disruptive technology and the board: The tip of the iceberg. Econ. Bus. Rev. **3**(1), (2017)
14. Gami, M., Jetly, P., Mehta, N., Patil, S.: (2019, April). Robotic process automation–future of business organizations: A review. In: 2nd International conference on advances in science & technology (ICAST), (2019).
15. Gotthardt, M., Koivulaakso, D., Paksoy, O., Saramo, C., Martikainen, M., Lehner, O.: Current state and challenges in the implementation of smart robotic process automation in accounting and auditing. ACRN J. Financ. Risk Perspect, (2020)
16. Aguirre, S., Rodriguez, A.: (2017, September). Automation of a business process using robotic process automation (RPA): A case study. In: Workshop on engineering applications, pp. 65–71. Springer, Cham (2017)
17. Schatsky, D., Muraskin, C., Iyengar, K.: Robotic process automation. A path to the cognitive enterprise. Deloitte Consulting, New York (2017)
18. Jovanović, S.Z., Đurić, J.S., Šibalija, T.V.: Robotic process automation: overview and opportunities. International Journal Advanced Quality **46**(3–4), 34–39 (2018)
19. https://www.lateetud.com/the-evolution-of-process-automation-technology Web. 29 Jan 2019
20. Das, D., Pandey, I., Chakraborty, A., Banerjee, J.S.: Analysis of implementation factors of 3D printer: the key enabling technology for making prototypes of the engineering design and manufacturing. Int. J. Comput. Appl. **1**, 8–14 (2017)
21. Das, D., Pandey, I., Banerjee, J.S.: An in-depth study of implementation issues of 3D printer. In: Proceedings of MICRO 2016 conference on microelectronics, circuits and systems, pp. 45–49 (2016)
22. Met, İ., Kabukçu, D., Uzunoğulları, G., Soyalp, Ü., Dakdevir, T.: Transformation of business model in finance sector with artificial intelligence and robotic process automation. In: Digital business strategies in blockchain ecosystems, pp. 3–29. Springer, Cham (2020)
23. https://www.kofax.com/Blog/2018/august/robotic-process-automation-rpa-past-present-and-future Web. 29 January 2019
24. Banerjee, J., Maiti, S., Chakraborty, S., Dutta, S., Chakraborty, A., Banerjee, J.S.: Impact of machine learning in various network security applications. In: 2019 3rd International conference on computing methodologies and communication (ICCMC), pp. 276–281. IEEE (2019)
25. Chattopadhyay, J., Kundu, S., Chakraborty, A., Banerjee, J.S.: Facial expression recognition for human computer interaction. In: International conference on computational vision and bio inspired computing, pp. 1181–1192. Springer, Cham (2018)
26. Guhathakurata, S., Kundu, S., Chakraborty, A., Banerjee, J.S.: A novel approach to predict COVID-19 using support vector machine. In: Data Science for COVID-19, pp. 351–364. Academic Press (2021)
27. Guhathakurata, S., Saha, S., Kundu, S., Chakraborty, A., Banerjee, J. S.: A new approach to predict COVID-19 using artificial neural networks. In: Cyber-physical systems, pp. 139–160. Academic Press (2022)
28. Saha, P., Guhathakurata, S., Saha, S., Chakraborty, A., Banerjee, J.S.: Application of machine learning in app-based cab booking system: a survey on Indian scenario. In: Applications of artificial intelligence in engineering, pp. 483–497. Springer, Singapore (2021)
29. Biswas, S., Sharma, L.K., Ranjan, R., Saha, S., Chakraborty, A., Banerjee, J.S.: Smart farming and water saving-based intelligent irrigation system implementation using the internet of things. In: Recent trends in computational intelligence enabled research, pp. 339–354. Academic Press (2021)
30. Mandal, J.K., Misra, S., Banerjee, J.S., Nayak, S. (eds.).: Applications of machine intelligence in engineering: Proceedings of 2nd global conference on artificial intelligence and applications (GCAIA, 2021), September 8–10, 2021, Jaipur, India. CRC Press (2022)
31. Guhathakurata, S., Saha, S., Kundu, S., Chakraborty, A., Banerjee, J.S.:.South Asian countries are less fatal concerning COVID-19: a fact-finding procedure integrating machine learning & multiple criteria decision-making (MCDM) technique. J. Inst. Eng. (India): Ser. B **102**(6), 1249–1263 (2021)

32. Guhathakurata, S., Saha, S., Kundu, S., Chakraborty, A., Banerjee, J.S.: South Asian countries are less fatal concerning COVID-19: a hybrid approach using machine learning and M-AHP. In: Computational Intelligence Techniques for combating COVID-19, pp. 1–26. Springer, Cham (2021)
33. Chakraborty, A., Singh, B., Sau, A., Sanyal, D., Sarkar, B., Basu, S., Banerjee, J.S.: Intelligent vehicle accident detection and smart rescue system. In: Applications of Machine Intelligence in Engineering, pp. 565–57. CRC Press (2022)
34. Nilsson, N.J.: Principles of artificial intelligence. Morgan Kaufmann Editors (2014)
35. Das, K., Banerjee, J.S.: Green IoT for intelligent cyber-physical systems in industry 4.0: A review of enabling technologies, and solutions. In: Applications of machine intelligence in engineering, (pp. 463–478). CRC Press (2022)
36. Ribeiro, J., Lima, R., Eckhardt, T., Paiva, S.: Robotic process automation and artificial intelligence in industry 4.0–a literature review. Procedia Comput. Sci. **181**, 51–58 (2021)
37. Bahrin, M.A.K., Othman, M.F., Azli, N.N., Talib, M.F.: Industry 4.0: A review on industrial automation and robotic. Jurnal Teknologi **78**(6–13), 137–143 (2016).
38. Banerjee, J. S., Bhattacharyya, S., Obaid, A. J. & Yeh, W. C (eds.).: Intelligent Cyber-Physical Systems Security for Industry 4.0: Applications, Challenges and Management, CRC Press (2022)
39. Zheng, P., Sang, Z., Zhong, R. Y., Liu, Y., Liu, C., Mubarok, K., Xu, X.:. Smart manufacturing systems for Industry 4.0: Conceptual framework, scenarios, and future perspectives. Front. Mech. Eng. **13**(2), 137–150 (2018)
40. Ustundag, A., Cevikcan, E.: Industry 4.0: managing the digital transformation. Springer Editors (2017). Available from https://www.springer.com/gp/book/9783319578699
41. Banerjee, J.S., Mahmud, M. & Brown, D.: Heart Rate Variability-Based Mental Stress Detection: An Explainable Machine Learning Approach. SN COMPUT. SCI **4**, 176 (2023)
42. Chakraborty, A., Banerjee, J.S., Bhadra, R., Dutta, A., Ganguly, S., Das, D., Kundu, S., Mahmud, M., Saha G.: A Framework of Intelligent Mental Health Monitoring in Smart Cities and Societies. IETE J Res (2023)
43. Bhattacharyya, S., Banerjee, J. S., & Köppen, M (eds.).: Human-Centric Smart Computing: Proceedings of ICHCSC 2022, Springer. (2022)

Chapter 2
Role of RPA in Intelligent Auditing

Ganeshayya Shidaganti, Lokesh Ramdas, Kesevan Sekar Balaji, and Ahmed Bawazir

Abstract The age of automation has arrived, bringing with it new possibilities for incorporating innovative technologies into IT Audit functions. Various organisations, both major and little, have already started their work by taking a voyage into the world of automation. Departments, which have long been plagued by manual processes and tiresome duties, may now delegate most of the "grunt work" to digital employees who don't mind working long hours and repeating tasks. The practise of reviewing and assessing an organisation's IT infrastructure, policies, and processes is known as IT auditing. When it comes to auditing especially in the field of Information security, the auditors must go through various monotonous tasks of obtaining details of various workstations and creating checklists, so as to produce findings or artifacts as evidence for an organisation's compliance with a standard. This chapter will present the concept of robotic process automation (RPA) and its effect on the domains of accounting and auditing, with a focus on the necessity for Intelligent Auditing. An explanation of how intelligent audits is done in light of numerous data and security issues, as well as an overview of automation and AI in finance and auditing. Finally, consider how Intelligent Auditing may be expanded with the help of RPA in the future.

2.1 Introduction

Human creativity is the root of ingenuity, as it allows us to offer new thoughts and ideas; Its worth, however, is decided by its scalability and application, it will not be capable of functioning and will be a mere proposition or idea. Human advancement and evolution drive corporate innovation and creativity, allowing firms to advance and keep up with human life. The sustainability and growth of economic success are both dependent on innovation. It ushers in the execution of creative ideas, service

G. Shidaganti (✉) · L. Ramdas · K. S. Balaji · A. Bawazir
Department of Computer Science and Engineering, M. S. Ramaiah Institute of Technology, (Affiliated to VTU), Bangalore, Karnataka, India
e-mail: ganeshayyashidaganti@msrit.edu

S. Bhattacharyya et al. (eds.), *Confluence of Artificial Intelligence and Robotic Process Automation*, Smart Innovation, Systems and Technologies 335,
https://doi.org/10.1007/978-981-19-8296-5_2

upgrades, and the creation of new goods. These act as catalytic agents, assisting in market adaption and corporate expansion. Progress in any element of the business must be incremental (which implies adoption of new technologies) in order to derive the maximum value from them. Aside from that, the main goal is supposed to be equipped for such interruptions in the future.

In today's world, computing technology innovations happen at a breakneck speed, and keeping up with them is critical but challenging. Robotisation is a broad and dynamic phenomenon that has been increasingly prominent in recent years. As the economic world becomes more globalised and the Internet advances, information flows more freely, and the world is witnessing a swift process of digitalisation of complete humanity. The digital age has brought about a number of significant changes in the labour market, including positions in which employees are no longer desired such specialised procedures for which robots have been used to replace humans, as well as fresh job opportunities that require the expansion of innovative digital skills, as well as the necessity for an integrated method to information systems. Since the 1970s, robots have been used in a variety of sectors and manufacturing processes. Subsequently, Robots began to be utilised in a variety of service industries, including tourism, financial services, and, more recently, healthcare, auditing and accounting.

An audit is a thorough study of a company's balance sheets. Investors and other stakeholders rely on audits to ensure that a company's performance reports are accurate. Regulators can also rely on audits to ensure that a corporation is following all applicable legal and regulatory requirements. Auditing is applicable for various sectors like Financial, Economy, IT, Energy, Government, Medical, Supply Chain and so on. The expansion and acceptance of information technology have had a direct impact on the field of accounting services due to the nature of its unique task. Starting with moving on to cognitive technologies like Robotic Process Automation (RPA), computer vision, computer visualizations, Artificial intelligence (AI), cloud-based data storage, and huge data set management are all included in the development of ERP(Enterprise Resource Planning) systems.

Many accounting and auditing duties are repetitious and entail interactions with several people. High levels of transaction processing are present in these systems. When involved in prompt decision-making, there is the potential for the use of RPA in these areas [1] (Fig. 2.1).

RPA refers to the software that is incorporated into a company's operations. IT infrastructure that already exists RPA can be programmed to do a variety of tasks,do repetitive chores, therefore relieving staff of their responsibilities in industries such as bills and transactions, they bear a heavy weight. Processing, filling out a variety of forms, spreadsheets (online or offline), reporting, creation and database maintenance, data verification and validation, concatenation of databases and data reconciliation [2] (Fig. 2.2).

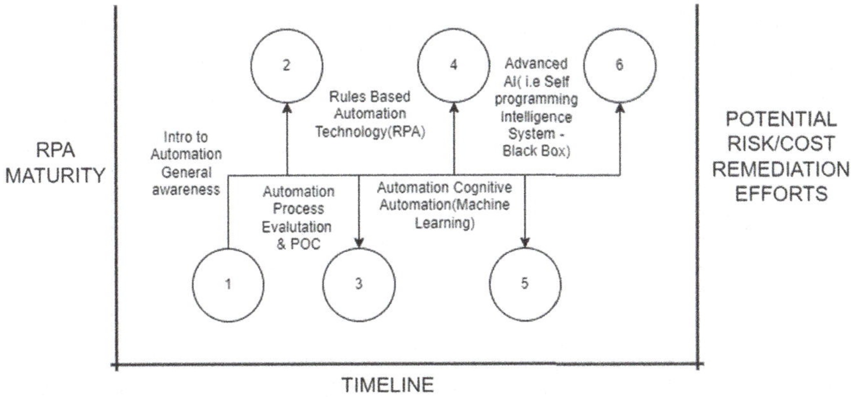

Fig. 2.1 Evolution of RPA

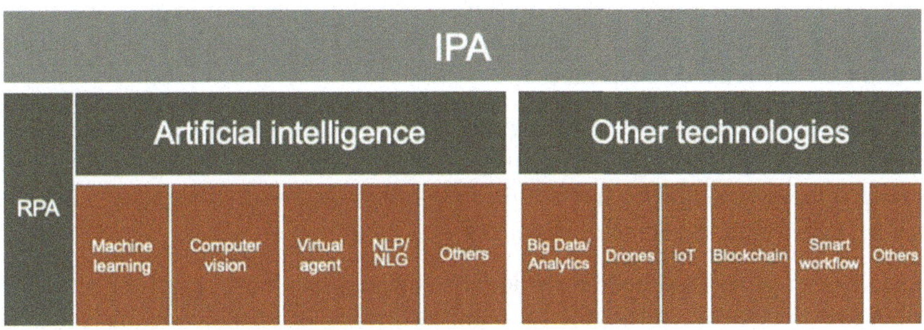

Fig. 2.2 An ecosystem of emerging technologies makes up IPA

2.1.1 Comprehending Robotic Process Automation (RPA)

The buzzword in the corporate world at the moment is 'automation', a word used by specialists in the field of business who believe is significantly transforming every aspect of business. With so many cutting-edge technologies available today, it's only logical that all organisations will benefit. RPA is something we have, which has been found to be extremely efficient in standardised, recurrent and rule-based operations, alongside emerging technologies like Big Data, Artificial Intelligence (AI), Blockchain, Internet of Things and so on. "As a virtual worker, RPA duplicates user activities to decrease or eliminate human intervention in dull, repetitive, and manually intensive operations," according to Capgemini. Deloitte had this to say about exponential technologies, according to them "RPA is computer-coded software, commonly referred to as BOT, that emulates human actions and is able to drive automation of rule-based processes". It's a great way to automate any operation that relies heavily on data entry, data manipulation, triggering reactions, or interfacing with other digital systems.“ As a result, Robotic Process Automation is pre-programmed software that is meant to do human-like tasks and is referred to as robots or software-based robots. Cloud accounting software is a wonderful example of RPA. In layman's terms, Robotic process automation (RPA) automates human activities and saves money and time [3].

2.1.2 *Objectives*

Better Governance: This begins with identifying which tests and processes are the most viable candidates for automation by defining roles, responsibilities, and structures. Processes for approving designs and deployment methods, as well as producing standardised documentation, should all be covered by a governance structure.

Change Management: Is a term used to describe the process of Change as an unavoidable reality. That is why it is critical to have mechanisms in place for monitoring and correcting changes to automated tests and processes, as well as for dealing with the downstream consequences.

Continuous Testing and Monitoring: Testing and monitoring are done on a regular basis. Periodic quality assurance testing is required due to the changing nature of business operations. Testing and monitoring should also be done on a regular basis to keep up with the changing environment.

Handling and Processing of Exceptions: To triage issues that may arise, a structure and methodology should be built, distinguishing between operational and technical exceptions and channeling them accordingly.

Sets of skills and training Using automation and cognitive intelligence tools frequently necessitates IT and data science expertise that isn't always present in a traditional internal audit organisation. Continuous capability assessments should be conducted by programme leaders, who should either augment gaps with roles-based training or onboard new resources as needed [4].

2.2 Literature Survey

From a methodological perspective, we planned to conduct a literature review by first thoroughly examining the papers that addressed the topic of RPA in accounting and audit. In this regard, we used the Google Scholar database and conducted a search without specifying a time limit using the terms "RPA in accounting and audit" and individually, "Robotic process automation" OR "RPA" AND "accounting" OR "audit".

A total of 40 articles that contained these keywords in the title, abstract, or keywords section were found in the first round of search results. The second stage involved downloading and scanning the papers using the search terms audit, accounting, and RPA after the duplicates had been eliminated. Then, the papers that only indirectly addressed one of the three formulated research topics were eliminated (i.e. either mentioned RPA only in a small number of sentences or only used the term robot, automatisation, without developing the idea of RPA). 19 articles made up the final sample as a result. These publications were categorised and grouped in the third stage based on factors such as the year of publication, the number of authors per article, the place of origin of the authors, and the methodology employed. To come up with the answers to the three research questions, we carefully read and analysed the articles that made up the final sample. The papers under analysis address one or more of the established questions listed in Table No. 1 in whole or in part (Table 2.1).

Table 2.1 Literature survey

Articles analyzed in terms of topics covered					
No	Authors	Year	RPA in accounting and accounting management	RPA in audit	RPA implementation
1	Anagnosle	2017	X		
2	Appelbaum and Nehmer	2017	X		
3	Tucker	2017	X		
4	Cooper et al.	2018	X	X	
5	Oevarajan	2018		X	X
6	Fernandez and Aman	2018	X		X
7	Moffittetal	2018		X	X
8	Vasarhelyi and Rozario	2018	X	X	X
9	Zhang et al.	2018	X		
10	Zheng	2018	X		
11	Ansari el al.	2019			X
12	Cohen et al.	2019		X	X
13	Jedrzejka	2019	X		X
14	Kaya et al.	2019	X		
15	Kokina and Blanchette	2019	X		X
16	Kruskopf et al.	2019	X	X	
17	Huang and Vssarhelyi	2019		X	X
18	Zhang	2019		X	X
19	Gotthardt et al.	2020	X	X	X

2.3 Influence of RPA on Accounting and Audit

Among the most important elements in the digital business transformation technique is automation. Employee task automation is dependent on people's ability to perceive and analyse information, whereas Robots and RPA technologies are required for process automation. RPA can thus be effectively adopted and used in any sort of organisation or department, including firms that provide specialist accounting and auditing services. Robotic Process Automation is an expression for automation of repetitious, organised, processes on the basis of rules and is a sort of software that imitates a human's actions while executing a task inside a process. It can speed up, improve, and "never fatigue" repeated tasks, freeing workers from a tremendous amount of work. To put it another way, RPA aids in improving the efficiency of corporate processes while also reducing human errors and costs. Furthermore, at the user interface level, it can interact with other software applications.

Robotic Process Automation is a computer program that enables firm coworkers to run current applications using installed software on a computer or "robot" such as data manipulation, transaction processing, triggering reactions, and additional digital systems interfacing. RPA is a software configuration that transfers information obtained from emails and spreadsheets which are used as input sources for registration systems like ERP and customer relationship management systems like CRM. RPA, in a larger sense, is a term that is a term used to describe a range of technology that includes autonomous systems, artificial intelligence, machine learning and robotics. RPA, for example, in order to execute activities utilising the company's key systems, accesses and manipulates spreadsheets, documents, and emails. Finance and accounting, Production, sales, acquisitions, supply chain management, customer service, and human resources are just a few of the departments involved where RPA can easily automate current processes and procedures.

In contrast to macros, RPA robots may communicate with numerous systems, work autonomously, and do repetitive jobs that require only binary decisions. As RPA moves toward automation or cognitive automation, it will be capable to do that tasks aren't expected to be done on a regular basis and which require judgement according to professional guidelines implemented to unorganised data(information).

RPA software may be installed on a real or virtual machine and is compatible with all client-inherited systems, including cloud, Java, Citrix, web applications, ERP, mainframe programmes, and other applications. RPA has a lot of potential in audit and accounting since many roles need interacting with several systems, doing large amounts of transaction processing, and making real-time decisions. RPA represents a chance to increase the quality of accounting services supplied, yet there are some concerns that its adoption will result in individuals being replaced by robots. In the real world, it will result in a shift in the professional accountant's position, with more time being devoted to analysis and forecasting at the expense of mundane tasks.

Accounting and auditing procedures have used a range of computer-assisted tools and processes throughout their history, which are frequently linked by multiple manual stages and "clicks" of affirmation. Robotic Process Automation may automate and take control of a complete set of repetitive procedures Few portions of the audit process, particularly those susceptible to the usage of workflows and those implicating repetitive judgments, can be automated with the use of RPA. RPA represents a significant and troubling shift from current primarily manual auditing practices, Nonetheless, it has the potential to enable auditors to work at a far greater level. However, because of the strictly regulated scope of audit and accounting services, the Robotic Process Automation for audit services is still in its early stages. RPA to help with audit automation and to create frameworks for using Robotic Process Automation in audits.

Robotic Process Automation may be utilised in the automation of the auditing process and jobs with repeatable and deterministic judgments that require process and time/motion improvements. Reconciliations, internal control testing, and detail testing are instances of audit tasks that Robotic Process Automation is able to automate. RPA can also be utilised to help with audit testing by automating processes. Consider implementing RPA in the Employee Benefit Plan Audit's substantial testing. Routine operations, such as forwarding a client's data from the previous year inside the accounting organisation's audit structure, might be performed by a "bot" (a computer script programmed to complete a certain action).

Accounting is the systematic and chronological recording of transactions and economic activity over time. Invoicing, salaries, and settlements, for example, are repeated processes with a huge number of transactions that occur inside a period and from one period to the next, organisations can streamline the entire activity by automating these procedures, lowering costs and risk of error. Additionally, because the whole accounting activity comprises following well-defined processes via procedures for work, organisations that provide accounting services might entirely profit from Robotic Process Automation. Because software updates can be done extremely fast, RPA can deal with changes in legislation, which are normally rather frequent in the tax industry.

Those pursuing products and services inputs through purchase afterwards payment, frequent closures, and monthly period or external broadcasting are the easiest to automate since they are regular and don't need significant large-impact judgement making or specialist reasoning. The use of "task-oriented" robots that only perform computations and keep special cases to humans could automate the operations. To put it another way, the human job of "extracting data from one system, conducting data processing, and transporting the changed data to another system" is going to be automated. The "bank reconciliations" operation is one such example. Accountants used to have to manually analyse and correlate transactions, find inconsistencies, and create journal entries before RPA. RPA is used by accountants to find, analyse, and resolve the sources of discrepancies.

The automation of the habit of reimbursing distributors and retrieving from clients in or by digital transfer or physical money would comprise the following. Robots utilise their credentials to sign in, search for fresh receipts, "couple" them with connected orders, demand and await for authorisations, undertake financial documents and other private logistics, and eventually, process and complete the transaction and notify an alert notification along with an official statement that the operation is finished. The method is reiterated for as long as there exist outstanding and unpaid receipts. It is faster and more accurate to extract, validate, and enter transaction data from/into various computer systems than it is to do so manually. An accountant can supervise a group of robots and only intervene if there are any problems. Data that isn't in the right format prescribed, network difficulties, and also other system failures are examples of these concerns.

By maintaining databases for clients, suppliers, debtors, and creditors, issuing/receiving and processing invoices, approving, validating, and paying debts when they are due or sending notices of impending payments for amounts owed, and confirming the relationships between issued invoices and the products or services to be delivered, as well as invoices regarding the purchased goods or services, RPA can be remarkably successful in managing payables and receivables. Periodic closure and reporting is an example of a procedure that can be automated. It entails managing the collection and validation of large volume of data from multiple sources, that is accomplished using spreadsheets, tables, and distinct records lists. Errors and wasted time are caused by data purification and issues with concatenating, connecting, or updating data from several sources that require human data transfer. For large organisations' accounting departments, current regulatory and legal reporting obligations are growing increasingly rigorous.

Ending balance sheets, merging problems at the group level, and preparing financial statements while adhering to time constraints imposed by legislation or at the base scope demands good synchronization, qualified personnel, and solid corporate management. The closing method towards the conclusion of a timeframe has an undeviating influence on the reporting's result since the entire nature of the report's utilisation originates from the correctness, completeness, and reality of the information. RPA can be deployed in the following areas:

- Closing activities are completed on an annual, quarterly, and monthly basis, along with the publication of the accounting journal, registers on various types of operations, and specific examinations of the purchases (sales), documents (spreadsheets), and balances.
- Financial and operational performance information are included in monthly, quarterly, and yearly reports to management, as well as external reports that are either compulsory and regulated or optional, depending on the organisation.

RPA can also be used to perform accounting functions such as Short-term control and forecasting planning and budgeting different activities or procedures in the long term, medium term, and short term taking into consideration many eventualities within a brief duration of time, resulting in a more efficient decision-making process [5].

2.3.1 Automation in Finance

All company departments, inclusive of every department have used automation. Accounting and finance being a key pillar is a vital support column for every organisation, has resisted changes a little more. Until recently, they relied on traditional techniques, which are time intensive, laborious, monotonous, and prone to errors. However, businesses and singletons who used automation in the start saw significant advantages, and these encouraging indicators encouraged additional people to follow suit. It is self-evident that deploying software like RPA provides numerous benefits to any company or a single entity dealing with large amounts of financial data.

RPA has a number of advantages, including the ability to:

- Make the procedure more efficient.
- Reducing time.
- Reduce costs by empowering employees.
- Reduction in errors.
- Save time and effort.
- Assist in beating the competition.
- Become cutting-edge and forward-thinking.
- Boost scalability and flexibility.
- Improve staff morale [6].

With so many advantages, it's no surprise that RPA is gaining momentum and popularity and is growing ever so quickly in these changing times.

2.4 Intelligent Auditing

RPA is rapidly being utilised by prominent audit organisations to replace different human operations in financial audits of a set of yearly financial records. Audit companies, particularly the bigger ones, are increasingly using software and computer technologies to increase rigour and efficiency in their work. Even today, a considerable number of manual, repetitive, and tedious duties based on well-defined processes take the auditors' time. The auditor performs a number of tasks throughout the financial account audit mission, some of which can be automated: data preparation, file organisation, data integration from various files, basic audit tests, data copying and pasting, and manual comments.

Audit procedures can be automated because they are well-defined, based on professional rules and regulations, subject to quality control by professional and supervisory bodies, and have significant implications, which saves time and money. On the other hand, audit procedures are well-defined, based on professional rules and regulations, subject to quality control by professional and supervisory bodies, and have significant implications. Large audit companies have started to rethink their entire business process by integrating new technology into their current operations in

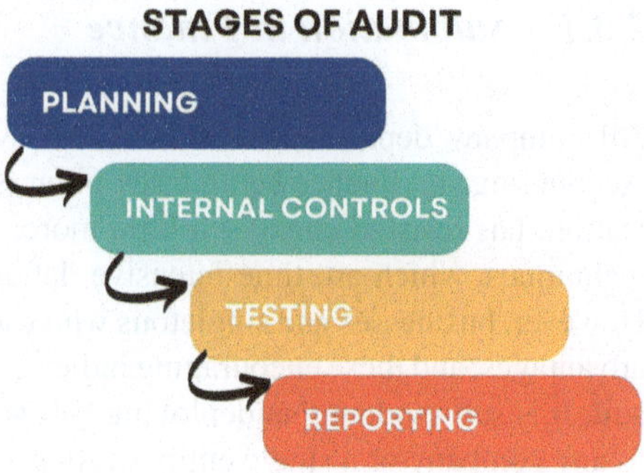

Fig. 2.3 Stages of audit

order to increase the effectiveness and efficiency of audit procedures. RPA has been utilised in a variety of auditing processes, including revenue audits, stock audits, pension plan audits, loan audits, document auditing, e-commerce website auditing, and continuous auditing [7].

There are several stages involved in auditing as shown in Fig. 2.3:

- The first phase is the planning stage. A corporation meets with the auditing firm at this point to discuss features such as the extent of the engagement, techniques, and aim.
- The next stage is to put internal controls in place. At this step, auditors gather financial documents as well as any other pertinent information needed to complete their audits. The data is needed to determine the correctness of the financial accounts.
- The third step is the testing stage. Auditors apply a variety of tests to ensure that the financial statements are accurate at this time. Verifying transactions and controlling procedures may be required, as well as requesting extra information.
- The fourth step is the reporting stage. The auditors write a report when they've completed all of the examinations, providing their opinion on the financial accounts' correctness [8].

The papers you utilise to substantiate your audit conclusion are referred to as audit evidence. When conducting an audit, you will come across a variety of evidence. Documents may be prepared by client workers or by third parties. You must grasp the four concepts of evidence in order to appropriately evaluate the strength of the evidence you gather:

- The nature of the evidence, such as whether it is vocally, visually, or via writing.
- Appropriateness refers to the evidence's quality, relevance, and trustworthiness.
- Sufficiency: The amount of trial data–sufficient empirical proof to assess the assertions of the management through the clients end.
- Evaluation: Determining whether the data is convincing enough for you to create an opinion.

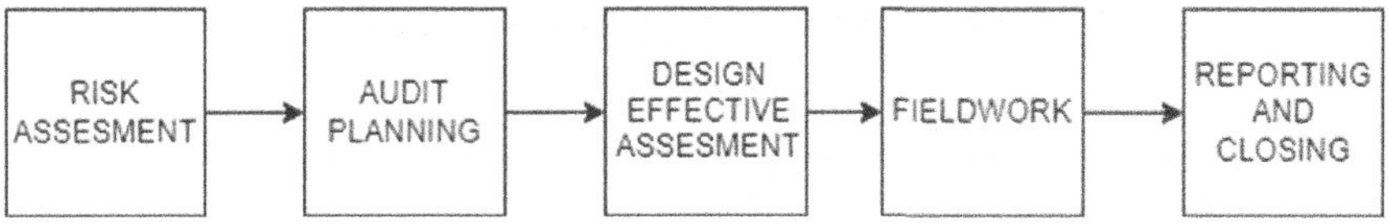

Fig. 2.4 Leveraging advanced analytics and automation across internal audit

Possible to present an algorithmic technique as a Four-Stage progression (Figs. 2.5 and 2.6):

- Establishing the auditor's duties and terms of employment.
- Preparing for the audit, which includes details such as deadlines and the departments that the auditor will examine.
- Assembling audit data, or, more accurately, gathering and cleansing data from various sources like financial accounts and statements.
- Putting together the report of the auditor [9].

2.5 AI in Audit

Newer auditing research has mostly focused on critical issues.

- Developing cognitive help for auditors using a virtual assistant and cognitive technologies: In an audit brainstorming session, information retrieval and risk assessment are made easier with interactive decision support. In the accounting/auditing workplace, Intelligent Virtual Agent (IVA) could be useful as shown in Fig. 2.6.
- Using natural language processing (NLP) to process textual material connected to audits, such as social media and contracts: It is used in contract audits, for example, to give auditors enough data to assess audit risk and provide audit proof.
- Textual analysis of tweets can augment analytical tools in the audit of a company's revenue account by providing data regarding consumer interest and satisfaction.
- Extracting sentiment features from documents following a combined theory of deep learning and NLP: For example, it can give relevant and reliable information to auditors by extracting sentiment elements from company communication documents and social media, such as forecasting internal control weaknesses, procurement irregularities, financial misstatements, and audit fees. Offers a blueprint for deep learning data identification and classification functions, and judgement capabilities for a range of audit procedures in various trials, with a focus on deep learning.
- Performing predictions and analytics using machine learning algorithms: For example, machine learning can enhance accounting projections. As a result, it may be applied to verify estimations. In addition, a mathematical ratio-based machine learning-based peer selection approach is offered as a potential alternative for risk assessment methods and analysis.

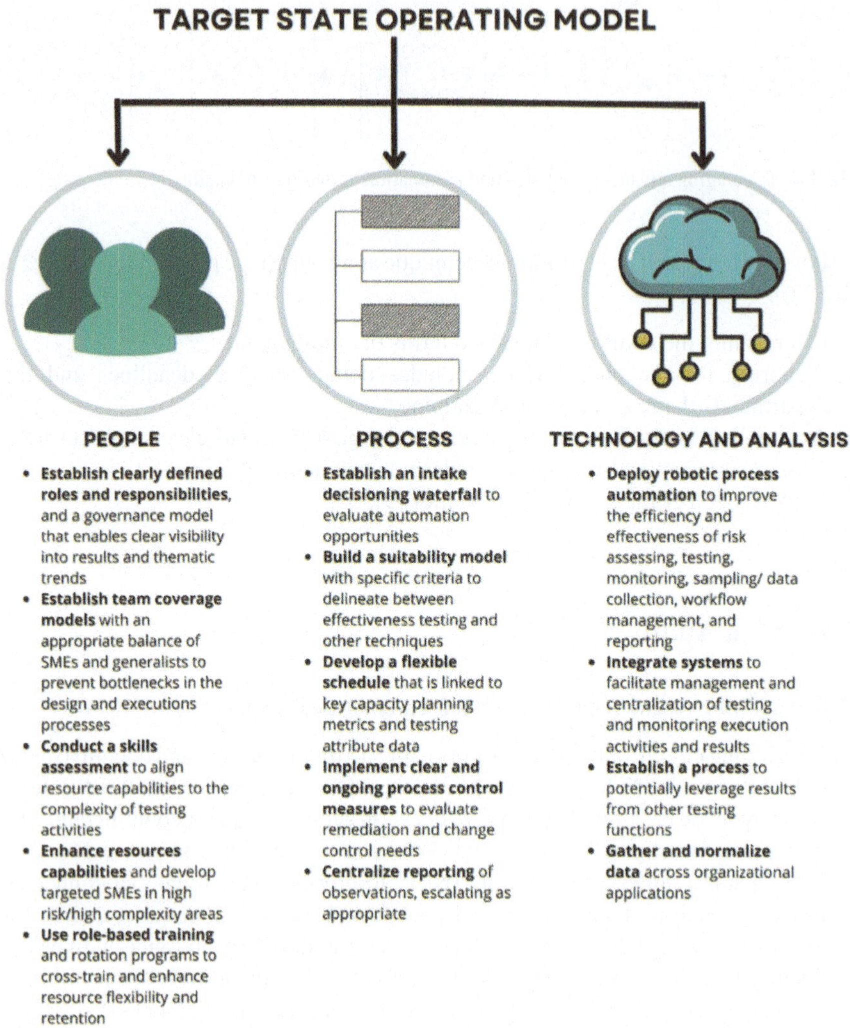

Fig. 2.5 Envisioning the target dtate

- Using image recognition in inventory audits: Inventory audits can be conducted with the help of drones and image recognition software.

Prominent financial companies are using artificial intelligence in their auditing services. KPMG's audit platform incorporates cognitive and predictive technology, and the firm has collaborated with intelligent agents to build naturalistic automation tools for chartered auditors. To obtain audit evidence, EY is utilising Artifical Intelligence to extract data from documents that have unstructured data (e.g., contracts, invoices, and photos). Machine learning is also used by Ernst & Young to analyse massive datasets in order to discover, evaluate and respond to the dangers of material

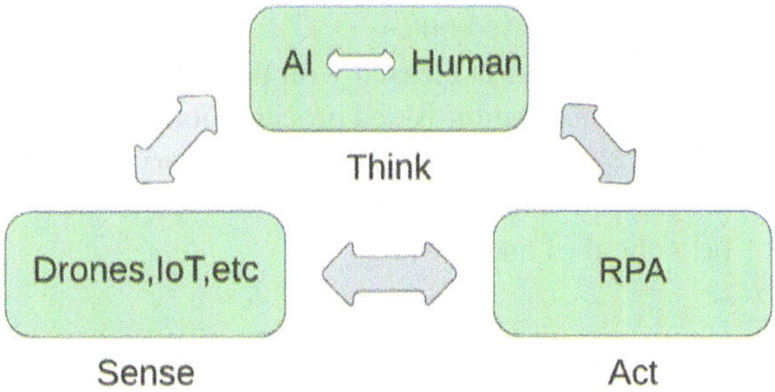

Fig. 2.6 The "Sense-think-act" loop of IPA

misrepresentation related to scams. Deloitte is analysing contracts with the help of analytics and cognitive technologies.

PwC is collaborating with H2O.ai, an artificial intelligence company, to develop an agent which utilises AI to analyse transactions and discover discrepancies in the statements. Machine learning functions are also embedded in the products of companies that provide common audit software, such as Galvanize 9, MindBridge Ai 10 and CaseWare 8 [10].

2.6 Task Organisation and Workflow in Auditing

2.6.1 Context on Workflow Analysis

A workflow is a sequence of actions that must be completed in a specific order to complete a business operation. It specifies the sequence in which tasks must be invoked, as well as task synchronisation and information flow. A business process is a set of procedures taken by a group of people to accomplish a specific goal. Tasks include things like generating a bill, sending an email and modifying a file. The use of software is required to complete a task.

Humans and/or systems Workflow management entails the following:

- Modelling of the process to encapsulate and characterise a process.
- Re-engineering processes to improve the process.
- Workflow automation and execution involve using both humans and information technology to implement, arrange, carry out, and control workflow tasks in accordance with the workflow specification. The definition of the workflow is built on top of the process modelling [11].

One way to represent the process is to use a communication-based strategy. Every stage in a process is abstracted into four phases,communication-based strategy is built on conversation between a consumer and a performer:

- A client demands a specific task to be carried out.
- The client and the worker confer and agree on the course of action to be taken.

- The task is carried out.
- If the consumer is pleased with the output, the action is considered a success. In this communication-based process model, a business process may be regarded as a network of workflows, where secondary workflows can be engaged to speed up progress toward the primary workflow, and the performer of one workflow could be a client of another [12].

2.6.2 *Audit Workflow*

Workflow theories are founded on business processes, but they may also be applied to the audit process since it is made up of a series of actions that turn inputs (auditable data) into outputs (audit opinions) for a specific goal. As a result, an audit partnership may be thought of as a web of audit processes, with the core workflow being the completion of the audit work and the four key audit stages being the subsidiary workflows (audit planning, internal control testing, substantive testing, and audit conclusion). For example, in the Single Audit planning approach, the finished planned working paper is a key procedure which entails answering a range of questions in order to determine if the auditee is at high risk and which funds must be validated. After that, a secondary workflow is started, either to undertake analytical procedures or to request professional judgments from the auditors in order to provide answers to inquiries in the planned working paper. Lower-level processes may be included in each secondary process. Individual audit jobs are the most fundamental operations that can't be broken down any further [13].

2.6.3 *Audit Task Structure*

Tasks for individual audits can be classed as semi-structured, unstructured or structured after the audit engagement (main process) has been split down into individual audit tasks. "The problem can be well-defined with a very restricted number of possibilities in structured activities, requiring very little judgement to make a final choice". Structured audit tasks can be done using components from the "Realm of RPA" like Desktop Robotic Process Automation, according to the automation continuum. Unstructured activities, on the other hand, involve "ill-defined problems with many viable solutions" that necessitate "great judgement and insight to choose among possibilities".

Unstructured audit jobs are located in "Realm of Cognitive Automation," according to the automation continuum. Semi-structured jobs are "tasks with a limited number of potential solutions, requiring a medium level of judgement to make a choice, and can be matched to the scope between RPA and Cognitive Automation." For semi-structured activities tools like Natural language processing, computer

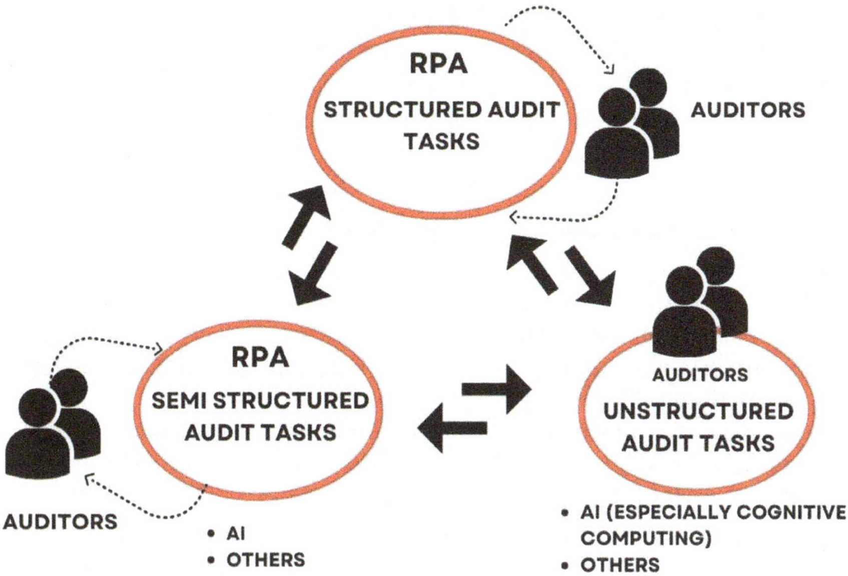

Fig. 2.7 RPA audit task structure

vision, and virtual agents are examples of technologies that can be employed [14] (Fig. 2.7).

2.6.4 "Auditor-in-the-Loop" IPA Environment

Following the classification of audit jobs as structured, semi-structured, or unstructured, and mapping them to the automation continuum, relevant automation technologies from the IPA portfolio can be selected for each work. Using technology from the "Realm of RPA," structured tasks may be automated with minimum human participation, however, auditors are still required to validate the outcome and manage with deviations. Frameworks from the "Realm of Cognitive Automation" offer useful information to auditors for unstructured tasks, allowing them to make better decisions. By interacting with the auditor, cognitive tools can learn to provide more relevant and accurate information over time. Auditors will be aided by appropriate technology in semi-structured duties so that they may focus on more critical responsibilities. Auditors will be brought in when the AI is confused about how to complete a job. The AI will learn from the event and increase its capacity to deal with similar circumstances in the future when the auditors have completed their duty. When various types of audit assignments are automated or complemented using tools from the IPA toolkit they are organised into a work process that includes auditors to control exceptions, review results, and make a contribution to their professional judgments, resulting in the "auditor-in-the-loop" IPA environment. As a consequence, even if IPA incorporates AI, auditors are not prohibited from conducting audits. Instead, it aims to cut down on the amount of time individuals spend on time-consuming and repetitive chores while also assisting them in making better decisions [15].

2.7 Implementation of Auditing Agents

When deciding on the implementation of an Intelligent auditing agent, we must focus on a task in the IT auditing stages which is a suitable case of automation. As discussed in audit workflow, it is vital to split the auditing tasks into individual indivisible audit tasks as described in (Figs 2.8 and 2.9).

In the initial stage of Identification, the IT auditing process will be disassembled into individual indivisible tasks. The audit tasks structure as discussed above the tasks must be well-defined with a very restricted number of possibilities in structured activities, requiring very minimal judgement to come up to a conclusive decision. In such conditions, the task is a prime candidate for automation [16] .

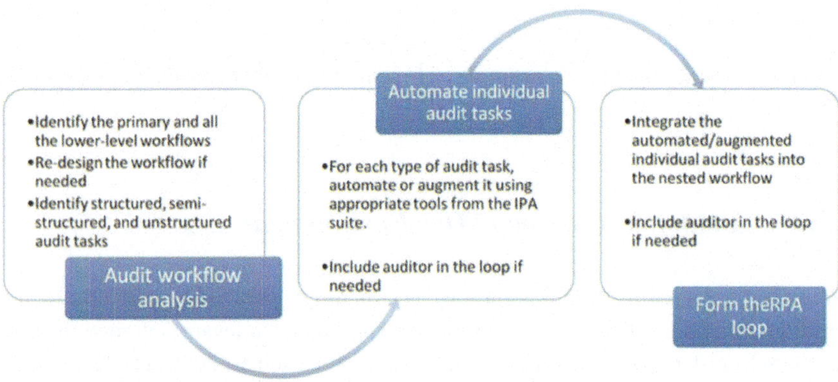

Fig. 2.8 Workflow of tasks

Fig. 2.9 Stages of implementing intelligent auditing

Fig. 2.10 Factors in developing RPA bot

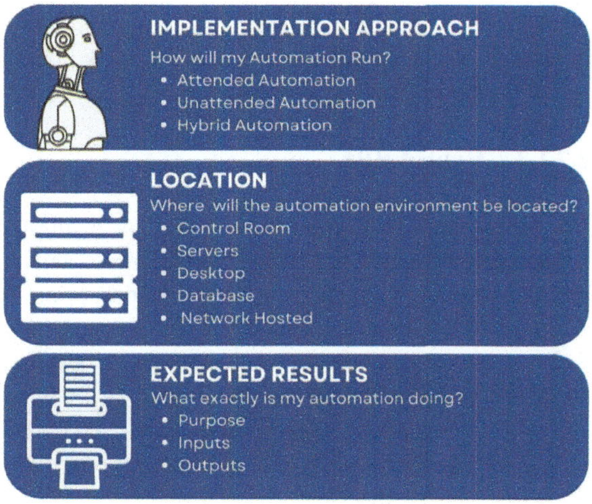

The second stage involves designing an automation bot. To design and develop one some factors are to be considered as shown in the diagram below (Fig. 2.10).

2.7.1 How Will the Automation Execute?

2.7.1.1 Attended Automation

Attended robotic process automation, like a virtual assistant, works alongside employees on their computers. By delivering real-time coaching and automated procedures, it optimises their repetitive routines, increases their productivity, and enhances the quality of their work. Before an employee may move on to another activity, the attended bot must complete each task.

Attending RPA happens when a customer service representative instructs a bot to carry out an address change request in a matter of seconds rather than making the changes manually, which might take a long time and involve many processes and clicks [17].

2.7.1.2 Unattended RPA

Bots perform group processes (a sequence of tasks) without the need for human involvement. They generally follow a predefined schedule or are the product of logic programming.

For example, an unattended bot can use a specified mechanism to recognise vendor bills provided by email, read such emails, record the accompanying invoices on the local system, and prepare them for payment. This sort of task may be done by unattended bots 24 h a day, seven days a week [18].

2.7.1.3 Hybrid RPA

Bot using natural language processing, for example, may listen to a conversation between two people, evaluate it in real-time, and then deliver information to the attending human to help them take the appropriate next actions. Alternatively, the AI-assisted bot might be talking with a person, and when certain demands in the dialogue arise that require the assistance of a human worker, they will enter the process to add value to the contact.

Hybrid RPA bots are most commonly utilised in client-facing environments, where they assist humans in responding to consumer requests [19].

2.7.2 Where is My Automation Environment Located?

The RPA process can be deployed onto various locations for operations. Ranging from a centralized management of automation like a control room or on-site or remote servers etc. It can even be executed in the desktop of the user mainly when it comes to attended robots. In terms of IT auditing, it can be executed on a network or a database, this enables direct access to the status and details of such entities.

2.7.3 What Exactly is My Automation Doing?

It is vital to know what the automation performs considering factors like its main purpose, its input and expected output. This clarity is needed well in advance in the designing and identification stages to avoid major risks of failure to occur in later stages in terms of its functionality and performance.

When it comes to the implementation stage, a designated staff has experience in any of the platforms to design and develop RPA bots to name a few, UiPath, Blue Prism, Automation Anywhere etc. The developer then ensures the perfect translation of the individual auditing tasks to be automated with full prior knowledge of the task with aid of the documentation provided by the auditors.

Consider a use case where when auditors perform a specific IT auditing process to certify an entity to be compliant with the standards a phase of evidence collection occurs. This phase involves gathering of data and creating reports for every entity such as servers or the management systems etc. used. One of the individual tasks is to collect and create a report on the server's status and performance. All the operations handled and performed by the server are stored as server logs. These logs essentially gives the entire history of the server execution and provides essential information.

But parsing through several data logs manually and generating a report for the same is a daunting task which requires repetition. Hence, creation and design of a RPA bot capable of parsing through rows of server logs extracting important terms and statements and generating a report based on the data collected is something which needs less judgement. With this case, a developer is assigned to create the respective bot with the platform of his choice and finally deploy the bot in a location specified in the initial stages of development and implementation.

After development and deployment, it is crucial to maintain and monitor the performance of the bot regularly and provide required modifications based on new needs.

Software developers' top priorities are frequently:

- Make it happen.
- It should be lucrative.
- Make it safe.
- That's the order.

The issue is that if you have to re-engineer your entire application/bot after it's been developed due to a security flaw, you've failed at priority 1 and are now eating into priority 2's profitability. It is quintessential to consider the risks and security considerations in every stage of development. Given below are the considerations to be concerned about when considering risk factors of a RPA bot [20].

2.7.4 RPA Security Considerations–Cybersecurity

Both systems and data should be protected from outside threats and viruses. It is vital that all the sensitive information and evidence collected by the bot be secure from such threats.

- **Encryption:** This ensures that the data managed by the bot is secured by encryption of data in transit and at rest utilising modern encryption technologies.
- **Firewalls:** Appropriate Firewalls must be set up and checked on a regular basis.
- **Authorised Access:** zones must be implemented in the server room to limit public access to the bot execution environment.
- **Antivirus:** To protect against the entry of a malicious entity into the bot operation environment, procedures should establish deployment and monitoring of antivirus software.
- **Intrusion Prevention/Detection Systems (IDS/IPS):** must be utilised to prevent high-risk external breaches. These systems' logs must be inspected and acted upon appropriately [21].

2.7.4.1 RPA Security Considerations–Logical Access

Individuals in charge of bot access management should utilise robust authentication techniques and review bot user access on a regular basis.

- **Authentication:** Robust authorisation techniques should be used to protect access to bots and bot data. For access to sensitive information present in servers remotely, multi-factor authentication should be employed.
- **Privileged Access:** The focus is on reducing the danger of bot data loss and corruption in the botting environment. Authorised users should have access to privileged accounts, which should be evaluated on a regular basis.
- Controls should be in place to mitigate the risk that user access credentials given are not permitted.
- Controls should handle the danger of user access not being updated or revoked in a timely manner after a change in employment status. This might result in unauthorised access to the bot's executing system and data [21].
- On a regular basis, access reviews should be done on all user access and permissions.

2.7.4.2 RPA Security Considerations–Physical Security

Environmental controls and physical access restrictions should be used to mitigate any risk to the physical servers present in which the bot executes.

- **Personnel Access:** To prevent data integrity from being compromised, physical access to the areas where the bots are housed should be limited to authorised persons.
- **Access Records:** Management should save and monitor access logs for anyone who enters the bot's physical location. It is necessary to investigate and resolve unidentified access.
- **Facility Monitoring:** Security staff should keep an eye on the facility and utilise surveillance techniques to guarantee that only authorised people are allowed in.
- **Environment Monitoring:** To safeguard the equipment from fire damage, power outages, and overheating, controls ensuring the safety of the equipment must be applied [21].

2.7.4.3 RPA Security Considerations–Operations

Backups of the bot, bot's data, and bot's system environment must be done on a regular basis, restoration testing must be done, and task failures must be rectified.

- **Backups:** In order to make sure that the essential bot execution can be recovered in the case of a disaster, incremental data backups of the bot execution environment should be done on a regular basis.

- For the sake of safety of the bot operational environment in case of a disaster yearly data recovery exercises must be conducted.
- **Processing:** Essential jobs that interact with the bot execution space should be set to run at predetermined periods and intervals. Job history records should be examined on a regular basis, and job failures should be resolved as soon as possible.
- **Schedules:** To assist assure data integrity, access to the bot's work scheduler must be confined to authorised persons.
- **Exception Handling:** Problems and failures in task processing should be tracked, investigated, and addressed [21].

2.7.4.4 RPA Security Considerations–Monitoring

To track and rectify detected faults, those in charge of maintaining the bot execution environment must use and configure health monitoring technologies (Fig. 2.11).

- **Triggers:** The company's personnel should look for triggers in the bot software that indicate any violation of certain defined rules or constraints.
- **Alerts:** Monitoring and alerting features must be enabled so that the team is automatically notified if an indicated trigger occurs.
- Monitoring tools should be set up to notify management if a critical bot system health issue arises that affects hard drive space, scheduled duties, performance, and the bot's software/hardware.
- **Capacity:** Monitoring mechanisms must be set up to notify the team if the bot execution system's systemic memory capacity is reached or approaching [21].

2.7.5 RPA Data Considerations: Confidentiality

If the bot uses confidential data, special care must be taken.

- **Disposal:** securely discarding confidential data
- Disclosure: preventing unwanted disclosure
- **Retention:** in compliance with regulations
- Access is restricted to only authorised personnel and is protected by a secure space.
- Similar to security consideration, with pipelining the data flow between nodes of the bot data must also be protected and secured in every stage and transmission of it given below are elaboration of main factors [21].

2.7.5.1 RPA Data Considerations: Integrity

A critical success factor is maintaining data integrity throughout the process.

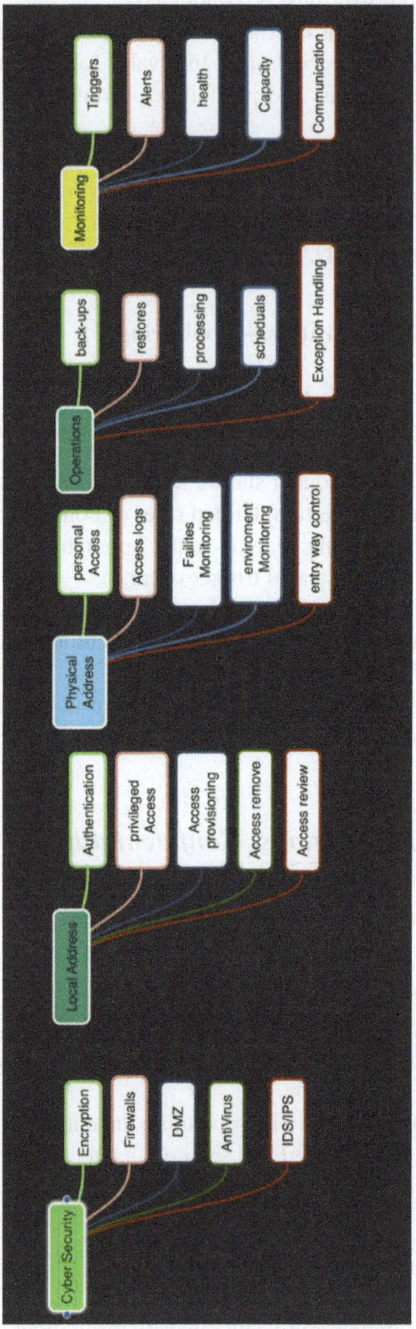

Fig. 2.11 RPA security considerations overview

- **Detect:** To detect issues with data concerns, reconciliations, and reviews.
- **Correct:** Appropriate handling of issues must be established.
- **Accuracy:** To validate whether all the transactions processed are accurate.
- **Completeness:** Verify all the items are been processed [12].

2.7.5.2 RPA Data Considerations: Availability

The data needed for processing must be available.

- Processing Capacity: Resource utilisation.
- Monitoring: Constant monitoring of the stack for issues.
- Backup: The data utilised by the system in which the bot executes.
- Restore: The ability to receive the data from the backup.
- Service Level Agreement: For giving clear information to handle issues [12].

2.7.5.3 Privacy Issues with RPA Data

Additional concerns must be made if the bot uses data that has privacy obligations.

- Notices: how is your data used?
- Collection: secure methods and storage
- Consent: obtain consent for usage.

2.7.5.4 RPA Data Considerations: Tracking and tracing

User must be in a position to determine during the processing, what happened to the data.

- **Measure/Track:** an event in which you follow someone or something's path or trail (historical).
- **Tracing**: a method of determining where something came from (current).
- **Capture:** any data usage-related action conducted by and to the bot.
- **Responsibility:** Unique IDs and access controls are used to ensure accountability.
- **Storing:** the logs and crucial data points are kept [12] (Fig. 2.12).

2.8 Future Scope

The RPA area is still in its early stages, and future research is needed to provide a more accurate view of the general picture, as well as a more thorough picture of the RPA phenomenon's use in accounting and auditing. The most difficult aspect of RPA implementation in general is figuring out how will users work alongside with RPA as well as the various technologies that are still developing go along with it.

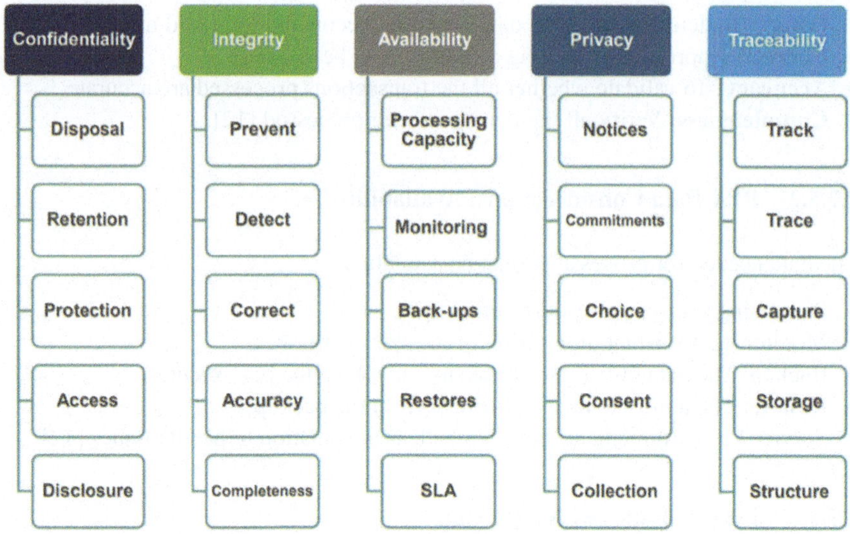

Fig. 2.12 RPA data considerations overview

Generally speaking, technology has a substantial influence on the accounting/audit occupation throughout the duration of time, but nowadays, more so than ever, it is having an even greater impact. Accounting and auditing professions will undergo considerable changes as a result of disruptive technology, both in terms of actual activity and potential specialised training. RPA has a lot of potential for automating accounting operations, and robots will eventually substitute expert accountants for a lot of their work, particularly monotonous and ordinary tasks. It's possible that some people will lose their jobs as a result of this process, both entry-level or entry-level-experienced roles in accounting and auditing businesses, on the job market, new employment may become available. Future accountants' responsibilities will expand beyond traditional accounting and financial reporting to include complicated data analysis, forecasting, and consulting. This shift necessitates the development of new talents and competencies in areas such as new technology, data processing, and effective integration, as well as skills in analysis and synthesis, critical thinking, and communication. Exchange of information and teamwork talents, sentimental awareness, conceptualising and addressing difficult problems, as well as creative thinking, adaptability, and patience in continual learning, will all be essential. New talents are gaining ground and will necessitate the formation of modern vocations, such as professional accountant specialising in huge data set analysis (data scientist), cybercrime expert or database analyst, data security professional or systems integrator, and blockchain or cloud accounting.

As a result, more investigation is required to look into the unfavorable impact of RPA adoption on professional accounting personnel's behaviour, firm organisational culture, the short- and medium-term costs and benefits. RPA also entails a shift in future professional accountants' education, which will require further research.

Accounting education should include cybersecurity training, RPA, cloud accounting, blockchain and big data analysis of huge datasets, as well as minimum programming abilities, in order to cope up with present technical problems and have success in developing the future in the field of RPA. Because the quality of accounting education has a direct impact on the employees and businesses' prosperity, It is critical that its content be rebuilt to reflect current technological advancements. Of doubt, the fundamental skills taught in today's auditing and accounting programmes will still be necessary in the future; they are the profession's main pillar. In addition, new courses are expected by technology advancement and must be incorporated in the course of study. They will aid future professionals in developing technical and social skills that will help them to participate in and integrate into the labour economy. Expertise of the software and its characteristics, in addition analysis, and security awareness are all technical talents. There are abilities that can assist an employee in interacting with programmes, artificial intelligence, and robots in general. Many duties will be performed by a combination of humans and robots. Although technical skills have long been regarded as important, they are becoming increasingly so as employees are required to bridge the gap between robots and humans.

2.9 Conclusion

We are experiencing a phenomenon that appears to represent, without saying so, the continuous and rapid progress of technology in all disciplines. Without a doubt, the robotization of services is our future. The robots' growing incorporation into our daily lives and as a result of one's professional efforts, a succession of unavoidable consequences arise, queries, to which a tiny portion, respectively, we attempted to provide some plausible answers in relation to RPA. Because of the large volume of information that needs to be handled, as well as the time savings, cost savings that Robotic Process Automation generates, RPA is gaining traction and becoming increasingly frequent in financial accounting duties at major corporations and firms that specialise in providing accounting and auditing services. Organisations that have started adopting RPA are seeing immediate benefits from the automation of specific audit and accounting tasks and procedures, particularly those that are well-organised, systematic, repeatable, and simple. Individuals interested in implementing Robotic Process Automation in their audits and accounting actions have to begin by categorising jobs based on their level of difficulty, then standardising and optimising processes, as well as adjusting foundations in the processes and business flow. Additionally, the addition of "digital personnel" as well as the automation of some functions via RPA would necessitate a rethinking of internal controls. By adjusting and mastering the necessities of new developing technologies, the accounting profession has the potential to become even more important in the future. Upcoming generations of successful certified public accountants will be required not only traditional knowledge but also advanced social skills in areas such as Information Technology, Artificial Intelligence and Robotic Process Automation.

References

1. Burgess, J.,Herman, B., Loriè, A.: Robotic Process Automation (RPA) and the auditor ISACA South Florida. RSM US LLP, (2020)
2. Stanciu, V., Rîndaşu, S-M.: Sustainable professional training—challenges and solutions in emerging European countries. Audit. Financ. **18** (4)(160), 771–784 (2020). https://doi.org/10.20869/AUDITF/2020/160/025
3. The future digital work force: robotic process automation (RPA). JISTEM— J. Inf. Syst. Technol. Manag. **16**, e201916001. TECSI Laboratório de Tecnologia e Sistemas de Informação—FEA/USP. (2019)
4. Willcocks, L.P., Lacity, M., Craig, A.: The IT function and robotic process automation. (2015)
5. Lacurezeanu, R., Tiron-Tudor, A., Bresfelean, V.P.: Robotic process automation in audit and accounting. Audit. Financ. **18** (4)(160), 752–770 (2020). https://doi.org/10.20869/AUDITF/2020/160/024
6. Asthana, S., Khurana, I., Raman, K.K.: Fee competition among Big 4 auditors and audit quality. Rev. Quant. Financ. Acc. **52**(2), 403–438 (2019). https://doi.org/10.1007/s11156-018-0714-9
7. Nunes, T., Leite, J., Pedrosa, I.: Intelligent process automation: an overview over the future of auditing. In: 2020 15th Iberian conference on ınformation systems and technologies (CISTI). IEEE. (2020)
8. Langer, M. et al.: Explainability auditing for intelligent systems: a rationale for multi-disciplinary perspectives. In: 2021 IEEE 29th ınternational requirements engineering conference workshops (REW). IEEE. (2021)
9. Brandas, C., Muntean, M., Didraga, O.: Intelligent decision support in auditing: Big Data and machine learning approach. In: 17th International conference on ınformatics in economy (IE 2018) education, research & business technologies. The Bucharest University of Economic Studies, Bucharest, Romania (2018)
10. Alpaydin, E.: Introduction to machine learning, 3rd edn. The MIT Press, (2014). Appelbaum, D., Nehmer, R. A.: Using drones in ınternal and external audits: an exploratory framework. J. Emerg. Technol. Account. **14**(1), 99–113 (2017). https://doi.org/10.2308/jeta-51704
11. Zhang, C.: Intelligent process automation in audit. J. Emerg Technol. Account. **16**(2), 69–88 (2019)
12. Moffitt, K.C., Rozario, A.M., Vasarhelyi, M.A.: Robotic process automation for auditing. J. Emerg Technol. Account. **15**(1), 1–10 (2018)
13. Chergui, M, Medromi, H., Sayouti, A.: Inter-organizational workflow for intelligent audit of information technologies in terms of entreprise business processes. Ed. Pref. Desk Manag. Ed. **5** (5), 98 (2014)
14. Abdolmohammadi, M.J.: A comprehensive taxonomy of audit task structure, professional rank and decision aids for behavioral research. Behav. Res. Account. **11**, 51 (1999)
15. Boersma, E.: Intelligent process automation framework: supporting the transformation of a manual process to an automation. MS thesis. University of Twente, (2020)
16. Alles, M., Kogan, A., Vasarhelyi, M.: Audit automation for implementing continuous auditing: Principles and problems. (2008). Retrieved from http://www.fdewb.unimaas.nl/irsais/pdf/IRSAIS2008papers/Alles_en_Vasarhelyipaper
17. Zhang, C., Thomas, C., Vasarhelyi, M.A.: Attended process automation in audit: A framework and a demonstration attended automation in audit. J. Infor. Syst, (2022)
18. Brandenburger, N.: Remote control of automation: workload, fatigue, and performance in unattended railway operation. Technische Universität Braunschweig, Diss (2021)
19. Guha, A., Samanta, D.: Hybrid approach to document anomaly detection: an application to facilitate RPA in title insurance. Int. J. Autom. Comput. **18**(1), 55–72 (2021)
20. Tripathi, A.M.: Learning robotic process automation: create software robots and automate business processes with the leading RPA tool–UiPath. Packt Publishing Ltd, (2018)
21. Cosner, J., Glenn, D., Okonkwo, C.C.: Are you ready for RPA?

Chapter 3
Impact of AI and RPA in Banking

Debanjana Dasgupta

Abstract This chapter talks about the impact of RPA and AI in the Banking Industry. Banking being one of the early adopters of RPA and now AI, in this chapter we analyze the current problems in banking industry and key disruptors and set the context to establish how Intelligent Automation with AI and RPA can help in addressing these disruptions. The chapter then focusses on how these technologies will transform the way the banks operate.The chapter then dives deep into real life use cases of Lending operations, Contact Center operations and fraud operations in a bank. For each of these areas, the chapter describes the business problem with a brief overview of the generic process and typical problems associated with that, what the levers of automation are that can be applied to address the problem and outline the solution with high-level components specified along with the benefits achieved through the solution. The chapter will end with stating what it takes for enterprises and what practices they need to build, what should be the operating model, delivery, support, and governance, to achieve the benefits that AI and RPA can bring to the enterprise.

3.1 Introduction

Banking industry is one of the leading industries that has always been early adopters of technology. They have embraced various technology innovations leading the way to other industries. Starting from the introduction of ATM or Automated Teller Machines to card-based payments in the previous century, and from there to the revolution of banking through online and mobile banking post-millennium, this industry has seen and experienced it all. In the last decade and throughout the pandemic, as we were forced to stay inside our houses with social distancing, digital payments have swept the world. From groceries to food deliveries, essential bill payments to medical tests, digital payments were our lifeline. Cardless transactions, Universal payment interface, and transfer of funds through mobile phones are services that

D. Dasgupta (✉)
Open Group Certified Distinguished Architect, New Delhi, India
e-mail: debanjanad@gmail.com

S. Bhattacharyya et al. (eds.), *Confluence of Artificial Intelligence and Robotic Process Automation*, Smart Innovation, Systems and Technologies 335,
https://doi.org/10.1007/978-981-19-8296-5_3

have become a part of our daily living. There are chatbots, virtual assistants, recommendation engines and many other technology innovations that are continuously helping, guiding, and suggesting how to manage our money.

Hence it is not surprising that when it comes to intelligent automation, banking industry takes a lead. Combining the power of artificial intelligence (AI) and robotic process automation (RPA) to automate business and IT processes, thus achieving higher operational efficiency and customer experience is the goal of top banks today. These technologies have immense potential if used in the right way for the right fit purpose. According to Mckinsey, "the potential for value creation is one of the largest across industries, as AI can potentially unlock $1 trillion of incremental value for banks, annually".

With that context in mind, in this chapter, we will discuss the impact of RPA and AI, the key levers of intelligent automation, in the context of banking industry.

Now in this chapter, we will refer to the term Intelligent Automation often as we speak about the applicability of RPA and AI in various process transformations. So let us first try to get a quick understanding of what Intelligent automation is.

In short, Intelligent automation leverages various levers for automating business and IT processes in an enterprise. This includes automating processes using various technologies like workflow, business rules, AI driven decisions, recommendations, conversations, and automated execution through Robotic Process Automation (RPA) along with traditional integration mechanisms like API integration, message-based integration, and others. It is a convergence of various technologies to achieve complete end to end automation of processes that form the core of the enterprise. AI and RPA are two of the key enablers of intelligent automation.

Intelligent automation driven by AI and RPA along with these other technologies can transform the way banking business is done. That is where the power of these technologies lies. AI led automation supplemented by RPA can drive up bank's revenues, lower operating costs, and improve compliance and customer satisfaction significantly.

In this chapter, we will try to understand this topic on the Impact of AI and RPA on banking along the following themes:

- Understanding the key challenges and causes of disruption in the banking industry
- Nature of transformation needed in these times to meet the challenges
- How AI and RPA are positioned to enable transformation
- What benefits can AI and RPA help banks to achieve
- Few Key use cases in banking that can be transformed with AI and RPA
- How do enterprises get ready to enable these technologies in their ecosystem?

We will discuss real use cases around Contact center automation, automation in Lending and Fraud Operations automation. We will briefly describe the processes and understand why RPA and AI are being considered good levers to resolve specific problems. We will discuss a high-level solution and what benefits can banking enterprises expect to reap with automation based on these levers for these use cases.

This chapter is divided into various sections. The next section will describe the challenges in the banking industry from a business perspective that we see today, followed by a section establishing the need of AI and RPA to address these challenges and enable the transformation. Section 3.4 is the core of this chapter, as we dive deep into three important use cases around lending operations, contact center operations and fraud management in banking. In this section, for each of these areas, we will describe the business problem with a brief overview of the generic process and typical problems associated with that, what the levers of automation are that can be applied to address the problem and outline the solution with high level components specified along with the benefits achieved through the solution. The last section of this chapter addresses the changes that need to be brought into the enterprise starting from operating model, methodology and tools, design and development life cycle of technology projects and the governance and support framework.

There are associated diagrams and references to substantiate and describe each of these sections as we progress through the chapter.

3.1.1 Literature Review

With respect to the challenges faced in the banking industry and the potential areas of transformation that can be achieved with AI and RPA, this chapter refers to various published papers and research articles, analyst briefings and other online scholarly articles. These are listed in the Reference Section for the readers.

There are numerous papers and research articles, analyst reports examining AI and RPA implementations in banking and assessing the impact. Such papers have studied data from banks around the application of artificial intelligence methodology. The result of such studies states that the banks are using various AI services to improve customer needs. There are also research done to discuss the changes that AI and RPA can bring to the banking industry and the impact that will have on human workforce and manpower [1], stating how AI has been rapidly transforming the dynamics of banking and financial services industry. Some of them address interesting points about the ethico-neutral character of technology and its attendant threats like cyber-crimes and macro-financial risks [2]. There is also a question on sustainability as technology tends to replace humans and the related personal touch which most often is the essence of financial services industry [3] thriving on the art of customization and customer delight. There is an abundance of research articles describing the various aspects of the impact of automation on banking, how AI and RPA can be adopted individually as points solutions or as a combination and what are the risks associated with the adoption of each. There are also references from various analyst research on where the industry is heading and what business foresee as their roadmap [4] in the context of Intelligent Automation. These articles have been analyzed and leveraged based on their relevance within the chapter. There is a complete list of references in the Reference section at the end of the chapter.

3.2 Challenges in the Banking Industry

Banks have undergone significant transformations over the years. From completely physical brick and mortar structures with strict and fixed hours of operation to digital banks of the future, there are significant changes that banks are making in their operating model to stay competitive. In these rapidly changing times with rapid technology evolution, banks need to rapidly transform [5] to keep up with the business and technology innovations and the changing customer dynamics [6]. With that context, let us first understand what are the key challenges that banks and financial institutions are facing today. The primary challenges that the banking industry is facing can be listed as follows as depicted in Fig. 3.1.

As listed in the diagram, you can see the challenges are as follows:

- Changing expectations of customer
- Fintech disruption
- High cost to income ratio
- Complex regulatory controls.

These challenges are spread across the various areas and domains of the bank—starting from customer experience to banking operations and external factors to controls and compliance. In the following section, we will discuss the challenges in detail and understand the impact.

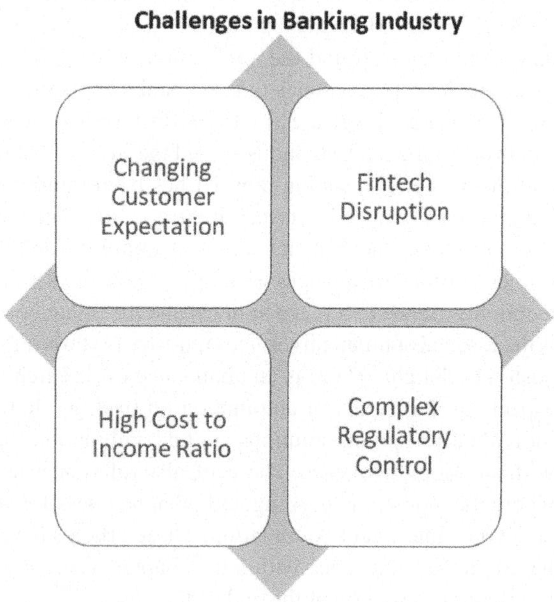

Fig. 3.1 Key challenges in banking

3.2.1 Changing Expectations of Customers

With millennials being a significant part of the customer base and the widespread adoption of pervasive technology, banks are experiencing a sea wave change in the customer expectations of the varying demographics. the expectation from the banks is now to provide a customer focused, engaging experience with real time virtual services and service interactions at a low cost.

With the rapid rise of digitization around us and Covid 19 pandemic accelerating this rise, customers have developed the mindset of "on demand" and "right now" when it comes to services. Banking service is no different from that. People want value for money, they do not care who the service provider is as long as they are getting the services based on their requirements, at the time when they need it. The concept of loyalty is diminishing, as more and more choices from banks to FinTech, from cards to digital wallets are offered to the customers. Customers are looking for a ubiquitous experience which is agnostic of the channel of communications. Be it branch banking or online or mobile app, customers are looking for continuous engagement and seamless transitions. A young home buyer can inquire about a loan through social media, or apply through his/her mobile app with the help of a chat and when they need any clarification beyond the assistance of a chatbot, a customer representative should be reachable. The transition should be smooth and seamless, the application should be submitted within minutes and traceability of status should be provided to the customer.

Apart from usability, experience, quality, and cost of service, nowadays customers are also looking for various value-added services like the ecosystem of partners and marketplace, data driven personalized insights and relevant advice based on life stages and other personalization parameters through the various channels of interaction. A combination of these services would significantly impact customer acquisition and impact.

So, we see that with the change in customer expectations, there is a need for banks to adopt rapid digitization, in a balanced way so that customers do not miss the human connection. Customers are engaged and are provided with the services they need, and they may not need but would be good to have. Providing a superlative customer experience with an optimum combination of human and digital experience is the key, and banks need to strategize and plan for that to enable them to grow in this changing landscape.

3.2.2 Fintech Disruption

FinTechs are financial service providers which typically provide various financial services, bundled together like a bank or individual unbundled services, both of which are enabled by digital technology. They usually offer innovative services and products like wallets, mobile payments, alternative financing etc. In recent times, fintechs have changed the way customers access and manage their finances. With gaining popularity, they have challenged the traditional banks. Many banks have been forced to adopt and accelerate their digitization programs because of the competition from fintechs.

With their novel business models, fintechs bring financial services right at the customer's doorstep, on demand. With large scale digitization, they enable customers to buy, change and transact on the services and products they offer, in an automated and virtual mode. These are very appealing, to not only millennials but any tech-savvy customer as well. With the pandemic, digital payments have gained larger adoption and many fintechs providing such services have seen significant growth in customer base. Many of us have become used to digital payments and mobile wallets and do not think twice before using them in our daily lives.

FinTech's are also characterized by high agility and lower operational costs. Supplemented by latest technologies they can aggregate and disaggregate their offerings based on changes in the market and customer needs. The adoption latest technologies including artificial intelligence, analytics, automation, and others make them flexible and highly efficient at the same time. Not only are they throwing challenges at traditional banks by causing customer churn, but also challenging the operating model of traditional banks by demonstrating optimized cost of operations through higher degree of automation adoption. In order to compete against the fintechs and provide comparable services and products, banks see a strong need to digitally transform and adopt technologies like AI and RPA to be relevant and stay profitable in the business.

3.2.3 High Cost to Income Ratio

Cost to income ratio is one of the key KPIs (key performance indicator) of any bank. It is defined as the ratio of the cost of operations to the bank's operating income and is an indication of the profitability of the bank. The higher the ratio, it means that the bank is incurring more cost in operating the bank than its income and is not as profitable as intended. One of the ways of improving cost to income ratio is by optimizing operational efficiency and thus reducing the operating cost. With limited growth and steady competition from fintechs, banks are looking for strong cost improvement measures to lower the operating cost. Headcount reduction is one of the knee jerk reactions to reducing cost. However, just focusing on headcount reduction may work as a tactical measure to improve costs. In the strategic and long term, banks are planning to optimize and systematically eliminate waste by automating more end to end processes. This includes promoting straight through processing, adoption of self-service channels, creating mechanism for self-healing and auto resolving processes with costly human intervention as the last resort. A key lever for such transformation is through intelligent automation across the front, mid and back offices of the bank. Thus automation, standardization, and simplification of products, services and underlying processes can go a long way in reducing the operating costs of a bank. Be it in account opening or provisioning a loan, improvement in cycle time, reduction in rework or getting it right the first time all of these add up to improving and optimizing the operational cost. You can debate and say that the complexity of the processes in not new and was always there. That is true, but now what impacts the banks is that due to the complexity, the time to market and time to value becomes high and impedes digitization, and digitization is the trend of the hour [7]. A longer cycle time in basic services like account opening or loan approval not only cause customer dissatisfaction and customer churn but also incurs cost and revenue leakage. These are just a few examples, there are many other opportunities across the various business processes across the bank that have the potential for significant savings and optimization. All of these add up to result in significant cost savings. Additionally, with advanced technologies like analytics and artificial intelligence [8] to recommend relevant products to customers and automation driven digitized operations to operate the bank seamlessly, banks can improve their margins and return to growth. Thus, technologies like AI and RPA, the key enablers for intelligent automation can contribute significantly in optimizing the operating cost.

3.2.4 Complex Regulatory Compliance

Another important factor that is impacting profitability in banks, is high cost of regulatory compliance and penalties associated with it. Analyst's reports show that banks end up spending almost 10% of their operating cost in compliance. Post the global financial crisis in the previous decade, banks and financial institutions face very strict laws and regulations. To adhere to these, banks need to invest larger amounts in human resources, technology, and outsourced services to manage the regulatory compliance. In cases of non-adherence, the banks face large amounts of penalties by the regulatory authorities. Top global banks like UBS, Barclays, NatWest (formerly RBS) and others have faced large penalties. Closer to home, Reserve Bank of India has imposed large fines on banks like Punjab National Bank, and ICICI Bank [9] for deficiencies in regulatory compliance. These costs add up to the overall expenses and dent the cost to income ratio further.

Maintaining a large pool of skilled compliance officers who will read through every document and contract and raise alerts to remediate on time so that compliance is ensured, is a very costly and time inefficient process. On top of that, there is a good probability of errors creeping in, as the volumes increase, and the officers are loaded with work. Development of compliance programs with the help of technology, that improve efficiency in regulatory compliance operations [10] and improve accuracy, can help banks in significantly improving their operating cost margin. Using technologies like artificial intelligence and RPA can not only help in ensuring the right controls are enabled in the business processes but can also help in accelerating responses to regulators which often draw penalties for delay.

3.3 Need for Artificial Intelligence and RPA in Banking

From the above discussion, it is apparent that to counter the key challenges faced by the banking industry today, technology adoption to enable digital transformation is important. Digital transformation with automation as one of the key levers is one of the key measures that banks must take. Technologies that can help automate processes, mimic labour-intensive human activities, and augment human decision taking abilities are of extreme value in managing and countering the challenges that they are facing today. Technologies like artificial intelligence and RPA, along with workflow, decisions and others that constitute intelligent automation need to be adopted by banks and financial institutions.

Thus, using the challenges of cost reduction, changing customer preferences and fintech competition, as catalysts, transformation in operations and transformation in customer experience becomes a strong imperative for banks and financial institutions. Through a strategic focus and structured approach, banks can promote and achieve cost reduction, profitability and growth enabled by technologies like artificial intelligence and robotic process automation [11]. The adoption of these breakthrough technologies have an immense potential to automate end to end processes [12] with intelligent automation and generate substantial cost savings and promote growth. In the following sections, we will discuss, how artificial intelligence and robotic process automation are key to achieve end to end automation in automating the banking processes and what benefits do they bring to the bank.

To start with, let us first understand, what are the key characteristics of these technologies and how these can be beneficial. The following two sections will talk briefly about RPA and AI, and discuss their characteristics and capabilities that make them key in achieving intelligent automation in an enterprise. This will form the basis of our discussion in the later part of the chapter, where we will discuss real life use cases in banking that can be automated by RPA and artificial intelligence and the impact they have on the enterprise.

3.3.1 Robotic Process Automation (RPA)

Robotic Process Automation is a way of automating business and IT processes through programmable software components called 'bots'. The bots [13] are programmed to perform a set of tasks and activities mimicking the activities of a human worker.

The key characteristics of RPA are:

- Programmable
- Rule-based
- Mimic human activities
- Can work in existing IT ecosystem.

These characteristics make RPA, a technology that has now become a quick fix to automate business and IT processes. They are not too complex to code and work well when the process execution follows standard rules and activities. They can mimic human activities like matching data between screens, data entry, copying data from various systems and creating an output and others. They need to be explicitly programmed for any activity that it is made for and do not have intelligence or self-learning capabilities. One of the important characteristics [14] is that they are non-invasive and can work with the existing IT systems without the existing systems having to undergo any change.

Now when it comes to automation, not all processes are best automated with RPA. There are certain characteristics of processes that make them fit for RPA based automation. These are:

- Processes are repetitive and rule-based
- Underlying data the process operates on is structured
- Process is operated in a digitized environment.

While selecting processes for automation by RPA, these characteristics should be evaluated carefully, to ensure the feasibility of automation. This is important because choosing the best process, which is fit for RPA based automation, ensures that the maximum benefits of RPA [15] based automation is achieved.

Now let us briefly touch upon the key benefits that can be achieved through RPA based automation.

The key benefits are:

- Saving manual labour and cost
- Increase in accuracy and consistency of outcome.

Improvement in cycle time and customer satisfaction

These benefits are intertwined and related. As you can see, they directly contribute to the overall business benefit [16, 17] that can be achieved through RPA.

For more information on RPA and how it is positioned in the context of Intelligent Automation, refer to the references cited at the end of the chapter.

3.3.2 Artificial Intelligence (AI)

Artificial Intelligence is a very broad area, where we develop computer systems to imitate and augment human intelligence. This can include learning, problem solving, decision making, image recognition, language interpretation and many others. We will not discuss artificial intelligence in detail in this chapter, rather focus on some of the key characteristics and benefits of the technology [18, 19].

The key characteristics of artificial intelligence are:

- Augments human intelligence though recommendation, interpretation, decisioning and learning

- Works on structured, unstructured, and semi-structured data
- Can be costly and complex to build and operate
- Need to be trained but not explicitly programmed for all conditions and variations of the data.

These characteristics make artificial intelligence, which includes natural language processing, image recognition, recommendations and decisions, emotion analysis etc., a technology that has now become a key lever to automate business and IT processes end to end. The benefits of applying artificial intelligence to make the process intelligent can be significantly high if done appropriately i.e., choosing the right process, having good, relevant, clean data, optimized model and skilled resources. However, implementing artificial intelligence solutions to automate processes can be complex and costly [20], due to the inherent complexity of the technology [21], necessity of good quality of data and skilled resources needed to build the models.

Just like we saw in RPA, not all processes are best automated or made efficient with artificial intelligence. There are certain characteristics of processes that make them fit for automation based on artificial intelligence solutions. These are:

- Underlying data is unstructured which needs to be interpreted and understood to process and derive decisions
- Requirement of a system that can learn from data and derive insights
- Identification of patterns, image recognition, and language interpretation are key requirements
- Need for conversation and disambiguation through man–machine interaction.

There are other characteristics of AI as well, the ones listed above are important in the context of intelligent automation. While selecting processes for automation by artificial intelligence, these characteristics should be evaluated carefully, to ensure the feasibility of automation. This is important because choosing the right process which is fit for artificial intelligence and its wide range of capabilities for automation ensures that the maximum benefits are achieved. The benefits from artificial intelligence are almost similar to the benefits of intelligent automation [22] improvement in efficiency, accuracy, customer and employee experience and cost reduction. For more details on AI in the context of Intelligent Automation [23, 24], refer to the references section. Let us now analyze, why these can play important roles in resolving some of the problems of banking industry, that we discussed in Sect. 3.2.

3.3.3 Role of Artificial Intelligence and RPA in Banking

Based on the above discussion, it is evident that adoption of large scale digitization and automation programs with RPA and AI(supplemented by other technologies like workflow and integration) can help in resolving the core problems that banks are facing to a considerable extent. Beating the Fintech's in the digitization game or providing supreme customer experience to millennials and beyond are some of the things that can be made possible by the combined power of RPA and AI enabling Intelligent automation. Having a robust compliance management framework that can automatically embed and check the controls in the business process, trigger alerts in violation, and monitor frauds and thefts will not only optimize the regulatory operations [25], but also help in improving the brand value, and reduce churn and improve customer, employee, and partner experience for the bank.

From onboarding a customer to completing the KYC, applying for a loan to the credit decision, most of the processes in a bank are complex, long running, and need manual intervention and a lot of data analysis and validation, to put it very simplistically. Many of these processes and subprocesses need SME (subject matter expert) knowledge, human intelligence, and analysis to decide the best course of action. Others might need complex data manipulation, consolidation and reconciliation from multiple sources making the task immensely labour-intensive and prone to manual errors. Take the example of a reconciliation process in banks. The reconciliation process [26] is a daily end of the day activity for banks to reconcile all the transactions. At a high level, this process will involve checking all the transactions of the bank—withdrawals and deposits, matching them with the ledger of payables and receivables and finally evaluating and matching the balance. This process is typically done by a back-office team, who spend their time matching ledger entries with the transactions, flagging any discrepancies and checking the balance in the books. This is an extremely repeatable and rule-based process and automating this with help of RPA [27] and integration between various systems is feasible. Not only will this reduce the manual effort and reduce the probability of manual errors, but also lift the morale of the employees and improve their experience. They can now be deployed to more value-added tasks like analyzing the discrepancies flagged that cannot be auto resolved.

Think of the scenario of the customer care representatives. Most common enquiries to the bank would typically be around their balance, withdrawals, and deposits or complaints about a product or service. The customer service representatives respond to queries and resolve them. However, the performance is dependent on their skills and knowledge. A senior and experienced representative will resolve a query within minutes over someone who is new to the job. A voice enabled chatbot with intelligence built in, can converse with the customer on call or through a portal/mobile application, understand and interpret the customer's ask and fetch relevant data from the back end to display or read out to the customer. The chatbot once trained will have uniform performance and will be available at any time of the day and can be scaled to handle peak loads as and when necessary. A voice enabled chatbot will thus help to ease the bulk of the customer representative's workload and enable him/her to focus on the more difficult customers and complex problems.

Even complex processes like lending, retail and commercial, have a lot of potential to use AI and RPA to help make important decisions like deal recommendation, automatic credit analysis etc. These processes typically involve analysis of a lot of data and documents from the commercial or retail borrower, complex evaluations on funding decisions and credit worthiness of the borrower and the project/property/item to be financed before finally deciding on the credit approval. Additionally, commercial loan amounts are large amounts, has higher risk and are of high value for banks. Automating such processes to enable straight through processing [28] for even a fraction of the total applications and flagging of any irregularities appropriately can help banks highly optimize their operational efficiency [29].

Thus, the combination of AI and RPA, to completely automate processes end to end, front to back, have the potential to completely transform the banking operations. RPA can be leveraged to automate the rule-based and labour-intensive repeatable activities while artificial intelligence with its wide-ranging complex processing capabilities [30, 31] can provide conversational AI in the form of chatbots, recommendation engines, automated decisions (like risk assessment), detecting fraud and money laundering activities. Artificial Intelligence can augment the human intelligence and analysis and do the thinking on behalf of the human resources and RPA can be the arms and legs and do the execution. Both together, glued with an orchestration engine has the power to address most of the challenges that banks are experiencing today.

These are just a few examples and the possibly is huge if you think of the whole banking ecosystem with business and operations. Realizing the importance of the potential of this combination, banks are now considering large scale intelligent automation deployments in their enterprises, powered by AI and RPA, as a game changer in their transformation.

Let us now dive into the details of a few use cases, to understand the power of AI and RPA in banking. Through these use cases, we will attempt to give an idea about the complexity of the business problem and how AI and RPA can be applied to resolve it.

3.4 Artificial Intelligence and RPA Use Cases in Banking

In this section, we will discuss three important use cases where RPA and AI can be applied to improve the process metrics by automating it and thus improve the efficiency of operations. For each of the use cases, we will briefly describe the problem statement and define a high-level solution with AI and RPA. We will also touch upon the potential benefits of the solution.

3.4.1 Use Case 1—Contact Center and Usage of Chatbots

Contact center is one of the key areas of customer connect for any business. A good experience at the contact center reflects on the NPS score and overall customer experience. This is also an area where due to the direct contact with the customer, there are opportunities to create leads for additional business and upselling products and services. The area is operated by customer service representatives who usually

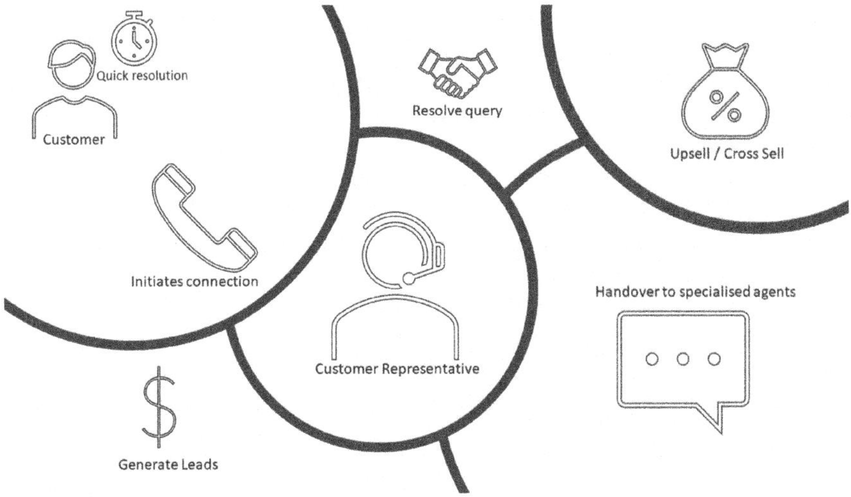

Fig. 3.2 Customer service representative ecosystem

handle high volumes of calls. The work can be repetitive and monotonous, and they need to access the right information very quickly to provide value to the customers. So, it comes as no surprise, that one of the key pain points in this area is how to get the query resolved at the first interaction with the customer, how to provide the same quality of service to all customers - day or night and how to provide personalized service and prevent employee churn despite the high workload? That is where a chatbot or a virtual assistant comes into play (see Fig. 3.2).

Do you know what is common between Amy, Erica, Cora and AmEx [32]?

Well, they are the chatbots of some of the top global banks. Amy is the chatbot for HSBC, Erica for Bank of America, Cora is for RBS and AmEx is for American Express. Being early adopters, many global banks have already deployed chatbots and virtual assistants successfully to manage customer interactions [33] and provide superior customer service. The below diagram is a thematic representation of a life of a customer agent who interacts with the customer typically over a phone call or may be text chat/email, and resolves queries with a potential to generate leads and earn revenue for the bank. A chatbot, who is a digital customer representative [34] in this context, works in the same ecosystem. It is a part of the digital workforce, can scale, does not need holidays, or fall sick and there is no churn.

Chatbots are one of the prevalent applications of artificial intelligence, especially in the banking industry. Imagine how many times we have called the bank for simple issues like lost card, blocked account, simple account balance or how to change my name in the account and similar other activities? To get such queries resolved, it needs time from the customer end as well as from the bank's perspective. You might have to wait till you get to talk to a representative. Within the bank, there is a team of customer representatives who are responding to such queries day in and day out. A system that could automatically respond to the customer and guide and direct them to the appropriate action would have manifold benefits. It puts the customer in control. He/she can get the matter resolved according to their time and take appropriate actions, thus enabling self-service. It also allows to take off significant load from the customer representatives so that they can focus on the real complex negotiations with customers.

So chatbots are one of the efficient means of automating services in contact center operations. With conversational AI, chatbots can converse and communicate with customers, interpret the intent of the conversation, and resolve their queries or direct to the appropriate resource–human or digital. Digital assistants and chatbots have the potential to enhance the omnichannel experience and can transform customer service to a large extent.

Let us now focus on the technology behind a chatbot that imparts the intelligence to conduct human like conversations and how the chatbot solution will look like.

Technology Enabling a Chatbot

Chatbots can be of two types–simple rules-based ones and AI enabled ones. Rule-based chatbots typically operate based on a menu of fixed options and can execute a set of pre-defined tasks. Any question or option that is not within the pre-defined menu cannot be addressed by these chatbots (see Fig. 3.3). These are typically key word based and are usually meant to address a limited set of queries for a specific set of users around very defined and limited topics, they do not scale in terms of usage and applicability across larger domains of service.

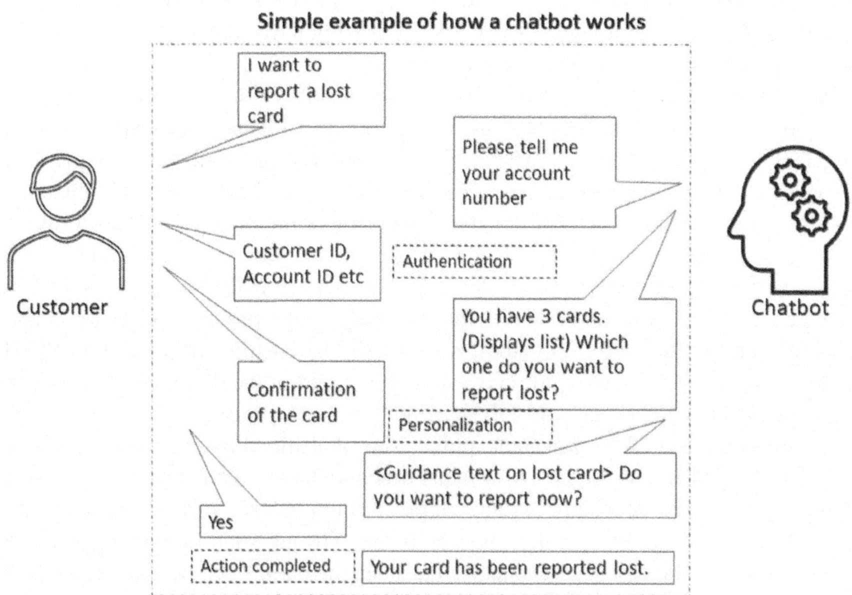

Fig. 3.3 Communication between a customer and chatbot

The other kind of chatbots are based on artificial intelligence. These are the ones that can potentially achieve the benefit and impact we discussed in the previous section. These chatbots derive their intelligence from machine learning models which allow them to interpret and understand the queries, extract relevant data from the conversation and apply them to find the best answer to resolve the query.

The key characteristic of a chatbot is natural language-based conversation. We do not have to use domain specific keywords or specific codes to converse with the chatbot. It is as natural as a human-to-human conversation. It can ask questions to reduce ambiguity and interpret the customer's query perfectly.

Let us now understand how artificial intelligence makes this possible.

Here the intelligence in the chatbot is achieved through natural language processing capability. Using natural language processing algorithms, the intent of the query or conversation can be understood and through entity recognition algorithms, relevant data form the conversation can be extracted. A combination of these data will give the exact query that the customer wants to ask and relevant data that is necessary to execute the query and find the best resolution. The data can further be used to trigger transactions that will resolve the problem customer wants to resolve. The following diagram indicates the main high-level components of a chatbot followed by a brief description of the component.

Let us now describe the components one by one (see Fig. 3.4).

Chatbot UI

This is a user interface through which the conversation happens between the customer and the chatbot. There are various scripting languages and frameworks that you can use to build the chatbot UI.

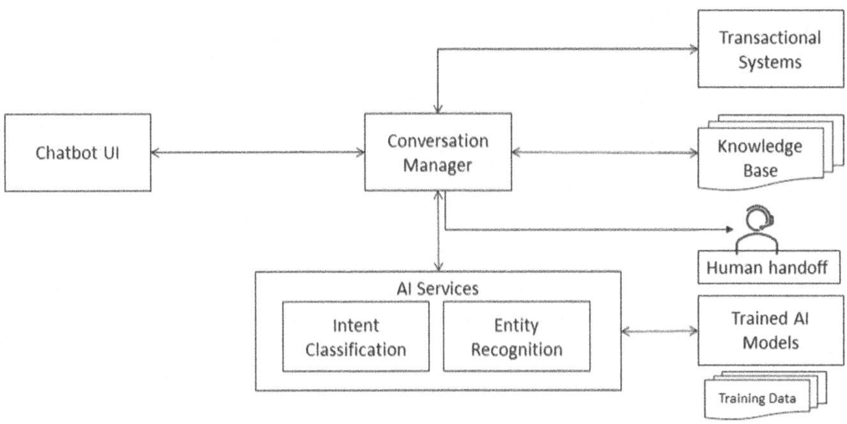

Fig. 3.4 Components of a chatbot solution

Conversation Manager

This is the component that controls the conversation flow and orchestrates the interaction. It receives the inputs from the user, invokes the AI services, processes the output from the services and uses it to fetch the right resolution data. The output from the AI services are the 'intent' and 'relevant data' associated with the intent. For e.g., from the above conversation example, the intent is "report lost card" and data is "Account No" and "Card Number". Based on this information, it goes to the knowledge base or the corpus to find the right resolution and displays it to the user. So, the user will see a guidance text as to how to report a lost card. In case, the user wants to proceed with the action of reporting, the transaction related to the resolution execution (reporting a lost card), the chatbot will invoke the bank's transactional services to complete that.

AI Services

There are two main services that are shown here. One is intent classification whose purpose is to identify the right intent of the customer query and the second one is entity recognition, which will extract the right entities from the conversation. Typically, the services are the components that encapsulate the core algorithms for artificial intelligence.

Intent classification works to automatically classify the text and categorizes it into intents like account closing, lost card etc. The service uses a machine learning model to identify the intent used in the customer conversation and train it based on the data. There are various Natural language processing algorithms that can be leveraged to create the intent classification models. Natural language processing uses computational linguistics which is the rule-based modelling of the human spoken language with intelligent algorithms such as statistical, machine, and deep learning algorithms. It typically uses keyword extraction, topic modelling, knowledge graphs, entity recognition, sentiment analysis, lemmatization etc. to correctly interpret and identify the intent. There are many service providers that provide such pre trained language specific and domain specific models and related AI services which can be leveraged. There are also many chatbot frameworks that can be leveraged to build a chatbot very quickly. Additionally, custom models can also be built and retrained with enterprise specific data. The custom models may have higher precision and accuracy, but it also involved cost and time to build.

Entity extraction is a text analysis technique that can extract relevant data from unstructured text and can classify it into pre-defined categories. Similar to intent classification, there are prebuilt services for entity extraction in the market, which can be leveraged, or custom models can be built with enterprise data for better accuracy of the model. Named entity recognition plays a significant role in natural language processing and can automatically provide the syntactic analysis of the conversation.

Knowledge Base

This is the corpus and information source from where the answers to the customers queries will come. These could be documents or text extracts with procedural guidance to resolve the queries. This needs to be built and collated from existing resources in the bank, diligently, for the chatbot to be effective and revised frequently as new queries come in or the procedures change.

Transaction Systems

Transaction systems are typically the systems or services that are leveraged to initiate and execute transactions in a bank.

Human Handoff

There is finally a handoff to connect to a human agent in case the chatbot is not able to respond to the customer query. In such cases, the query is logged and revisited for evaluation of any refinement to the model or the corpus.

Benefits and Impact of a Chatbot

So why should banks invest in chatbots? What are the benefits it can bring to the bank?

There are a few key benefits that can be achieved through chatbots in banking operations.

Customer Experience Improvement

With chatbots, customers do not have to wait in line and can access them from anywhere anytime to resolve their queries and address their issues. The enablement of self-service through chatbots puts the customer in control of the interaction and improves the cycle time of the issue resolution. There is also a certain degree of personalization that is usually brought in through the chatbots. The actions and recommendations are based on the customer's data and profile which enables the bank to offer a tailored customer experience based on the requirements of the customer. This genuinely impacts the customer experience and promotes he brand value of the bank positively.

Improve operational efficiency with Cost Reduction

This benefit is obvious. The chatbots can reduce the workload of the customer service agents by addressing the customer queries and proving guidance to resolve them or triggering the resolution action itself. Chatbots can interact with customers and assist in all stages of the customer lifecycle–be it onboarding, account maintenance or special events in the customer journey. Deploying a chatbot with relevant and accurate corpus and knowledge base can result in significant reduction of customer services agents, who can now focus on resolving the complex queries and negotiations, thus adding revenue to the bank.

The Future with Chatbots

So why should banks invest in chatbots? What are the benefits it can bring to the bank?

Chatbots as virtual assistants are gaining ground and many banks have already implemented chatbots or are in the process of deploying one soon. However, chatbots can be taken beyond just conversational AI in further innovative modes to reap the maximum benefit. These are some of the additional potential uses of chatbots that banks can leverage.

Extreme personalization: Chatbots can greet customers (in their language) and based on their publicly available data (from social media), they can detect life events and recommend various products and services. Based on the mood of the customer, that the chatbot can detect through conversations, it can roll out offers that are completely customized (movie ticket, dinner voucher or larger items like retail loan discount, tips on entrepreneurship etc.).

Virtual Reality with Chatbots: Taking personalization further, through virtual reality, it can show the impact of a Fixed Deposit or a Residential Loan within the customer's context.

Completely touchless transactions: Chatbots can authenticate customers based on biometrics and enable completely touchless transactions–through voice enabled chat and biometric authentication.

So as banks ride on the trend of digitization, chatbots are one of the most beneficial areas they can focus on to improve the KPIs around customer experience and operational efficiency very quickly. As the technology advancement continues, chatbots with advanced capabilities will become key components in driving operational efficiency and customer satisfaction.

3.4.2 Use Case 2—Loan Processing

Let us now dive into the details of another key banking process which is Loans. Loans are one of the key products that banks provide and is one of the areas which generates a lot of revenue for the bank.

Loan Process Overview

Loans are a key business of the bank, and it is very important that the operations around loans–starting from origination to disbursements are done with supreme efficiency (see Fig. 3.5). This is one area that generates revenue for the bank and hence it is imperative to have high operational efficiency along with compliance adherence while providing excellent customer experience.

Let us start with understanding the loan process. The figure below gives a high-level view of the loan process in a bank. This is a generic process and there may be specific variations when it is realized for a specific bank. The idea here is to give an idea about the basic activities that happen during loan processing in a bank.

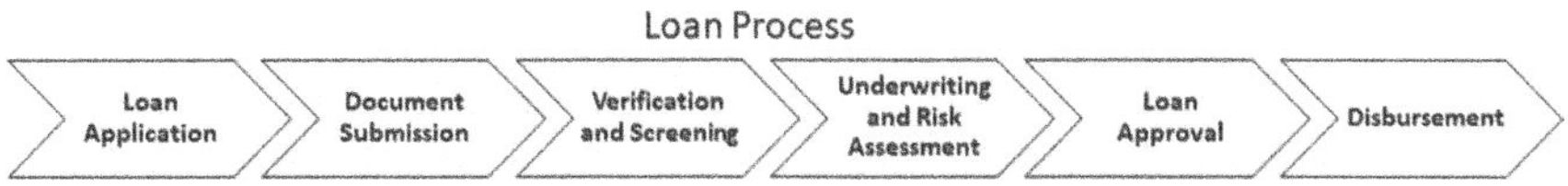

Fig. 3.5 A loan process flow

Typically, the loan process includes filling in a lot of forms by the customer, which then gets verified by the bank. The bank does a screen and risk assessment to evaluate the credit worthiness of the customer, post which the loan is approved, and funds are disbursed.

As discussed in the previous sections, intelligent automation enabled by AI and RPA are key levers to automate processes–front to back and transform business operations. In the following Fig. 3.6, we have indicated the potential levers that can be used to transform the loan process. These are some of the common mechanisms and there may be many other ways to transform the process with other technology levers based on the requirements and capabilities and constraints of the customers technology landscape.

The application of technology, like AI and RPA can radically transform the process. At each stage of the process, these various technology levers have the potential to improve the key KPIs of the process like cycle time and improve overall efficiency if used correctly and supported by good, clean, relevant data.

Technology Solution for Automated Loan Processing

Let us now traverse the process from start to finish and understand the applicability of the technology and the reasons around it.

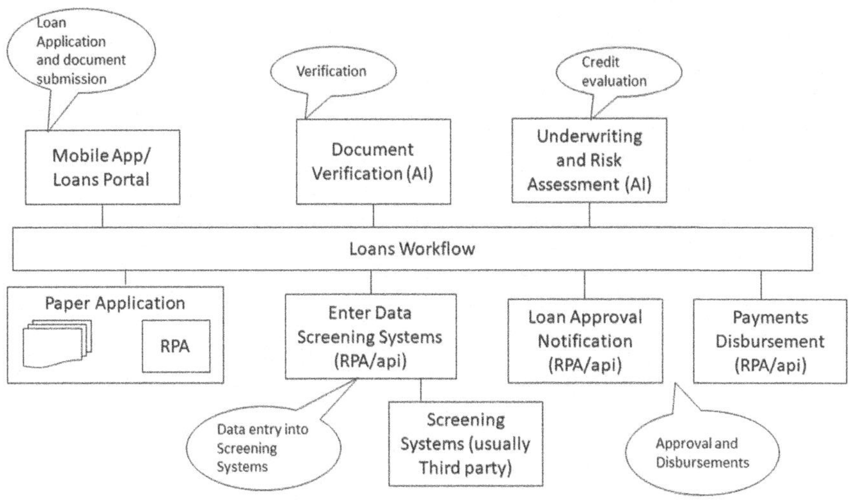

Fig. 3.6 Levers of automation for the loans process

In this figure, we have shown a workflow which orchestrates the flow and manages the states of the loans process. That is the recommended way in which a typical loan process should be implemented. In the following discussion, we will not delve into the workflow and consider it as an assumption. We will focus on the applicability of AI and RPA in this process automation.

Loan Application and Document Submission

In loan application and document submission subprocess, the customer submits the loan application along with the necessary documentation. This a document heavy process. In many banks there are online portals or mobile apps through which you can submit a loan application and upload digital documents. However, there are still many banks, where only a part of this process is digitized, and they still reply on paper submission of forms and documents. In those scenarios, there is strong applicability of RPA, where the paper document data can be extracted by OCR (Optical character recognition) technologies and then copied into the bank's loan systems by an RPA. The data entry is a manual process, prone to errors and applying RPA on this process will not only improve accuracy but also the processing speed of data entry for the applications. Since the application documents have a standard format, there may not be a need to use artificial intelligence here. Instead, a combination of OCR and RPA can be sufficient to convert the paper applications into digitized data and recorded into the systems of record. However, that needs to be assessed during a feasibility study. Additionally, some of the top RPA products available in the market have built in OCR capabilities that can also be leveraged.

Validation and Screening

In the next subprocess, the purpose is to validate the documents submitted and verify the claims the customer has made regarding their identity, income and source of funds, purpose of loan and other validations (see Fig. 3.7). To validate data provided via documents, firstly we need to extract the right data from the documents and interpret what it means. For e.g.–if the document attached is a driver's license, then we need to extract the name, driving license number, expiration date etc. from there and then validate all these data against a trusted source–like the motor vehicle database of the transport department in the government. For this, there is a need to employ artificial intelligence to do the document processing and entity extractions before we can validate the data. Typically, documents will be in the form of images or pdfs. For the data extraction, we need to extract the text from these source images and pdf files. Keeping in mind the variety of documents and not so standard formats, we will need to first use OCR, and then use segmentation and text processing techniques like text retrieval, paragraph retrieval, key value pair extractions and other specialized techniques. This will give us a text version of the documents submitted as pdfs and images. Then we will apply text analysis techniques like named entity recognition, key word matching and custom annotations to identify the meaning of the texts. For those parts of the text that cannot be matched to entities directly, we can use custom annotations to help with the manual processing. The figure below shows a component view of the AI component that can be leveraged here.

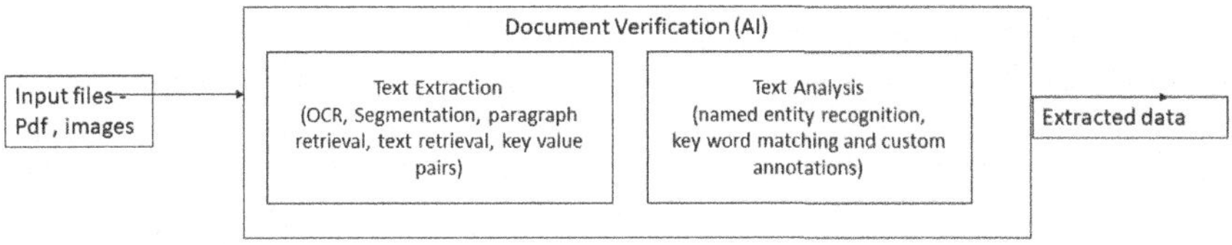

Fig. 3.7 AI component for document verification

Post the data extraction, the extracted data from the documents are verified against trusted sources like the bank's internal systems, government databases and other trusted third-party systems (see Fig. 3.8). Usually, these are system integrations—real time or batch depending on the provider systems. Screening of customer profile and transactions is slightly different. Typically, there are specialized products and systems which do screening by consolidating the data across various public and private sources—some online, some offline. Bank agents must enter data into the screening systems and submit the data. The screening systems get back with a consolidated report on the customer, highlighting any alerts or outliers in the customer's profile review. The activity of entering the data into these screening systems, (which can be multiple) is tedious and labour-intensive. This is where RPA can play a critical role. Being third-party systems, they may or may not provide integration capabilities though apis or otherwise and RPA becomes an ideal choice. Also, these systems are typically hosted inside the client's network, on premise in most cases because of the data sensitivity. So, RPA is the most non-invasive and cost-effective way to automate the tedious data entry into the screening systems. This reduces the manual errors as well as speeds up the process significantly. The adjoining figure depicts the components for Verification and Screening sub process.

Underwriting and Risk Assessment

The next subprocess is one of the very important sub processes of the loans process. It is the evaluation of the credit worthiness of the customer and involves analyzing a lot of data, from both internal and external sources, to assess the risk associated

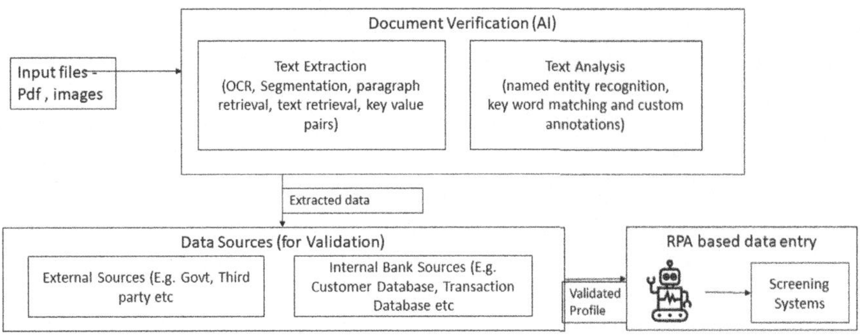

Fig. 3.8 Data extraction and validation solution

with lending money to the customer. The data is consolidated from various sources and analyzed to find any patterns of fraud or money laundering or other criminal activities, The data is also analyzed to predict whether the customer's source of wealth and finance over the loan duration is sufficient to make him/her an eligible borrower and various other dimensions of analysis to assess the credit worthiness of the customer. Usually, a risk rating is also associated with the customer. Thus, to transform this process, a combination of AI and RPA could be leveraged. RPA bots can consolidate the data from various external and internal sources, pulling related and relevant data across various screens, spreadsheets, or crawl portals to fetch the relevant data. This data can then be cleansed and used to predict a risk assessment score. Some of the common AI algorithms to build a risk assessment model are K-Nearest Neighbours, Logistic Regression, Decision Trees, and Neural networks. There are also various service providers that encapsulate the complexity of these algorithms and provide you with services that you can use to create and train your model for risk assessment and credit evaluation. The RPA and AI combination greatly helps in assisting the underwriters to do the credit evaluation.

Loan Approval and Disbursement

The next two subprocesses are loan approval and payment disbursement. These are simpler subprocesses which need recording the approval based on all the analysis done so far, notifying the customer, and disbursing the payments. They can typically be automated through the workflow and api integration or RPA, depending on the requirements.

There can be a debate on what is recommended, between api integration and RPA. We have seen that many banks have started to consider RPA as a tactical approach–something that is a quick fix and temporary whereas the API based integration is more strategic and long term. Generally, API integrations to automate data transfers between systems are recommended but they may need longer time for analysis and build, especially in a federated IT environment. The decision to go for RPA or API integration should be decided on a case-to-case basis, considering the factors of speed, effort, cost and benefit derived.

So, from the above discussion, as we traversed the loan process, we saw that by applying AI and RPA along with workflow and integration, the loans process can be radically transformed and there is a strong potential of significant improvement in the KPIs. The intelligent automation as described above will promote straight through processing, reducing cycle time and improving customer experience. The impact of applying these technologies thus is very apparent and directly influences the key KPIs of the loan business.

With that let us now dive deep into the third use case around fraud operations in a bank.

3.4.3 Use Case 3–Fraud Operations

The third and last use case we will discuss in this chapter us around fraud detection (see Fig. 3.9). This is another core area of banking operations and is very cost intensive. Financial frauds and scams are unfortunately widely prevalent these days. Most of us have received messages or phone calls asking for OTPs and passwords in the context of account expiry or mobile deactivation. Fraudsters use customers' accounts or cards or online banking credentials to authorize payments from the customer's account to themselves without the knowledge of the customer [35]. Examples: Identity theft, stolen debit, or credit card credentials etc. Scammers trick or convince customers to authorize payments from his/her bank account to the fraudster's account or make customers compromise the security of the account. Examples: Phishing, Phone Scams, Social Media Scams etc.

As digitization is on the rise and there are high volumes of digital payments, frauds and scams are posing a new threat to banks and their customers. There is a report on key findings and insights from PwC's Global Economic Crime and Fraud Survey 2020, which states that a whopping $42 Billion [36] has been lost to fraud related crimes in 2 years. Another report from ukfinace.org states "Unauthorized financial fraud losses across payment cards, remote banking and cheques totaled £783.8 million in 2020". Now, that is a lot of money and that can impact the performance of financial institutions hard. The report also goes on to say, that the earlier you invest money in tackling fraud, the better is the return. However, despite this imminent threat to financial institutions, not enough is being done. A KPMG survey found that the costs associated with fraud are increasing at a far greater rate than the spends associated to prevent them.

When it comes to Fraud operations in a bank, there are two dimensions. One is responding to a fraud attack and the second is prevention of fraud. In both, one of the key enablers in efficient fraud operations is data and technology. This combination can effectively battle fraud by detecting patterns, identifying outliers, and establishing traceability between the money trails and can significantly improve and augment the human intelligence while detecting and handling frauds [37].

Fraud operations in banks have well defined functions and sub-functions in line with the type of frauds that occur. Fraud has many sub-functions such as Scams, Card Fraud, Non-Card Fraud, delayed payments etc. Typically, frauds are identified and reported through channels like impacted customers, detected and alerted by bank systems (manual or automatic) or are detected by third parties like audits and whistle blowers. Out of these, most of the cases are typically reported by customers.

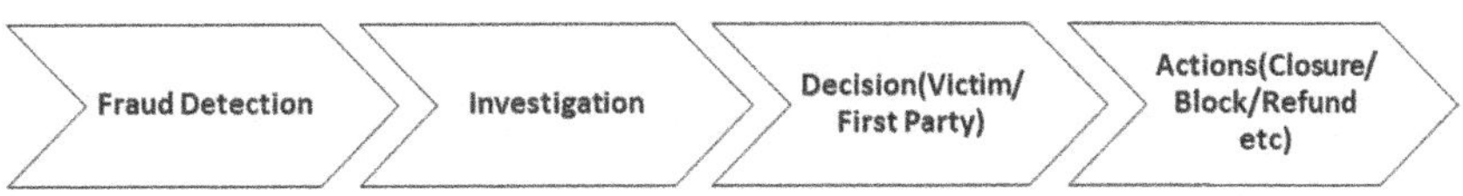

Fig. 3.9 Fraud operations process

The fraud operations process at a high level includes preparing and collating data from various source systems and external agencies and then analysis to find patterns and relationships between the data, and identify anomalies and outliers to detect the fraud. It also includes responding to alerts and investigating transactions and profiles for abnormal activities and identifying any financial crime. Investigations can be extensive, involving internal and external parties and processing and analyzing data to establish the financial trail. Post the investigations, there may be various outcomes, positive and negative. There may be reversal of transactions or refunds given to customers, or there may be closure of account and initiation of arbitration against fraudulent parties. The following diagram shows a high-level flow of the fraud operations.

In this overall process, the maximum potential of improving efficiency and KPI with technology are around the fraud detection, investigation, and decision. This is a good case of combining the power of RPA and AI together (see Fig. 3.10). The RPA bots can be very effective in opening various systems, external and internal and preparing the data for investigation. RPA will be very effective in copying data between screens, matching data between systems, cross referencing, and reducing noise in the data which are essential to prepare the data to build the predictive models. Once the data is prepared, we can use artificial intelligence–machine learning models to identify patterns and relationships to help aid the investigation and arrive at the decision of the fraud investigation. This is where Machine learning with clustering techniques, classification, and regression algorithms can be used to find patterns, associations and relationships, anomalies, and outliers. In this case as you can see, a combination of both supervised and unsupervised machine learning algorithms works well, to detect unknown patterns and relationships in the data.

In this use case, we are not getting into the details of the components, you can refer to the books and references mentioned at the end of the chapter to get more information.

Benefits of AI and RPA in the above three use cases.

By now, you have got an idea of the immense power of AI and RPA in banking processes. They have the potential to transform the processes and radically improve the efficiency and speed. Since technology is driving the process operations, the scalability will also significantly improve.

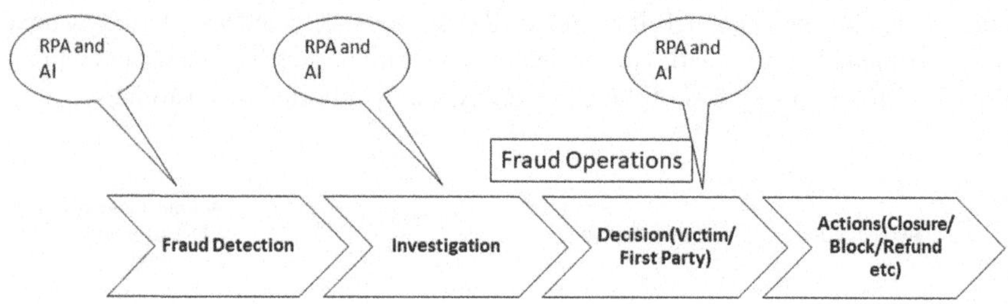

Fig. 3.10 Levers of automation in fraud operations process

Let us now summarize the key impacts and benefits those financial institutions can achieve by leveraging the power of Intelligent automation enabled by artificial intelligence and RPA.

- Speed—Using RPA and AI with machine learning to detect frauds or assess risk during underwriting and credit assessment will increase the speed of operations. Machine learning will aid in detecting patterns and relationships in minutes as opposed to long periods of human analysis. Fraud detection systems which employ machine learning with supervised unsupervised learning algorithms to detect known and unknown patterns will cast a very wide net to detect most fraudulent activities. This will be way more efficient and faster than rule-based fraud detection that most mature banks have today. Conversational AI in chatbots will enable complete self-service in contact centers.
- Scalability—Employing technology like AI and RPA in banks improve their operations scalability. There will be obvious improvement in speed and accuracy, banks would be able to process more workload by leveraging these technologies than by using human workforce alone.
- Efficiency—Efficiency improvement will also happen as there will be an increase in accuracy, decrease in rework, improvement in first time right cases and greater straight through processing. All these will improve the efficiency of operations and positively impact the cost to impact ration.

Let us now summarize the three use cases we have discussed and compare the before and after scenarios in the transformation journeys of customers and bank employees. This will help us understand the impact that these transformations that AI and RPA have brought about as depicted in the following table.

Impact of AI and RPA on the use cases discussed

Use case	Before transformation	Post transformation with AI and RPA
Contact center	1. Manual handling of calls 2. Fetching of data from various sytems to resolve customer query 3. Call handling time and resolution is dependent on the knowledge and skill of the service representative	1. Automated handling of calls by the chatbot (speech capability can be added) 2. Chatbot can automatically fetch data from various source systems based on the customer's identity 3. Uniform and standard experience and cycle time 4. Self-service and quick resolution

(continued)

(continued)

Use case	Before transformation	Post transformation with AI and RPA
Lending	1. High volume of documents to be processed 2. Manual data entry into various systems 3. Complex analysis of variety of data to identify risk and credit assessment by underwriters 4. Lack of uniform notification at important milestones of the life cycle 5. Manual reconciliation	1. Automated document processing with AI enabled data extraction and processing 2. Automated data entry by RPA 3. AI led risk assessment to augment human analysis 4. Uniform and standard notifications 5. Higher % of straight through processing 6. Automated reconciliation
Fraud operations	1. High volume of data to be collated from variety of external and internal sources 2. Tedious work of identifying anomalies, patterns, and relationship between data 3. Rule-based fraud detection (limited events can be detected) 4. False alerts 5. Laborious investigations 6. High cost of operations	1. Automated data consolidation and preparation by RPA 2. AI led pattern and relationship identification 3. AI led fraud alerts with wider net of detection 4. Reduction in false alerts 5. AI augmented investigation 5. Reduction in cycle time and improved efficiency

With that, we will end this section on the use cases and now try to understand, that what kind of changes that bank need to adopt, so that they can enable these technology adoptions to automate end to end processes with RPA and AI.

3.5 Impact on the Enterprise to Enable AI and RPA

So far, we have discussed what kind of benefits AI and RPA can help banks achieve, and how they can improve the operational efficiency and customer experience. Now in the following section, we will understand how can banks enable AI and RPA in their ecosystem, what are the changes in operating model they need to implement, what are the different capabilities and resources they need to invest in to make this real and achieve the kind of benefits we have discussed above.

To understand this let us first briefly touch upon the lifecycle of AI systems and RPA bots.

3.5.1 AI Systems

Broadly artificial intelligent systems have the following broad phases.

- Define the objectives, kpi to be improved, problem statement and business requirements: In this phase, the objective and problem are elaborated, and the business requirements are laid down. Along with that the business dimensions that will measure the success of the AI system should also be finalized. This establishes traceability of the technology solution with the impact on business.
- Data identification, acquisition, and preparation: This is the most critical and time-consuming phase in the lifecycle. This phase will include identifying the right data, their sources, data cleansing and feature engineering. Feature engineering is the activity in which we select the relevant features of the data that will help to build the model. In this phase, the data pipeline needs to be defined, designed, and built to create a clean and relevant data set for the AI model.
- Define and develop the AI model: In this phase, the right algorithms are selected for training the AI model based on the dataset. There will be various experiments done with multiple models and the best and most optimized model will be selected for production deployment.
- Testing and refinement of model: Typically, there are several experiments that are run to select the best model based on accuracy and precision parameters. To scale such model development, training, and testing, typically a training and testing pipeline is recommended.
- Deployment, monitoring, and support: The model can be deployed on the selected environment which can be a server or a container, on premise or on cloud based on the requirements and the deployment model defined. During monitoring the performance of the model can be observed and tuned both in training and the production pipelines.

3.5.2 RPA Systems

For RPA systems, these phases are similar in purpose and can be described as:

- Define problem statement and business requirements: This is the usual inception phase where the problem is elaborated, and requirements are established. The process or activity that will be automated by the RPA bot is also identified and finalized.
- Process Design and RPA bot Design: This is a key phase, where the process or activity that is to be automated is documented step by step and the bot design is created. During this phase, best practices of design like reusability and modularity should be applied to design the bot.
- Bot Development: This is the core development phase, where the bot is programmed based on the design created in the design phase.

- Testing and Deployment: This phase includes testing the bot in a pre-production like environment on production like systems before it is deployed to production.
- Monitoring and Support: Bot support will include various levels of support to detect bot failure and then resolve the failure.

For additional details on these lifecycles, you can leverage the references and also my book on Intelligent Automation.

3.5.3 Enterprise Changes

Based on the above discussion on the lifecycle, we now have a fair idea on the various activities that need to be undertaken in the design and development of an AI solution or a bot. Now let us understand from an enterprise perspective, what we need to enable the existing IT and business teams for creating such solutions or how to build a new team to enable these transformational technologies.

We have seen that intelligent automation with AI and RPA are becoming a key imperative in the enterprise transformation journey, considering the disruptions that are impacting the businesses today. Enabled with the power of these technologies, businesses will be able to deliver the outcomes customers want and achieve their business goals and we have seen the kind of business benefits AI and RPA will deliver through a few key use cases.

In this section, we will discuss the other side of the impact of these technologies. We will understand what changes or practices enterprises are needed to follow to successfully adopt these technologies and deliver the desired outcomes.

The key dimensions of bringing in new technology in a large organization are primarily around the awareness and adoption of nature of tools and technology platform that is coming in, the processes and methods that need to be introduced or modified, what kind of talent and skills are needed to be developed, how will the governance and support look like. Finally, what will operating model look like and how does it work with the existing business and IT ecosystem.

As depicted in Fig. 3.11, we will discuss these dimensions in detail in the following sections.

3.5.4 Operating Model

This is a very key dimension that defines how the new technology group will operate within the enterprise. The stakeholders for each phase in the lifecycle are decided and interactions are guided based on the operating model. The lifecycle is typically comprised of opportunity identification, design, and build, deploy and manage and governance and support. The operating structure regarding ownership, management and staffing of each of these lifecycle activities are defined within the operating model.

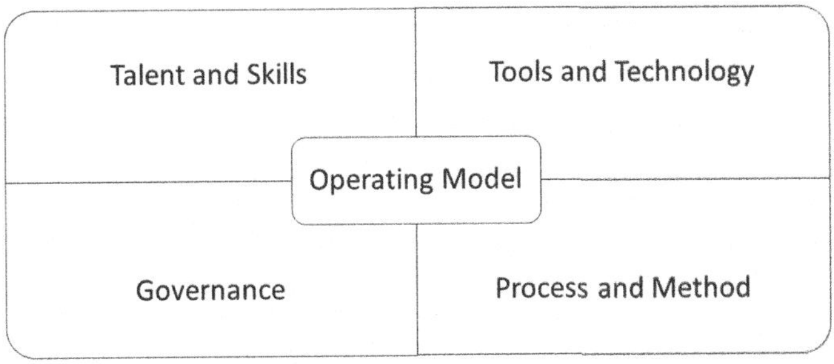

Fig. 3.11 Dimensions of change

The operating model also includes the roles and responsibilities of the new group and how they will interact with the rest of the organization. In many organizations, rolling off a strategic technology like AI or RPA can also be done through a Center of Excellence. The same dimensions as above would apply in case of a Center of Excellence as well. The structure and organization of the Center of Excellence or the new technology group (however it will be called within the enterprise) are strongly influenced by factors like who owns the CoE, whether it is IT or business and where the funding of the CoE comes from. The operating model whether it is a CoE or termed as new technology group is also responsible for increasing awareness, and knowledge about the capabilities of the technology and working towards increasing adoption in the enterprise.

3.5.5 *Tools Technology and Platform*

This is another core dimension as a new technology is rolled out in the organization. The choice of tools and platform is evaluated based on the objective, purpose, and requirements of the technology. A basic tool chain to adopt the technology from design, development, deployment, and monitoring needs to be defined to accelerate the adoption in a uniform way. In more mature organizations, a reference architecture is defined, and components are aggregated through build, buy or reuse decisions, eventually building a platform. In case of RPA, the tool, or tools of choice for the enterprise should be evaluated based on the type of use cases that are prevalent in the enterprise and how well they fit into the overall ecosystem. Decisions on how to host the RPA, whether to use a cloud hosted model or on premise, should also be decided under this dimension. Similarly for AI, the service provides, frameworks, open source, and standards should be defined along with the hosting considerations. For both RPA and AI, design, development and deployment best practices should be formulated and adhered to, to ensure maintainability of the solutions.

3.5.6 Process and Methods

This dimension defines the methodology for opportunity management, design and development, deployment, monitoring, and support for solutions enabled by AI and RPA. How these systems will integrate with other workstreams, and technologies and the respective ownerships are also defined under this dimension. The tool chain that we mentioned in the technology dimension, needs to be enforced under the process and methods to ensure uniformity in design and development practices.

On the above discussion on the lifecycle, we now have a fair idea on the various activities that need to be undertaken the design and development of an AI solution or a bot. Now let us understand from an enterprise perspective, what we need to enable the existing IT and business teams for creating such solutions or how to build a new.

3.5.7 Governance and Support

RPA and AI are not standalone technologies and will work together along with the other applications, platforms, and technology groups in the enterprise ecosystem. Establishing a strong governance structure and defining the responsibilities and dependencies between the key groups is thus essential. The support structure also needs to be well thought of. If the support organization becomes resource heavy, the cost benefit of the transformation may be negated. Hence what is recommended is that, once you have a steady pool of bots or other AI components deployed in production, look towards automating the support. For Eg, Level1 tickets should be targeted to be addressed by self-healing systems, and Level2 tickets by chatbots. Only if it comes to code changes, it should go to the pool of technical resources. This is just an example, but a detailed strategy to handle support is essential to scale these technologies.

3.5.8 Talent and Skills

This is a very important dimension, since this is the one that will drive the new technology adoption practice and the success of the enablement largely depends on how skillful and efficient the resources are. They can be insourced or outsourced based on the structure and requirements of the enterprise. If skills are built inhouse, there needs to be a training strategy and learning plan to reskill and upskill the employees. There should be strong encouragement and backing for trainings and certifications so that the skills of the employees can be validated. "Train the instructor" programs can help scale such learning programs, when the skills need to be built fast and for a larger pool of resources.

The five dimensions discussed above will help initiate, nurture, and mature the new technology group and help the enterprise to transform their processes using these technologies. Finally, such a new group needs to be staffed with skilled resources and the group needs to have very strong sponsorship from executives in the enterprise to be successful.

3.6 Conclusion

Banking being the early adopter of technology continues to lead this space. A report from Business Insider quotes that $199B is the size of savings in Front Office, $217B in mid-office and $31B in back office in core banking operations. Intelligent Automation with AI and RPA are featured among the winning strategies that are being employed by banks. However, there are a few pitfalls. Random or ad-hoc initiatives may backfire in this space. There is a strong need to have a holistic intelligent automation strategy and roadmap should be created at enterprise as well as group levels to identify the best opportunities and iteratively build the intelligent automation capabilities with AI and RPA. This should bring in all the stake holders along the lines of businesses, partners, and third parties and enable banks to truly reap the benefits of these technologies and transform into the "intelligent bank" of the future.

References

1. Smith, A., Nobanee, H.: Artificial intelligence: In banking a mini-review. (2020)
2. Geetha, A.: A study on artificial intelligence (AI) in banking and financial services
3. Sharma, M., Kaur, N., Sahdev S.L How AI is changing the face of modern-day banks
4. Mehrotra, A.: Artificial intelligence in financial services – need to blend automation with human touch. In: 2019 International conference on automation, computational and technology management, (2019)
5. Navleen, K.,Sahdev, S.L., Sharma, D.: Banking 4.0: -The influence of artificial intelligence on the banking industry & how ai is changing the face of modern day banks‖.(2020)
6. Vaishak, B.: RPA in banking sector. Int. J. Res. Appl. Sci. Eng. Technol. (2022)
7. Königstorfer, F., Thalmann, S.: Applications of artificial intelligence in commercial banks—A research agenda for behavioral finance.(2020)
8. https://www.bcg.com/publications/2018/four-ways-banks-can-radically-reduce-costs
9. https://www.firstpost.com/business/rbi-imposes-penalty-on-pnb-icici-bank-for-deficiencies-in-regulatory-compliance-10214511.html
10. https://internationalbanker.com/technology/the-cost-of-compliance/
11. Rahman, M., Ming, T.H., Baigh, T.A. and Sarker, M.: Adoption of artificial intelligence in banking services: an empirical analysis, (2021)
12. Fares, O.H., Butt, I. Lee, S.H.M.: Utilization of artificial intelligence in the banking sector: a systematic literature review. J. Financ. Serv. Mark, (2022)
13. Rameshbabu, M.: Rise of robotic process automation to revolutionise key banking processes. (2021)
14. Shende, A., Roy, S.: RPA implementation in banking. Int. J. Adv. Res. Sci-Ence, Commun. Technology

15. Wang, Y.: Applying robotic process automation in the banking industry. (2021)
16. Amin, D.A.: The potential benefits and risks of adopting RPAin the banking sector. (2022)
17. Langmann, C., Kokina, J.: RPA in accounting. In: Robotic process automation, pp. 243-262 (2021)
18. Nayanajith, D.A.G.: Artificial intelligence (AI) in Banking. (2022)
19. King, B.: The role of AI in banking, pp. 219–251. (2018)
20. Jaiswal, AK., Akhilesh, K.: (2020). Tomorrow's AI-enabled banking. In: Smart technologies, pp. 191-200. Springer, Singapore. (2020)
21. Dasgupta, D.: Intelligent automation simplified. BPB Publications (2020)
22. Dasgupta, D.: Intelligent automation simplified. BPB Publications (2020)
23. Biswas, S., Singh, S., Thomas, R., Carson, B.: AI-bank of the future: Can banks meet the AI challenge?. (2020). https://www.mckinsey.com/industries/financial-services/our-insights/ai-bank-of-the-future-can-banks-meet-the-ai-challenge
24. Garvey, J., Alexander O, Babczenko K, and all.: Customers in the spotlight: How FinTech is reshaping banking. https://www.pwc.com/gx/en/industries/financial-services/publications/fintech-is-reshaping-banking.html
25. Sheth, J.N., Jain, V., Roy, G., Chakraborty, A.: AI-driven banking services: the next frontier for a personalised experience in the emerging market. (2022)
26. Vijai, C., Suriyalakshmi S.M., Elayaraja, M.: The future of robotic process automation (RPA) in the banking sector for better customer experience. J. Commer. **8**(2), 61-65 (2018)
27. Yarlagadda, R.T.: The RPA and AI automation. Int. J. Creat. Res. Thoughts (IJCRT), 2320-2882 (2018)
28. Burgt, J.V.: Explainable AI in banking. (2020)
29. Aitken, M., Ng, M., Toreini, E., van Moorsel, A., Coopamootoo, K.P.L., Elliott, K.: (2020). Keeping it human: A focus group study of public attitudes towards AI in banking. In: European symposium on research in computer security (pp. 21-38). Springer, Cham (2020)
30. Xue, M., Xiu, G., Saravanan, V., Montenegro-Marin, C.E.: Cloud computing with AI for banking and e-commerce applications. (2021)
31. Jaiswal, A.K., Akhilesh, K.B.: Tomorrow's AI-enabled banking. In: Akhilesh, K., Möller, D. (eds.) Smart technologies. Springer, Singapore (2020)
32. https://enterprisebotmanager.com/7-of-the-best-chatbots-in-banking-and-what-to-watch-for-next/
33. https://www.smartbots.ai/use-cases-for-chatbots-in-banking/
34. https://thefinancialbrand.com/71251/chatbots-banking-trends-ai-cx/
35. https://assets.kpmg/content/dam/kpmg/xx/pdf/2019/05/global-banking-fraud-survey.pdf
36. https://www.pwc.com/gx/en/services/forensics/economic-crime-survey.html
37. https://www.businessinsider.in/finance/news/the-impact-of-artificial-intelligence-in-the-banking-sector-how-ai-is-being-used-in-2020/articleshow/72860899.cms
38. Ahmed, S., Alshater, M.M., El Ammari, A., Hammami, H.: Artificial intelligence and machine learning in finance: A bibliometric review. Res. Int. Bus. Financ. **61**, 101646 (2022)

Chapter 4
Robotic Process Automation: The Key to Reviving the Supply Chain Processes

Gowri Rajagopal and Raghuraman Ramamoorthy

Abstract Businesses are very cautious about their supply chain. Any delay or disruption may lead to unimaginable loss. The COVID-19 scenario has taken this concern to an epic hyper. Businesses could not meet their customer's needs due to the lack of Human Resource during the prolonged lockdown. Companies had to either downsize or shut down. Businesses started searching for means to operate their processes with as low resources as possible. Thus, many started adopting Robotic Process Automation though it is still evolving. There are many literatures that discuss the implementation of RPA in varied streams like accounting, finance, HR, and others. Yet there is a very limited study that has focused on RPA implementation in supply chain processes. The proposed chapter showers light on the adoption of RPA in the varied stages of supply chain. A complete real-time automation use case on invoice processing is also discussed. This chapter aims to help researchers, RPA enthusiasts, and entrepreneurs to realize the time, effort, and economic benefits of implementing RPA. Thus, encouraging them to adopt RPA in their supply chain processes.

4.1 Introduction

Business continuity and profitability are the major aspects concerning any organization. It is the capability to continue performing its core Supply Chain (SC) processes successfully and efficiently. Though there exists a pressing competition in the current dynamic business environment, not all businesses manage to thrive. A hundred percent business continuation can only be in the realms of an assumptive ideal world. processes involved in any SC are prone to be affected by disasters or unexpected events. Disruptions are a leading cause of delays in the process execution and completion.

This directly impacts the utilization of resources and largely affects profitability. Now, as the world has seen a digital boom in past few years due to industry 4.0, the

G. Rajagopal (✉) · R. Ramamoorthy
Rapid Acceleration partners Pvt. Ltd, Chennai, Tamil Nadu 600 026, India
e-mail: gowri.rajagopal@gmail.com

S. Bhattacharyya et al. (eds.), *Confluence of Artificial Intelligence and Robotic Process Automation*, Smart Innovation, Systems and Technologies 335,
https://doi.org/10.1007/978-981-19-8296-5_4

supply chain has benefited from data-led technologies. However, the rampant spread of COVID-19 has left many supply chains broken irreparably.

The competitive dynamics of the business environment are pressing organizations to explore new solutions for overcoming such process disruptions and strengthening their SC to its normal best. While many are struggling to cope with the shortages and keep their businesses afloat. Many businesses face a lack of human resources, raw materials, etc., while many could not bring the level of productivity while working from home online. As the organizations look forward to a solution to continue with the regular processes many of them have resorted to automating their processes.

Gartner's claimed in one of their press-release in 2020 that "The decreased dependency on a human workforce for routine, digital processes will be more attractive to end-users not only for cost reduction benefits but also for insuring their business against future impacts like this pandemic" [1]. Adopting technologies like Robotic Process Automation (RPA) has helped organizations in keeping their grounds as they struggle past the COVID pandemic. Today many companies are resorting to RPA to fulfill their need for human resources and mend their broken link in the supply chain. and to sustain the optimal results to take advantage of RPA.

This chapter aims to shower light on the opportunities of implementing RPA in Supply Chain Management (SCM). The book chapter is segmented into four parts. The first part provides an introduction to the problem statement, the second segment introduces the supply chain and the processes involved. The Third segment introduces RPA and the fourth segment discusses the opportunities of automating the supply chain processes and introduces a use case. The fifth segment presents the discussion on incorporating RPA and IPA into SCM, and the pros and cons followed by the concluding remarks.

4.2 An Introduction to Supply Chain Management

4.2.1 Understanding Supply Chain

The terminology "Supply Chain" is a new term given to an old concept that has existed for ages. It first started as a military strategy to transport resources between places using optimal routes. As its efficiency eventually gained visibility, its applications increased in multiple dimensions. To the extent that, one could say where there is business there is a supply chain, this includes the IT domain as well. It was in 1983 that Keith Oliver from Booz Allen and Hamilton coined the term Supply Chain Management. Multiple researchers have defined supply chain in different terms. Wang [2] in his paper has listed several definitions provided by different sources [2]. Oxford University Press has defined SC with respect to production and distribution processes involved in a business.

Londe [3] views SC as a group of several independent firms that come together for manufacturing and deliver the product to the end-user [3]. Mostly these firms would be raw material and component producers, assemblers, wholesalers, retailer merchants, and transportation companies Pringle and Harris [4] point out the value addition that takes place through the involvement of all the different organizations in bringing a product or service from upstream to downstream to meet customer needs [4]. Lee and Billington [5] view the SC as a network of facilities that procure materials and convert the material to intermediate and finished products that are distributed to end customers [5].

While Quinn [6] described SC based on the activities involved in the transfer of goods from the procurement of raw materials to the completed goods reaching respective customers [6]. The interpretation of SC by Lummus and Vokurka [7] is around the network of entities involved in the transfer of material in the process of a product reaching its customer [7]. Lambert et al. [8] view a supply chain concerning the alignment that is required amid the various partners involved in the SC while manufacturing and pushing it to the market of firms with products or services [8]. Christopher [9] poses SC as a management approach for aligning the organizations and the processes involved in creating value or service to a customer [9]. Gaither and Frazier [10] explain the SC concerning the material flow between different organizations from the stage of basic raw materials to the finished goods and the delivery to the ultimate consumer [10]. Prakash et al. [11] have described a SC as a collection of firms that forward materials linearly [11]. It could be observed that until the 1980s the term SC and logistics were used interchangeably and even so later it was defined in different ways under different perspectives.

4.2.2 Overview of Supply Chain Stages

A SC in general, involves all the activities required in creating a product or a service. It contains many processes, from upstream raw material procurement to reaching downstream end-users. Organizations collaborate to achieve a competitive advantage from the value generated by the product. In short, a supply chain links different organizations to add value to the end-user. The value addition can be a service or a product. Raw material suppliers, transportation providers, and warehousing are all major players in SC. Not least say the need to meet the customer needs with the supply where logistics and demand planning play a huge role.

The exchange of information, finance, and material across SC entities is yet another challenge. Deckert [12] points out the extensive number of processes carried out in SC. Starting from setting out tenders, identifying economical quotations, partner selection, acquiring product orders, scheduling procurement of raw materials, timely manufacturing, packaging, dispatch, and the last mile delivery to name some [12]. A typical supply chain stage is shown below as discussed by Sunil and Petrer [13]. In Fig. 4.1, the point of origin relates to the source from which the raw materials are extracted [13]. Supplier segments procure raw materials from point of origin and supply the same to the manufacturers. The manufacturers in turn manufacture the product and push it to the distributors.

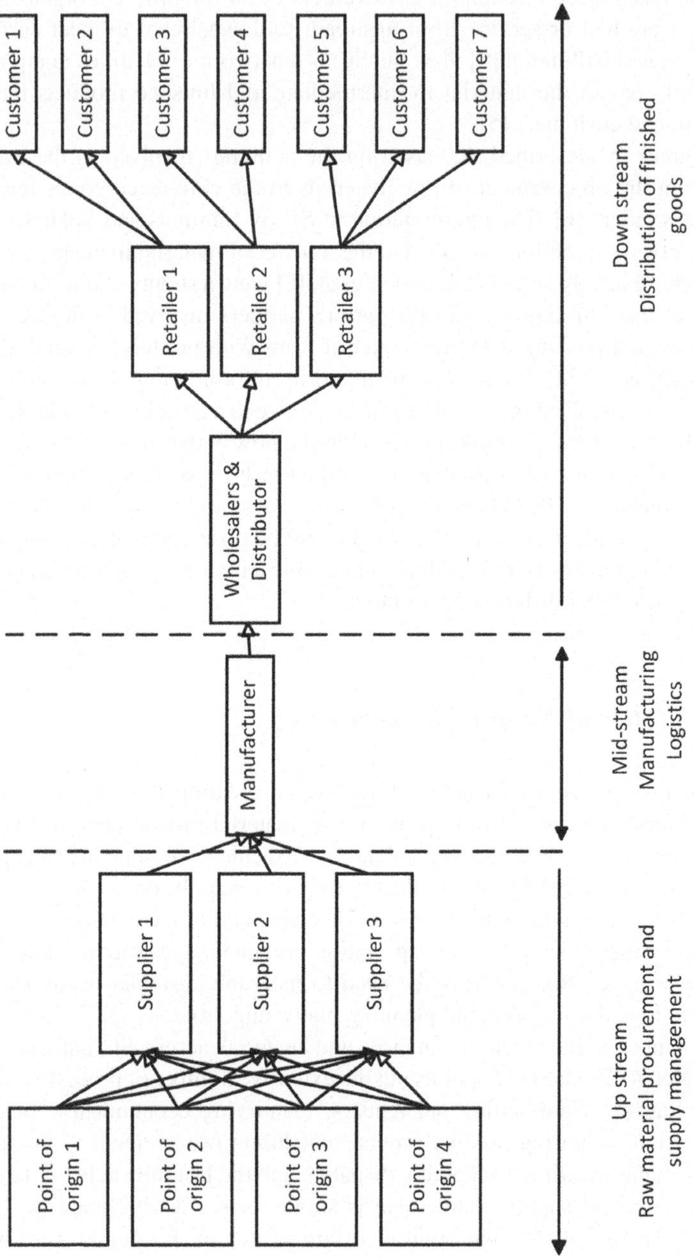

Fig. 4.1 Various stages of supply chain

The retailers purchase the products from the distributors and make them available to the customers. Here, the stages prior to the manufacturer from whom they procure the material inputs are termed as the upstream and those segments after the manufacturers are the downstream segments of any supply chain. Some also claim the stages involved in manufacturing, packaging the products, and making them available to the downstream as the midstream operations.

4.2.3 A Macro View of Selective Supply Chains

We already discussed that the stages in any business's supply chains are similar, yet the processes, the source materials and finished goods/products and services, and the flow of materials in and out of these stages would be different from business to business. We would like to give a macro view of some selective supply chain and their stages for better understanding.

Agriculture Supply Chain Consider an agriculture supply chain, what a consumer views are fresh produce at the grocers or supermarket. However, agricultural supply chain is an extensive one. There could be multiple sources of suppliers at varied tiers. The point of origin, i.e., the nth tier Supplier could be a source of seed, a chemical industry, paper industry, and metal industry. Tier n–1 could consist of package or carton producers, OEMs, etc., that supply the resources needed for the next tier. Tier1 suppliers are fertilizer manufacturers, farm equipment vendors, and automobiles like tractor retailers from whom they interact directly for their business needs. The midstream could be the farming, and collection of yield and packaging. The downstream consists of dealers that purchase their produces and supply it to the wholesalers who in turn supply to retailers like grocers/supermarkets, or even a petite shop from where a customer could purchase these vegetables, pulses, etc.

Pharmaceutical Supply Chain Similarly, when we look at a pharmaceutical industry the supply chain stages are the same but the type of material, information, manufacturing processes, and other flows vary. For example, the point of origin or the nth tier suppliers for a pharmaceutical company would be oil industry, paper manufacturers, and farmers, who supply the raw materials for the manufacturing of medicine and packaging them accordingly in the upstream and the midstream. In the downstream, the medicines are transported to the distributors from whom the local pharmacies obtain the medicine stocks. Hospitals and other consumers would purchase these medicines from pharmacies.

Automobile supply chain An automobile supply chain is also a network of enterprises that manufacture vehicles. It links the varied part suppliers to the manufacturers. these suppliers are in multi tiers. The material movement takes place from the raw material source to the car owners. Figure 4.2 shows that tier2 suppliers procure materials from multiple raw material sources. tier2 suppliers could be anyone from a bearing to rubber tube manufacturers. For example, a steel plate manufacturer would

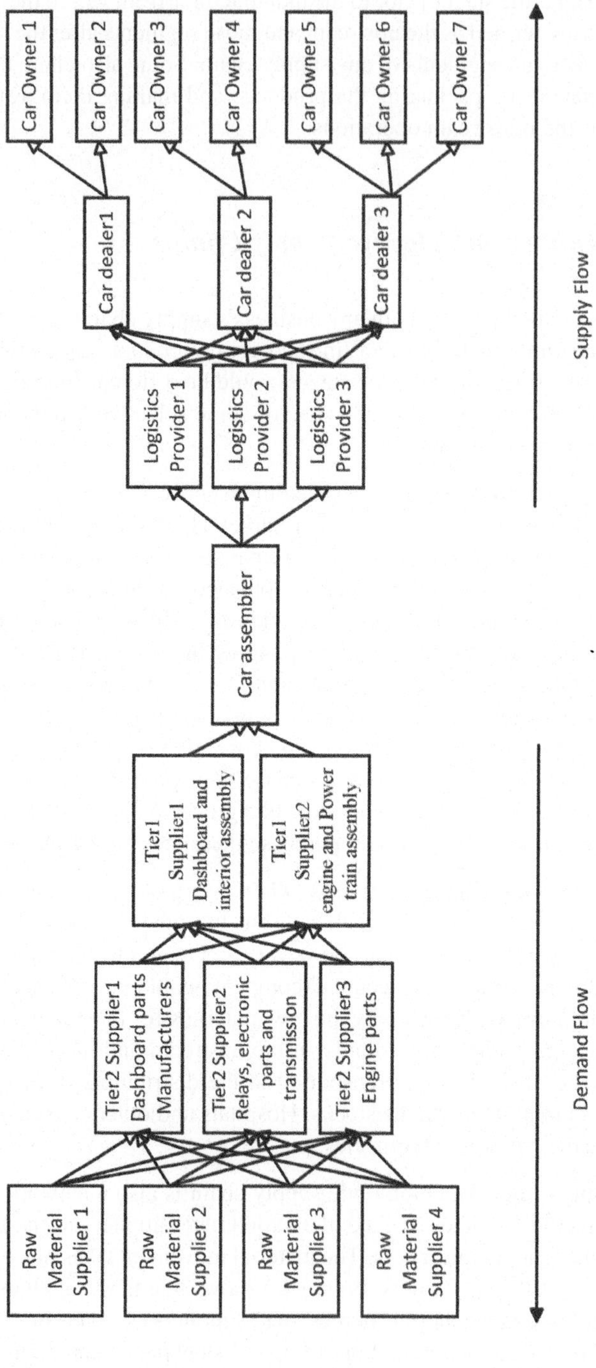

Fig. 4.2 A generic view of supply chain in the automobile industry

source from iron ore extractors. Similarly, a tier2 rubber tube manufacturer could procure rubber sap from the respective farmers. Tier1 tyre manufacturers would source from the tier2 suppliers. For example, tier1 suppliers can procure steel plates for making car chassis. Similarly, if tier1 supplier is a tyre manufacturer then they may purchase rubber tubes and steel plates for their tyres, frames, and nozzles. If tier1 supplier is a dashboard assembler then they would source for dashboard body parts like the varied measuring devices like odometer, fuel indicator, ABS indicators, airbags, and relays for assembly. The car manufacturer would in turn purchase these assembled parts like chassis, wheels, dashboard body, etc., for the final assembly. As the car is finally assembled it is pushed to the downstream dealers with the help of transportation or logistics providers.

The dealers, as any customer may come with the intent of buying a car facilitate in selection and purchase of the car through their services. The automobile supply chain process may seem very simple but involves varied levels of transactions in terms of material flows, information flows, and fund flow like any other SC. As the industry usually would source from multiple suppliers the distribution complexity increases tremendously. Salomon et al. [14] suggest that these concerns would eventually lead to problems like resource allocation and inventory maintenance. Prakash et al. [11] add to this list by pointing out the issues of maintenance, demand forecasting and cash flow-related problems.

As we explore the processes involved in various supply chain, we could observe that the Information transfer between the different stages is humungous. Accessing data maintained in multiple silos that too in different locations is a big concern. Security concerns also add up to it. Besides these, there are industries that still maintain their data by large in legacy systems which is another issue. To tackle these issues in this era of industry 4.0, many companies are adopting approaches like blockchain, IoT, AI, ML, and RPA. Multiple authors have proposed that the use of RPA is easier under several circumstances. Before we can discuss the benefits let's first understand the what's and whys of RPA.

4.3 Understanding RPA

4.3.1 What is Robotic Process Automation?

RPA stands for Robotic Process Automation, which is a software robot that mimics human activity to complete complex tasks. In a variety of industries, RPA is becoming increasingly common [3]. It requires the use of a programming interface to automate the repetition of routine tasks.

RPA refers to the use of specialized technologies and methodologies to automate routine human activities using software and algorithms. It is primarily guided by basic rules and business logic, and it interacts with a variety of information systems through existing graphic user interfaces. Its features include the use of a non-invasive software robot known as a "bot" to automate repeatable and rule-based tasks. RPA's meaning has recently been expanded to include its use in conjunction with AI, cognitive computing, process mining, and data analytics RPA can now be reallocated from repetitive and error-prone procedures in business processes to more complex knowledge-intensive and value-adding activities due to the adoption of advanced digital technologies [15]. There are multiple definitions for RPA.

Patrick Geary, Blue Prism's Marketing Officer, was the first to use robotic process automation in 2012. RPA was invented by Cyrille Bataller and Adrien Jacquot, according to the European Patent Office (EPO). Osman [16], defines RPA as "a technology that enables to automate the execution of repetitive and manually intensive activities" [16]. While Gartner's see RPA as a productivity tool, Sutherland views RPA as computer technology and technique that performs a Full-time equivalent (FTE) job rather than an individual to control existing application [1].

According to the Institute for Robotic Process Automation & Artificial Intelligence (RPAAI), It is the use of technology to program software robots to capture and interpret existing applications in order to process transactions, manipulate data, and communicate with other software systems [6].

4.3.2 Market for RPA

Zhang and Liu, [17], in their research, compiled the number of advisory and consulting companies and pointed out that the global value of the RPA tools market in 2018 was estimated to stand at USD 849 million [17], while Gartner, in 2019 projected USD 1.3 billion. According to Gartner, Inc.'s forecast for 2020, the global robotic process automation (RPA) software sales will hit $1.89 billion in 2021, up 19% from 2020 [1]. Despite the COVID-19 pandemic's economic impact, the RPA market is projected to expand at double-digit rates through 2024.

Fabrizio Biscotti, research vice president at Gartner points out that the ability of RPA projects to enhance process efficiency, speed, and productivity is a key driver for them, as organizations strive to meet the demands of cost reduction during COVID-19. Investing in RPA software allows businesses to make rapid progress on their digital optimization projects. Gartner's forecasted that its revenue from RPA software would hit $1.58 billion in 2020, up 11.9% from 2019 [1]. Average RPA prices are projected to drop 10–15% by 2020, with annual drops of 5–10% expected in 2021 and 2022. While Forrester's research (March 9, 2022) furthers these predictions to reach $6.5 billion by 2025 and that the curve would flatten due to the AI incorporation in the solutions.

There are multiple RPA development tools/products available in the market. Some renowned software is Automation Anywhere, UiPath, Blue Prism, Work Fusion, PEGA, Kofax, Redwood, and Laiye. As the market for RPA is increasing, many organizations are exploring their processes for automation. Thus, the spectrum of RPA applications is widening day by day.

4.3.3 Overview of RPA Life Cycle

The development of RPA and its implementation is segregated into six phases, namely the Discovery Phase, Analysis Phase; Design Phase; Development phase; Deployment phase; Control and Monitoring phase, and, finally, the Evaluation and performance phase [18].

Figure 4.3 represents the life cycle of an RPA. It includes the **Discovery Phase** involves the studying of existing practices and processes. From the observation, the processes that can be automated are identified as automation candidates. **Analysis Phase** involves evaluating and assessing the feasibility of automating a particular process using a thorough analysis of the effort needed to self-motivate the process while taking into account the process' execution characteristics. The **Design Phase** starts with the processes that passed the feasibility study. This phase defines the behavior, data flow, tasks, and other aspects of the RPA process that must be enforced.

This is followed by the **Development phase** that brings into effect the automatable components of each process defined during the design phase. In the **Deployment Phase**, the robots that are created during the construction process are executed in their respective operational environments. The **Control and Monitoring Phase** oversees

Fig. 4.3 RPA life cycle

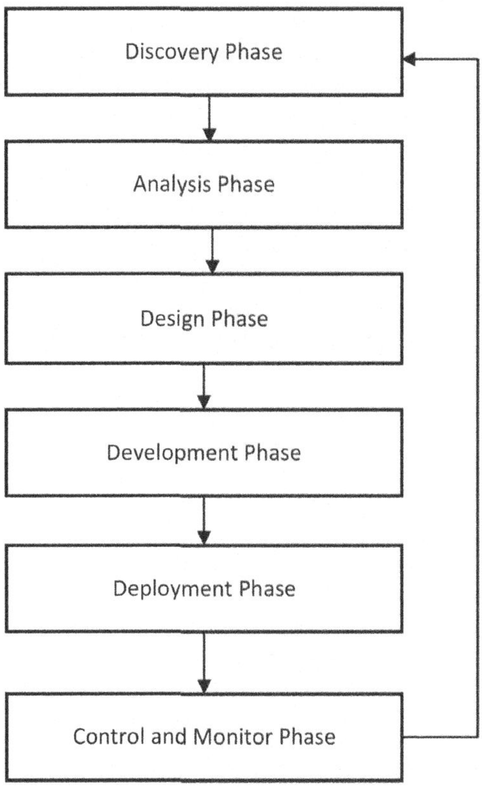

the supervision and monitoring of each robot's output after they have been deployed in their respective execution environments. The execution of robots is started in this process, and it is stopped in the event of serious errors, the execution status is monitored, and so on until the assigned tasks are completed. Finally, in the Evaluation and Performance Phase, the robots' performance is evaluated.

4.3.4 RPA Application

RPA offers several advantages, including increased efficiency with minimal process change, improved service quality, shorter delivery times, and the liberation of workers from mundane and repetitive tasks. RPA, according to Osman [16], fits well in any sector that devotes large amounts of labor, such as banking, healthcare, and insurance. RPA, works well in every sector, including banking, healthcare, and insurance, since they all devote considerable labor hours to tasks that software robots might manage [16]. Claims processing and account auditing (finance), order management, and inventory requirements preparation are only a few examples of RPA applications (Supply chain). Today's companies are actively searching for processes that can be streamlined using RPA to produce optimal results in order to take advantage of RPA. Unfortunately, there is no preset hard and fast rule that helps determine the suitability of automating a process by implementing RPA.

Many organizations walk the hard path of automating by trial or error and suffer from low to no ROI. According to Mühlberger et al. [19] finding the suitable process or activities for implementing RPA is a huge challenge [19]. They cite that the current approaches focus on profitability than assessing the viability of the process for automation. Scheppler and Weber [20] in their study discussed the importance of conducting a process assessment before implementing RPA due to the higher failure rate of RPA projects due to poor choice of processes [20]. RPA providers like UiPath and Power automate have brought about components like Process identifiers and process discovery that help in analyzing the automatable potential.

4.3.5 Understanding Process Automatability

As discussed above whether to automate or not is a big taxing question. Many automation specialists and organizations alike have to face it at one point or another and that too on a different scale [10]. Batakis et al. [21] have developed a process assessment model and process assessment formula using which they studied a process for its automatability and then using the formula they calculated the suitability of the process to be automated [21]. They also suggested that characteristics of automatable processes would expose the following characters:

a. High manual involvement

b. Repetitive processes
c. Process with standardized inputs and outputs
d. Adheres to preset rules
e. Structured data format
f. Accesses multiple data sources
g. High process maturity (in terms of ease of predictability of next state in the process)
h. High transaction volume
i. High time consumption
j. Less exception.

Christian et al. [22], elaborately added up to this list ten more characteristics and categorized them with their Process Characteristics Evaluation Framework (PCEF). PCEF segregated the process characters based on five perspectives, namely: Task; Time; Data; System, and Human [22]. The authors have explained the Task from the perspective of the execution of various activities that take place in a process. This involved standardization, maturity, determinism, and failure rate. Standardization is based on the number of varied activities and their different execution approaches in business. Maturity depicts variation in the number of deviation cases over time. The manual interaction and time taken for the interaction represented determinism. While the terminations and rework were taken as failure rate in the tasks.

Time Frequency is viewed according to the execution, duration, and urgency, where these values are based on the number of executions, the average completion time, and the average time to react. Data is explained with respect to its structured characteristics, while they also introduce the interfaces and stability of the system based on the exceptions and the number of systems involved. Finally, the authors have discussed the required human influence by the number of resources and erratic behavior. Yet one cannot say that all processes are suitable for automation. Hence, the user needs to be aware of the tenure in which RPA would start Breaking even and give Return on Investment.

4.4 RPA in the Supply Chain Processes

Supply chain management (SCM) utilizes RPA to automate monotonous, highly repetitive, data-intensive high time-consuming error-prone tasks. With the inclusion of Artificial intelligence, natural language processing and machine learning RPA are scaling up in terms of their application in SCM. It helps increase productivity and meet up demand fluctuations efficiently [23]. RPA provides the flexibility for faster scaling to meet the fluctuating demand. According to Vajgel et al. [24], the major benefits of introducing RPA into SC would be flexibility, scalability, and better control. Several authors point out RPA to be more reliable and speedy task completion [24]. This segment discusses the SC process that can be easily automated using RPA. Freeman [25] suggests payment processing, procurement, inventory management, warehouse management, and demand–supply planning as some of the key areas where RPA could be implemented [25]. Implementation of RPA in purchase order management and administrative tasks are also emphasized by Blume Global. They have also pointed out the use of NLP and text mining for responding to the questions raised during the tendering, proposal requests. These extensions to RPA provide the benefit of shorter turn-around time and faster responses.

Lawton [26] adds Data entry, Predictive maintenance, Logistics management, and Order management as other processes that can be automated [26]. Wollenhaupt [27] speaks about vendor payment invoices, inventory replenishments, purchase order management, and e-order auctions as tested processes for RPA implementation [27]. Viale and Zouari [28] in their paper have pointed out the implementation of RPA in Procurement [28]. UiPath in one of their RPA implementations journeys has successfully integrated ERP with their RPA for reordering inventory management for one of their Brazilian supermarket clients which eventually saved 1000 manual hours a month [29]. RPA also acts as a liaison between the different stages of the Supply Chain as it allows easy accessibility between different tools and systems [30]. This enables strong connectivity and information flow among the suppliers, manufacturers, distributors, retailers, and customers. The following segment would discuss the invoice processing approach with respect to the practical implementation of RPA.

4.4.1 Invoice Processing

With the incorporation of RPA, organizations can easily automate invoice processing activities. Irrespective of the multiple levels of audits, these processes still lack transactional visibility and are erratic [25]. Irrespective of the invoice being Accounts Payable (AP) or Accounts Receivable (AR) the processing of the documents has become easier with the help of RPA. The processes that are involved from the generation of invoices to payment fulfillment are what we term invoice processing. The process is small but the complexity arises due to the volume of documents that need to be processed and the need to extract them from different sources like physical documents, email, fax, etc. (UiPath).

Even if it is from one source says an email the documents can be of different file types. For example, if the document file type is a PDF, it can be a searchable PDF or a non-searchable PDF. A non-searchable PDF may require Optical Character Recognition (OCR) for the extraction of required data. Besides this, the design format of documents could be different where the coordinates of required data may be in different invoice documents. This may again use computer vision for extracting the details. From the use case in the following section, we could understand how the invoice is processed by an RPA better.

Use case: Invoice processing of Company XYZ Company XYZ is a Car manufacturer and imports parts from multiple OEMs like Engines, Dashboards, chassis, Wheels, and many other parts. It buys part from vendors across the world and from several of them it buys in credit. The vendors generate purchase orders and put together the ordered items. Once XYZ receives the ordered items the vendor supplies the invoice based on XYZ purchases. These invoices of accounts payable are paid within the mutually agreed time frame. Once the invoice is received, personnel from XYZ must verify the document scrupulously. The personnel had to verify and extract the Procurement Order number, the list of items received, the payable amount, ordered items quantity ordered, and the quantity for which the invoice has been generated, etc.

The extracted details were entered into a time-stamped excel workbook and forwarded to a superior for approval. In multiple scenarios, erratic details were entered, or duplicate invoices were received. Occasionally, multiple payments were done by mistake. Rectifying these issues led to hefty time delays from several hours to days. A sample invoice processing flow is shown in Fig. 4.4. Usually, the invoice document reaches Company XYZ via email attachments from a specified vendor as a pdf file or at times as a physical document. The flow depicts the processing of a pdf-based invoice received through electronic means.

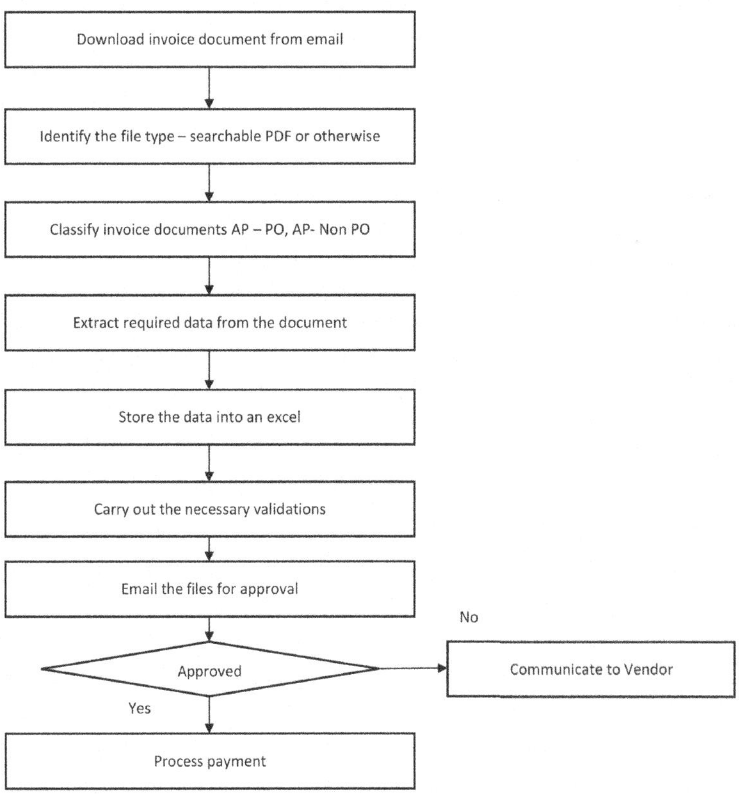

Fig. 4.4 Sample invoice process flow

The first bot selectively downloads the invoices in the pdf format attached to the emails and stores them in a local repository (folder). The pdf documents are further checked for their readability, i.e., is a searchable or non-searchable pdf. Once the bot confirms the type of pdf further actions are taken up, e.g., when the received pdf is a non-readable pdf it is sent to a specialized OCR-based extractor. Then the document is sent through a classification engine that extracts the purchase order number details. Based on the details extracted it is classified into a PO or non-PO. On classification, the specified document is segregated and stored in the respective folder.

As the file falls into the appropriate folder, the second bot picks it up and extracts all the key information. The extracted information is then inserted into excel sheets as per the customer requirement. The validations for checking the invoice could be carryout and the excel report along with the payment decisions would be sent to the concerned approval authority, who, then verifies the details and acknowledges the details. The third bot releases the payment to the vendor. A part of the process flow designed using the RPA software named RAPBot developed by Rapid Acceleration Partners Pvt., ltd, India, is shown in Fig. 4.5. This flow incorporates AI/ML-based classification of invoice documents with the support of Intelligent Process Automation to extract the key data. The data may be invoice number, date, part names, part details, number of part units, the total number of items, total amount payable, etc.

These bots would parse through the document check for the data and based on the results would classify the documents as PO or non-PO and push it into the appropriate directory for further processing. The documents are fetched from the local folder and the respective data are extracted from it and stored in a database. The major reason for involving IPA is that the documents are not uniform. There may be documents from different suppliers as discussed earlier and that too in different patterns of the same invoice forms. The coordinates of the respective metric may not be in the same location in all the invoices. The model must be trained to capture the metrics accurately which is always a challenge. Once the metrics are extracted, they are entered into another application that highlights the details extracted along with the original file. This mode is usually applicable when there is a human intervention required for cross-verification of the details extracted.

Another approach followed is where the data is collated into an excel workbook, the calculations and verification happen automatically, and on completion, the data is updated in the database. The updated data automatically gets loaded into Business Analytics tools like Tableau, Power BI, etc. The user is intimated by the bot via email once these actions are completed and the data are readily available on the dashboard. The compilation of the data analysis is shown to the user in the form of a dashboard for further the intimation process would usually be regarding the successful completion of the tasks or any error that may have occurred in the process or calculation.

Outcome of RPA implementation with the implementation of RPA by Company XYZ a tremendous level of human time and effort was reduced with the help of Invoice Processing bots. Eventually, as many business organizations adopt the hybrid workforce and implement RPA along with their employee many experiences

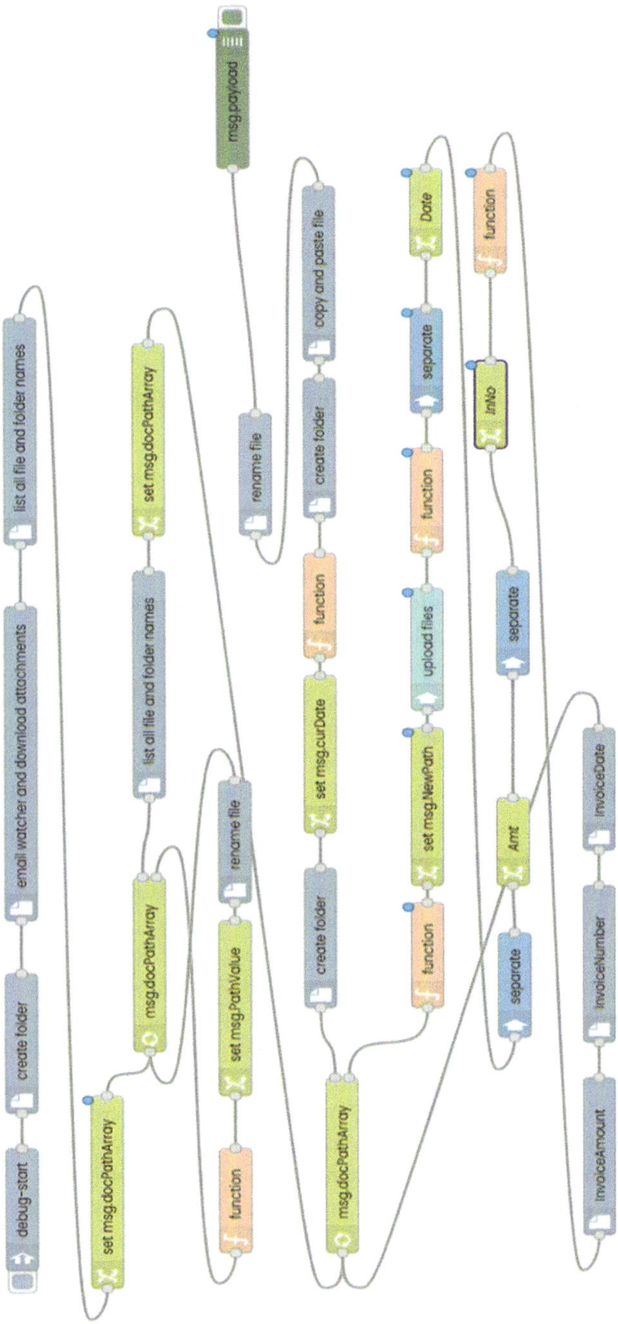

Fig. 4.5 PO invoice processing Bot—internal flow

increased time and economic benefit. Human assets could carry out more strategic and decisive tasks which are key to business success. Businesses started meeting their break-even on the RPA investment way faster and eventually increased the Return on Investment (ROI) year on year. Thanks to the fact that the bots could work 24/7 saving several hundreds of manual work hours within a short span of time.

As the number of errors reduced with the implementation of RPA the supply chain delays also reduced and cost-saving increased. Though there is so much improvement required in the prevailing supply chain practices, we as practitioners could see that RPA alongside AI and ML could lay a promising path to achieving the supply chain notion that the right product in the right quantity should reach the correct location at the correct time effectively.

4.4.2 Transformation of Legacy System with RPA

Legacy systems are the yesteryears' outdated technology and hardware. Organizations that have time-tested existence though are evolving continuously, legacy systems are still in use. Mainly due to the immediate adoption of new technology has practical difficulties. Sarpolis [31], in the article, explains that Supply chains rely a lot on legacy systems. Organizations have maintained inventory da-ta or executed planned maintenance that is tracked and stored in dumps of spreadsheets or property data that are siloed in varied legacy systems [29].

The main concern here is that they are so expensive and time-consuming to upgrade and rather bring down productivity. Besides, it is also difficult to make effective decisions when extracting data from varied sources is difficult. On several occasions, transferring these data from a legacy system to the cloud could seem to be a viable solution but getting it done is a tedious task in terms of the humungous data volume. Pendersen (2018) mentions that many organizations are evolving, and legacy systems are impacting their logistics and services [30]. Another issue with them is the security and compatibility with advanced systems [31].

This makes it difficult to maintain, support, or integrate it with new technologies and ultimately leads to customer dissatisfaction. RPA comes as a very easy and effective solution when it comes to transitioning from legacy [32]. RPA is also a faster and more economical option when compared to APIs that are used for the upgradation. Besides, API cannot be used for mainframe-based legacy systems but RPA can easily be implemented there. RPA makes the process of consuming the data way easier. RPA holds the upper hand in many ways as it allows easy integration, lower cost, less time of implementation, and it is comparatively secured.

In an article in 10xDs, it is pointed out that enhancing the legacy system with the deployment of RPA could save up 37% of cost [33]. While transforming a legacy system using RPA, it is important to identify the business operations, especially those with rule-based, repetitive, and mature processes. Integrating workflows using RPA reduces errors and increases speed and accuracy. Inventory management, CRM, finance systems, and many more that are used in the different stages of SC can be integrated. The incorporation of Intelligent Process Automation (IPA) involving Artificial Intelligence (AI) and Machine learning (ML) ensures a disruption-free execution of the legacy-based business processes. This also improves the efficiency and accuracy of the processes and reduces the cost and manual work to a greater extent.

4.4.3 Peek into IPA and Hyperautomation

Intelligent Process Automation enhances the capability of RPA as it adds the decision-making dimension to it. IPA has made cumbersome tasks like form filling, data extraction, and storing and retrieving files a quicker process. RPA is empowered in accessing and consuming unstructured data of more than 33 zetta-bytes with the help of IPA [35]. For example, consider that AI-powered chatbots when incorporated along with RPA would help in providing key customer support across all the segments of the supply chain. It extracts the required source of information, spins the appropriate answer, and presents it during any customer interaction that too with a personal touch. This saves tons of customer support personnel time in taking up simple and routine customer questions.

With IPA implemented in an automobile supply chain, one of the difficult tasks like customized quoting has been made possible by considering the model type, customer requirements, and time to manufacture, and that too within 30 min [36]. Besides these techniques like OCR have made accessing documents, data extraction, and making sense out of it in the manner of Document Understanding (DU) a child's play. In fact, with DU models' the sales cycle has become faster [37]. It has also improved the customer experience. When we combine AI, ML, and RPA, hyperautomation is born. This drives RPA to a higher level of application. Gartner, in 2021, has listed hyperautomation among the top ten technology trends and that 65% of organizations will transition into consuming this upcoming technology [38].

4.4.4 Democratizing Automation with Low Code No Code

As RPA and Hyper automation is taking the world on a spin every organization wants to try its hands on this upcoming technology. But implementing it in its raw coded form using python and java is not everyone's cup of tea. Hence, the No Code Low Code (NC-LC) approach has been developed. This approach enables anyone to develop an RPA and automate any level of flow with a very minimal amount of coding [36]. The NC-LC approach has democratized RPA and IPA to make it feasible for any RPA enthusiasts to try automating as per their needs. UIPath, Automation Anywhere, Blue Prism, Power Automate, Pega, and RAP Studio are all some of the NC-LC RPA providers. This has also brought down the developers' time from coding thousands of lines and increased the speed of project completion time [37]. Thanks to the No code approach of RPA development, it has made many solopreneurs possible and profitable as they can't afford to have more employees, nor can they afford to have a larger infrastructure footprint. Besides these solutions are also easy to deploy and they also bring down the Total Cost of Automation (TCA).

4.4.5 Advantages of RPA

With the implementation of RPA organizations could benefit multi-folds provided they choose the right process to automate. Multiple authors have projected that RPA gives the following advantages:

- Reduction in process execution time.
- Reduction in FTE cost with the reduction of execution time and administrative overheads.
- Better quality of service could be provided as the duplication and number of errors reduces.
- Easy extracting and consuming information from multiple systems/sources.
- Higher level of employee satisfaction through the reduction in monotonous and unproductive tasks.
- Enables utilization of time in more value-adding tasks rather than the mundane ones.
- Easy to scale up and meet up with the fluctuating supply chain demand.

4.4.6 Disadvantages of RPA

Though RPA has multiple advantages it also has its own practical challenges.

- The data that RPA would work upon need to be well structured else it will not help in decision making.
- Though RPA helps in digital transformation when implemented in the wrong process, it could cost significantly high operational overheads and difficult to revert the cost.
- RPA must be developed and deployed only after a thorough study of the process needs; otherwise, it could lead to erratic and unreliable outputs.

4.5 Conclusion

RPA is playing a major role in digitizing the supply chains. Many organizations are benefiting from transforming their business from legacy systems and eventually strengthening their processes. From making effective negotiations using the data extracted from different documents to simply downloading attachments from thousands of emails, classifying, and storing them in respective locations, RPA comes as an easy and economical option. It is also easy to increase the capability of RPA with the incorporation of AI/ML in order to make better decisions. Thus, RPA application spectrum is widening day by day as organizations are exploring ways to consume this advanced technology. With low code, no code approaches to RPA a lot of development and deployment hassles have been brought down. The recent advancements have enabled RPA as a scalable and affordable option. This chapter has brought to light the Supply chain processes, their stages, and the S.C. issues faced by various businesses. It has also emphasized the advantage of implementing RPA in a selective Supply Chain. With the invoice automation use case, where the tedious and time-consuming tasks of processing invoices could be carried out with ease and in an error-free manner. One could realize the capability of RPA in enhancing the Supply Chain efficiently. RPA adoption in Supply Chain is still at its nascent stage of growth at the moment. The cost and time of FTEs carrying out Supply Chain actions manually can be reduced drastically and that too with increasing the profitability. Hence, it is high time that industries emphasize on utilizing RPA to automate suitable processes in their Supply Chain.

References

1. Costello, K., Rimol, M.: Gartner says worldwide robotic process automation software revenue to reach nearly $2 Billion in 2021. https://www.gartner.com/en/newsroom/press-releases/2020-09-21-gartner-says-worldwide-robotic-process-automation-software-revenue-to-reach-nearly-2-billion-in-2021#:~:text=Worldwide%20RPA%20software%20revenue%20is,creating%20strong%20downward%20pricing%20pressure. Last Accessed 18 Oct 2021
2. Wang (John), Z.: Supply chain management for collection services of academic libraries: solving operational challenges and enhancing user productivity. Elsevier Publications, (2017)
3. Londe, L.: Partnerships in providing customer service: a third-party perspective. The Council of logistics management. Cincinnati (1989)
4. Pringle, A., Harris, E.: The risk adjustment of required rate of return for supply chain infrastructure investments. (1987)
5. Lee, H.L., Billington, C.: Material management in decentralized supply chains. Oper. Res **41**, 835–847 (1993)
6. Quinn, F.J.: What's the buzz? Logist. Manag. **36**(2), 43–47 (1997)
7. Lummus, R.R., Vokurka, R.J.: Defining supply chain management: a historical perspective and practical guidelines. Ind. Manag. Data Syst. **99**(1), 11–17 (1999)
8. Lambert, D.M., Cooper, M.C, Pagh, J.D.: What is management in supply chain management?—A critical review of definitions, frameworks and terminology. J. Manag. Policy Pract. **11**(4), (1998)
9. Christopher, M.: logistics and supply chain management: strategies for reducing cost and improving service, 2nd Edn. Prentice Hall, (1998)
10. Gaither, N., Frazier, G.: Operations management. South-western college Publishing, (2002)
11. Prakasha, A., Agarwala, A., Kumar, A.: Risk assessment in automobile supply chain. Mater. Today: Proc. **5**, 3571–3580 (2018)
12. Deckert, C.: Supply chain. encyclopedia of sustainable management. Springer Nature, Switzerland (2020)

13. Sunil, C., Petrer, M.: Supply chain management, strategy. planning and operation, 3rd edn. Pearson Publications, Prentice Hall (2016)
14. Salomon, V.A.P, Tramarico, C.L, Marins, F.A.S.: Analytic hierarchy process applied to supply chain management, applications and theory of analytic hierarchy process - decision making for strategic decisions. (2016)
15. Ivančić, L., Suša Vugec, D., Bosilj Vukšić, V.: Robotic process automation: systematic literature review. Lecture Notes Bus. Inf. Process. **361**, 280–295 (2019)
16. OSMAN, C-C.: Robotic process automation: lessons learned from case studies. Inform. Econ. **23**, 66–71 (2019)
17. Zhang, N., Liu, B.: The key factors affecting RPA-business alignment. In: ACM Proceedings of the 3rd international conference on crowd science and engineering, pp. 1–6 (2018)
18. Enriquez, J.G., Jimenez-Ramirez, A., Dominguez-Mayo, F.J., Garcia-Garcia, J.A.: Robotic process automation: a scientific and industrial systematic mapping study. IEEE Access **8**(March), 39113–39129 (2020)
19. Mühlberger, R., Bachhofner, S., Ferrer, E.C., Ciccio, C. Di, Weber, I., Wöhrer, M., Zdun, U.: Business process management: blockchain and robotic process automation forum. (2020)
20. Scheppler, B., Weber, C.: Robotic process automation. Informatik-Spektrum **43**, 152–156 (2020)
21. Batakis, N., Spanoudakis, N., Matsatsinis, N.: Decision support tool for ranking robotic process automation candidate projects. Univ, Crete (2020)
22. Christian, W.C., Stierle, M., Dunzer, S., Matzner, M.: A framework to evaluate the viability of robotic process automation for business process activities. Springer Nature Switzerland AG (2020)
23. How robotic process automation can streamline supply chain operations. https://www.blumeglobal.com/learning/robotic-process-automation. Last Accessed 18 Oct 2021
24. Vajgel, B., Correa, P.L.P., Tossoli De Sousa, T., Encinas Quille, R.V., Bedoya, J.A.R., Al-meida, G.M. De, Filgueiras, L.V.L., Demuner, V.R.S., Mollica, D.: Development of Intellgent robotic process automation: A utility case study in Brazil. IEEE Access. **9**, 71222–71235 (2021)
25. Freeman, O.: How RPA is revolutionizing supply chain networks, supply chain digital. https://supplychaindigital.com/technology-4/how-rpa-revolutionising-supply-chain-networks. Last accessed 20 Dec 2021
26. Lawton, G.: 7 use cases for RPA in supply chain and logistics, Techtarget. https://www.techtarget.com/searcherp/feature/7-use-cases-for-RPA-in-supply-chain-and-logistics. Last Accessed 18 Oct 2021
27. Wollenhaupt, G.: Contributor the emerging wave of procurement and spend-management technology. https://www.supplychaindive.com/news/emerging-wave-procurement-spend-technology-RPA/588897. Last Accessed 06 Oct 2021
28. Viale, L., Zouari, D.: Impact of digitalization on procurement: the case of robotic process automation. Supply Chain. Forum: Int. J, (2020)
29. UiPath.: Reordering produce with rpa frees mercadinhos são luiz employees to focus on store customers. https://www.uipath.com/resources/automation-case-studies/mercadinhos-sao-luiz. Last Accessed 18 Sep 2021
30. UiPath.: Robotic Process Automation (RPA) greatly improves warehouse efficiency at sf supply chain. https://www.uipath.com/resources/automation-case-studies/rpa-improves-warehouse-efficiency-at-sf-supply-chain. Last Accessed 12 Nov 2021
31. Sarpolis, N.: Five ways legacy systems could be hindering your supply chain. https://blog.flexis.com/five-ways-legacy-systems-could-be-hindering-your-supply-chain. Last Accessed 16 Sep 2021
32. Pedersen, M., Logistics, Y.: https://www.supplychainbrain.com/articles/28235-legacy-systems-the-problem-or-the-solution. Last Accessed 23 Mar 2022
33. Electroneek, Automation of legacy systems with RPA. (2021). https://electroneek.com/blog/rpa-for-msps/automation-of-legacy-systems-with-rpa. Last Accessed 23 Mar 2022
34. Behrens, K.: RPA as a solution to your legacy needs. (2014). https://www.uipath.com/blog/rpa/legacy-systems

35. 10xDS, How linking robotic process automation with legacy systems help enhance sytems integration. (2021). https://10xds.com/blog/how-rpa-enhances-legacy-systems. Last Accessed 12 Nov 2021
36. How intelligent automation enables humans to focus on higher value tasks. (2021). https://blog.rapidautomation.ai/how-intelligent-automation-enables-humans-to-focus-on-higher-value-tasks. Last Accessed 12 Nov 2021
37. Rapid Acceleration Partners ai, 4 benefits of hyper-automation for the automobile industry. (2022). https://blog.rapidautomation.ai/4-benefits-of-hyper-automation-for-the-automobile-industry, Last Accessed 12 Nov 2021
38. Automated Automobile Sales: The future of buying a car. (2021). https://blog.rapidautomation.ai/automated-automobile-sales-the-future-of-buying-a-car, Last Accessed 18 Sep 2021
39. Jack, M.: Robotic Process Automation (RPA) trends and predictions for 2022, https://wearebrain.com/blog/enterprise-automation/robotic-process-automation-rpa-trends-and-predictions-2022/#:~:text=Global%20Robotic%20Process%20Automation%20(RPA,hit%20%2413.74%20billion%20by%202028. Last Accessed 18 Sept 2021

Chapter 5
Intelligent Document Processing in End-to-End RPA Contexts: A Systematic Literature Review

A. Martínez-Rojas, J. M. López-Carnicer, J. González-Enríquez, A. Jiménez-Ramírez, and J. M. Sánchez-Oliva

Abstract Automating organizational processes typically involves document processing techniques for a large document set. For that purpose, the Intelligent Document Processing (IDP) paradigm has been studied for decades. With the fast emergence of Robotic Process Automation (RPA) in the process automation landscape, the industrial solution of IDP with RPA integration has risen significantly in the last few years. However, there is no up-to-date overview of the available knowledge in this area. Therefore, this chapter studies the current scientific knowledge about IDP and its integration into RPA through a systematic literature review that analyzed 77 primary studies. In addition, an industry review was performed, analyzing and characterizing 37 industrial tools. Although the results confirm the growth in the research interest in IDP in different dimensions, they also identify a lack of proposals that integrate IDP and RPA paradigms in confrontation with the industrial solutions that have increasingly led to its integration.

This research has been supported by the NICO project (PID2019-105455GB-C31) of the Spanish Ministry of Science, Innovation and Universities and the CODICE project (EXP 00130458/IDI-20210319 - P018-20/E09) of the Center for the Development of Industrial Technology (CDTI) and by the FPU scholarship program, granted by the Spanish Ministry of Education and Vocational Training (FPU20/05984).

A. Martínez-Rojas (✉) · J. M. López-Carnicer · J. González-Enríquez · A. Jiménez-Ramírez
Departamento de Lenguajes y Sistemas Informáticos, Escuela Técnica Superior de Ingeniería Informática, Avenida Reina Mercedes, s/n., 41012 Sevilla, Spain
e-mail: amrojas@us.es

J. M. López-Carnicer
e-mail: jlopez16@us.es

J. González-Enríquez
e-mail: jgenriquez@us.es

A. Jiménez-Ramírez
e-mail: ajramirez@us.es

J. M. Sánchez-Oliva
Servinform, S.A. Parque Industrial PISA, Calle Manufactura, 5, 41927 Mairena del Aljarafe, Sevilla, Spain
e-mail: jmsanchezo@servinform.es

S. Bhattacharyya et al. (eds.), *Confluence of Artificial Intelligence and Robotic Process Automation*, Smart Innovation, Systems and Technologies 335,
https://doi.org/10.1007/978-981-19-8296-5_5

Keywords Intelligent document processing (IDP) · Robotic process automation (RPA) · Systematic literature review · Industrial review

5.1 Introduction

The interest in the Robotic Process Automation (RPA) paradigm has increased in recent years [58, 79]. One of the most recurrent topics is its integration with cognitive elements [18]. More precisely, contemporary literature acknowledges that the field of Intelligent Document Processing (IDP) is of utmost importance as a complementary technology in RPA [58], since IDP is a common element in multiple business domains related to RPA, e.g., back-office and accounting. Furthermore, in the industrial scope, most RPA companies include IDP components in their tools. This fact is mainly due to the dependency on documents within their business processes.[1,2]

The IDP paradigm refers to the set of processes that transform structured or unstructured documents into usable information [72]. Similar to RPA, this area has been significantly active since 2017 regarding topics like the improvement in Optical Character Recognition (OCR) systems [55], the inclusion of cognitive elements [49], and its integration with RPA [18].

Although the current situation of IDP is explored in some studies [8, 19], they lack an in-depth study of the state of the art directly related to this topic. Meanwhile, other fields that are intrinsically related to IDP have been strongly studied, such as RPA [21, 79], Artificial Intelligence (AI) integration [6, 43], and specific areas inside IDP such as classification techniques [73] or OCR [52, 55, 87]. Nonetheless, they have been addressed as independent fields, thus lacking a broad vision in the area of IDP that shed light on their relationships and advances as a whole. Therefore, a comprehensive scientific study related to IDP in the context of RPA is necessary. Consequently, this chapter proposes a Systematic Literature Review (SLR) about IDP over the last years, which analyzes 77 documents and reports on the current state of this topic. In addition, the state of the industry related to this area is analyzed through the study of 37 tools.

The analysis of these studies and tools reveals that the scientific studies are focused on the use of new techniques that improve the performance in specific phases of the IDP pipeline, leaving aside integration with RPA. In contrast, industrial tools emphasize massive document processing and integration with RPA. Finally, a series of gaps and challenges have been detected, which open up future lines of work in the field of IDP. The rest of the chapter is organized as follows. Section 5.2 analyzes the related work that studies the importance of IDP in RPA contexts. Section 5.3 describes the scientific scope methodology and its application, and Sect. 5.4 describes the industrial research methodology and its application. Section 5.5 summarizes the

---

[1] https://www.automationanywhere.com/rpa/intelligent-document-processing.

[2] https://www.abbyy.com/flexicapture/.

limitations of the current work, discussions, gaps, and challenges. Finally, Sect. 5.6 concludes the chapter and describes future work.

5.2 Related Works

Reviews and surveys related to RPA, cognitive elements integration, or related fields of study widely appear in the literature. For instance, [36] investigates the definition of RPA by the community presenting an SLR. Enríquez et al. [21] analyze scientific publications and industrial tools in RPA by performing a systematic mapping study. In addition, Syed et al. [79] present an SLR proposing 15 challenges for future research. However, these works are just focused on RPA, and the relation with the IDP context is minimal.

A wide variety of studies are related to technologies associated with IDP and RPA, e.g., OCR. These studies compare industrial tools [80], study the integration of OCR with multiples technologies [77], and review technological OCR approaches [55]. Conversely to our work, these studies are just focused on one of the techniques that may be applied in IDP and RPA projects.

Although IDP is strongly related to RPA, not many studies investigate them together. Some studies analyze a particular part of the field, as Baviskar et al. [9] that presents a SLR that focuses on the processing of unstructured documents enhanced by AI, indicating future lines of research. In addition, Baviskar et al. [8] examine the influence and status of IDP from different perspectives, e.g., publications trend by year, publications type (i.e., conference and journal publication, among others), publications scientific discipline (i.e., computer science, engineering, and mathematics, among others), organized by authors, countries, organizations, and languages, and the most cited ones. Although this paper delivers a comprehensive statistical review of the literature, it does not analyze paradigms and techniques, unlike our study. Furthermore, Cristani et al. [19] develops a study about industrial IDP until 2018 that investigates the scientific publications and the approaches IDP companies implement. This study only considers certain business documents (i.e., invoices and orders). The authors also suggest an IDP architecture based on the study of the industries and scientific community perspective.

Unlike other approaches, the current chapter aims to study the IDP field and its relationship with RPA from a broad perspective for any scientific and industrial document domain. These analyses will answer a series of research questions, selecting only those studies and tools that satisfy specific criteria. Additionally, thanks to this methodological and in-depth scientific/industrial analysis, a series of untracked gaps and challenges have been identified in the field of IDP and the related RPA.

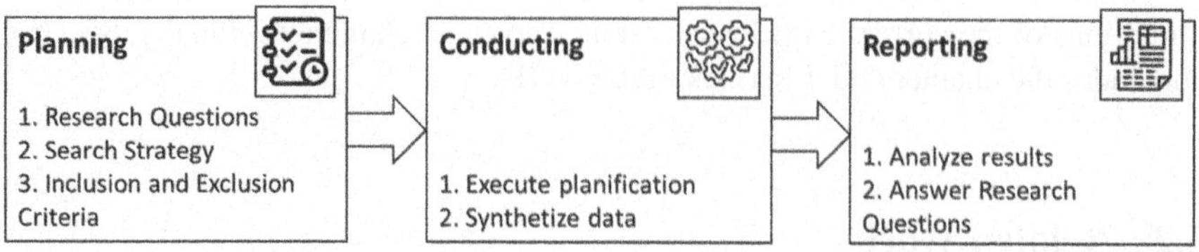

Fig. 5.1 Adapted methodology from Kitchenham and Charters [44]

5.3 Scientific Methodology

The SLR follows a widely accepted engineering methodology for conducting systematic reviews proposed by Kitchenham and Charters [44]. Figure 5.1 illustrates the method followed, including minor changes (i.e., reduce some steps) to adapt it to perform the industrial review.

5.3.1 Planning

This section describes the steps of the planning phase, i.e., the (RQs) to be solved, the search strategy, and the method's criteria selection.

5.3.1.1 Research Questions

In this study, a total of 7 RQs have been proposed to provide a comprehensive analysis of the state of the art in the context of IDP and RPA proposed (cf. Table 5.1).

5.3.1.2 Search Strategy

This step details how the studies were retrieved in the different digital libraries. Table 5.2 shows both the mandatory and the optional keywords that helped to build the queries. These words were combined in different digital libraries, i.e., IEEE Xplore, Scopus, Science Direct, and ACM. The concrete query strings executed for each digital library are shown in Table 5.3.

5.3.1.3 Inclusion and Exclusion Criteria

It is necessary to filter the results obtained from the queries to obtain the final list of primary studies to be analyzed in depth. The studies retrieved from the different libraries are filtered according to the inclusion/exclusion criteria described in Table 5.4.

Table 5.1 Research questions

RQ	Question	Aim
RQ1	What are the types of publications of the proposals?	Statistical analysis of studies based on the type of study (conference, article) and the number of publications
RQ2	How are IDP approaches being validated?	Verification of the types of validation in which the publications are included, whether scientific (running example, survey, or synthetic study), industrial (company case study, knowledge transfer, etc.), or without validation
RQ3	What are the proposals researched?	Analysis of the presented type of proposal (framework, algorithm, method, and theoretical analysis, among others)
RQ4	How integrated is task automation?	Analysis of integration with automated solutions or RPA. In case of not having it, evaluate its possible integration
RQ5	What are the types of documents addressed?	Analysis of the type of target document or specific application environment. Basic classification of the documents found, whether structured, unstructured, handwritten, or typed, and specify other relevant information about the documents
RQ6	What are the main application domains?	Context in which IDP proposals are developed. Domain refers to both the context of the problem being solved and the type of problem supported (banking, business, health, etc.)
RQ7	What are the main phases in which IDP proposals are framed?	Identification of the IDP pipeline phases more researched

Table 5.2 Proposed keywords

Mandatory words	Optional words
Document processing	Intelligent, cognitive, automation, pipeline, structured, unstructured, handwritten, typewritten, preprocessing, segmentation, classification, extraction, detection, RPA, robotic process automation, OCR, character recognition

5.3.2 *Conducting*

As depicted in Table 5.5, the initial search obtained 1,405 studies in total. The pool of studies ended up with 77 primary studies after applying the five inclusion/exclusion criteria, removing the duplicated entries obtained from the different digital libraries, and including an expert recommendation. Additionally, to ensure the reliability of C5, one of the authors that did not execute this criterion randomly selected 30% of

Table 5.3 Executed queries

Digital library	Query
IEEE XPLORE	("All Metadata":"Document Processing") AND ("All Metadata":"intelligent" OR "All Metadata":"cognitive" OR "All Metadata":"automation" OR "All Metadata":"pipeline" OR "All Metadata":"character recognition" OR "All Metadata":"automation" OR "All Metadata":"structured" OR "All Metadata":"unstructure" OR "All Metadata":"handwritten" OR "All Metadata":"typewritten" OR "All Metadata":"preprocessing" OR "All Metadata":"segmentation" OR "All Metadata":"classification" OR "All Metadata":"extraction" OR "All Metadata":"detection" OR "All Metadata":"RPA" OR "All Metadata":"robot process automation")
Scopus	TITLE-ABS-KEY (("Document processing") AND (("intelligent" OR "cognitive" OR "automation" OR "pipeline" OR "RPA" OR "robot process automation" OR "structured" OR "unstructured" OR "handwritten" OR "typewritten" OR "preprocessing" OR "segmentation" OR "classification" OR "extraction" OR "detection" OR "OCR" OR "character recognition")))
Science Direct[a]	("document processing") AND ("intelligent" OR "cognitive" OR "automation" OR "pipeline")
	("document processing") AND ("preprocessing" OR "segmentation" OR "classification" OR "extraction" OR "detection")
	("document processing") AND ("OCR" OR "character recognition" OR "RPA" OR "robot process automation")
	("document processing") AND ("structured" OR "unstructured" OR "handwritten" OR "typewritten")
ACM	Title:("document processing" AND ("intelligent" OR "cognitive" OR "automation" OR "OCR" OR "character recognition" OR "pipeline" OR "structured" OR "unstructured" OR "handwritten" OR "typewritten" OR "preprocessing" OR "segmentation" OR "classification" OR "extraction" OR "detection" OR "robot process automation" OR "RPA")) OR Abstract: ("document processing" AND ("intelligent" OR "cognitive" OR "automation" OR "OCR" OR "character recognition" OR "pipeline" OR "structured" OR "unstructured" OR "handwritten" OR "typewritten" OR "preprocessing" OR "segmentation" OR "classification" OR "extraction" OR "detection" OR "robot process automation" OR "RPA")) OR Keyword: ("document processing" AND ("intelligent" OR "cognitive" OR "automation" OR "OCR" OR "character recognition" OR "pipeline" OR "structured" OR "unstructured" OR "handwritten" OR "typewritten" OR "preprocessing" OR "segmentation" OR "classification" OR "extraction" OR "detection" OR "robot process automation" OR "RPA"))

[a]Due to search limitations, the 4 different searches were executed and pooled

the results and applied it to check that studies were correctly included/excluded. The resulting primary studies are summarized, synthesizing its content in Tables 5.6 and 5.7.[3]

[3] A complete analysis of the analysis executed can be found on the sheet with the title "Scientific" within the Excel document available at https://doi.org/10.5281/zenodo.6400519.

Table 5.4 Inclusion and exclusion criteria

Code	Description
C1	Mandatory and optional words must appear in the title, abstract, or keywords
C2	Studies must have a publication date of 2017 or later. Although IDP has been studied for decades, a new interest has arisen as new ways of employing emerging technologies in this sector have been found through the use of RPA and AI (cf. Baviskar et al. [8] and in Fig. 5.2), analyzing the publications in Scopus, one of the most prolific digital library between those used in the study This increase in the trend in IDP coincides in date with the increase in the number of publications in the field of RPA, as can be observed in the study presented by Enríquez et al. [21]
C3	Studies must be written in English and fit the area of engineering and/or computer science
C4	Only studies published in recognized rankings will be considered. More precisely, journal publications must belong to the Journal Citation Reports (JCR)[a], and conferences must belong to the GII-GRIN-SCIE (GGS)[2] ranking
C5	Studies must fit in the area of IDP and/or RPA. For this, the abstract and conclusions are analyzed

[a]Journal Citation Reports, https://jcr.clarivate.com
[b]Conference Rating GGS, http://scie.lcc.uma.es/gii-grin-scie-rating/ratingSearch.jsf

5.3.3 *Reporting Over the Research Questions*

The review methodology ends with the reporting step to answer the RQs.

5.3.3.1 RQ1. What are the Types of Publications of the Proposals?

This section is intended to reveal the statistic analysis of the primary studies analyzed in Tables 5.7 and 5.6. The origin is being investigated by means of the type of publication (i.e., journal or conference). In addition, the evolution of the trend in the number of publications is analyzed.

As shown in Fig. 5.3, the main type of contributions are conferences with 45 contributions (i.e., 58.44%) while 32 of them (i.e., 41.56%) correspond to journals. Additionally, as shown in Fig. 5.4, except for 2018, when there is a slight decrease in the number of publications, there is a clear upward trend up to the current date. Note that the number of publications in 2022 only considers the first month of the year and, therefore, it is expected that the trend will be kept.

5.3.3.2 RQ2. How are IDP Approaches Being Validated?

The type of validation proposed in the primary studies for IDP is analyzed regarding the context where they took place. In this sense, three contexts were observed: (i) scientific context, where validations are experiments or case studies with synthetic or real data; (ii) industrial context, where the proposed validation is related to real-world

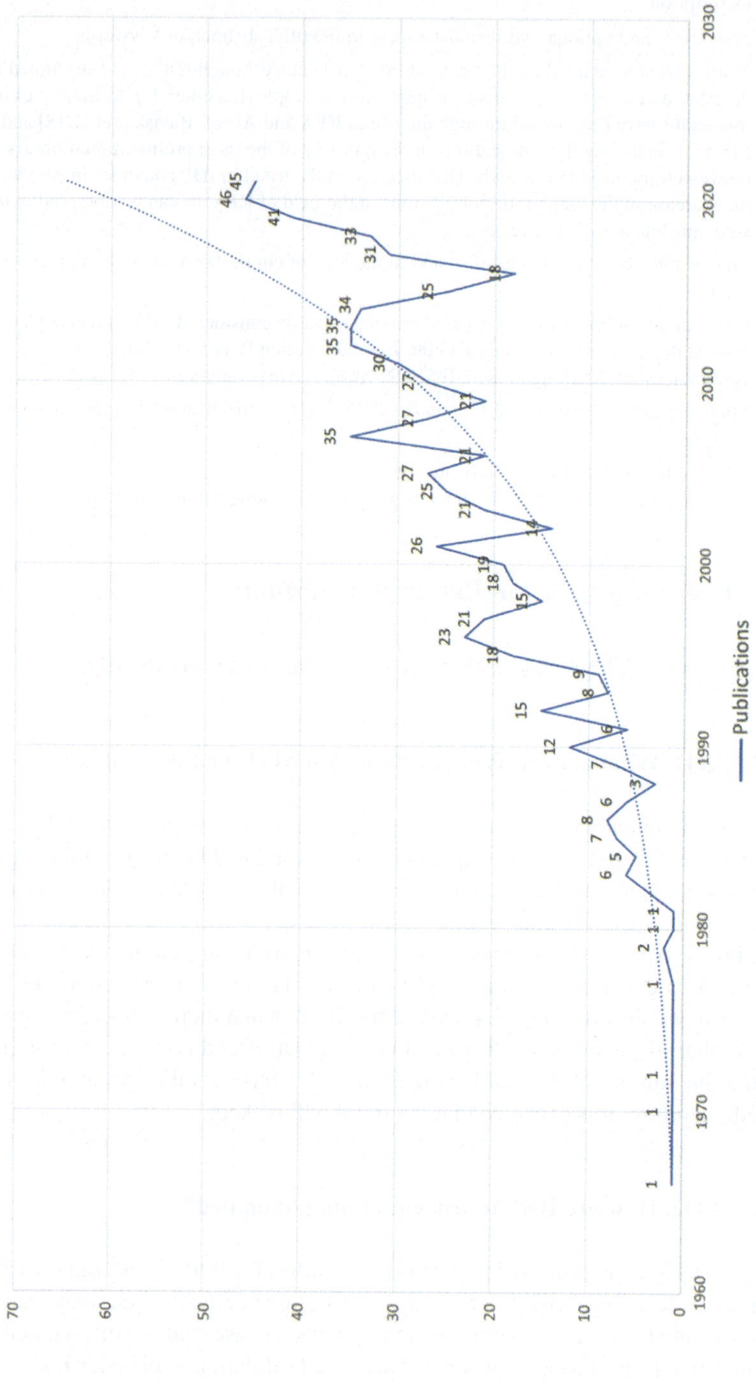

Fig. 5.2 Tendency of the number of IDP publications in the Scopus until December 2021

Table 5.5 Number of studies per digital library after applying the inclusion and exclusion criteria

Criteria	IEEE XPLORE	Science direct	Scopus	ACM	Totals
C1	339	176	843	47	1405
C2	54	27	203	10	1243
C3	50	19	194	10	273
C4	27	17	88	7	139
C5	22	12	68	5	107
Duplicates					76
Recomm.					77

Table 5.6 Journal primary studies

References	Titles	Years
[86]	Script identification of multi-script documents: a survey	2017
[33]	An efficient open system for offline handwritten signature identification based on curvelet transform and one-class principal component analysis	2017
[84]	Word graphs size impact on the performance of handwriting document applications	2017
[89]	DivaServices-a RESTful web service for document image analysis methods	2017
[54]	Sentence level matrix representation for document spectral clustering	2017
[3]	Blind document image quality prediction based on modification of quality aware clustering method integrating a patch selection strategy	2018
[10]	Text and non-text separation in offline document images: a survey	2018
[16]	Deep neural networks for document processing of music score images	2018
[90]	Paragraph vector representation based on word to vector and CNN learning	2018
[19]	Future paradigms of automated processing of business documents	2018
[65]	Text baseline detection, a single page trained system	2019
[4]	A new threshold selection method based on fuzzy expert systems for separating text from the background of document images	2019
[82]	CONFIRM—Clustering of noisy form images using robust matching	2019
[85]	Document representation and classification with Twitter-based document embedding adversarial domain-adaptation and query expansion	2019

(continued)

Table 5.6 (continued)

References	Titles	Years
[62]	Information extraction from text-intensive and visually rich banking documents	2020
[23]	A small samples training framework for deep learning-based automatic information extraction: case study of construction accident news reports analysis	2020
[42]	Holistic design for deep learning-based discovery of tabular structures in datasheet images	2020
[59]	Transformers-based information extraction with limited data for domain-specific business documents	2020
[40]	Additive angular margin loss in deep graph neural network classifier for learning graph edit distance	2020
[13]	A holistic approach for automatic deep understanding and protection of technical documents	2020
[78]	Forgery document detection in information management system using cognitive techniques	2020
[60]	Application of machine learning for document classification and processing in adaptive information systems	2020
[8]	A bibliometric survey on cognitive document processing	2020
[48]	TabCellNet: deep learning-based tabular cell structure detection	2021
[2]	Deep neural network-based contextual recognition of Arabic handwritten scripts	2021
[35]	Learning from similarity and information extraction from structured documents	2021
[57]	Blur2sharp: a GAN-based model for document image deblurring	2021
[73]	Document classification in robotic process automation using artificial intelligence—a preliminary literature review	2021
[11]	Beyond document object detection: instance-level segmentation of complex layouts	2021
[1]	TNCR: Table net detection and classification dataset	2022
[25]	TableDet: an end-to-end deep learning approach for table detection and table image classification in data sheet images	2022
[32]	Information extraction from scanned invoice images using text analysis and layout features	2022

case studies; and (iii) no validation, when the study does not propose any validation. As illustrated in Fig. 5.5, most of the proposals present scientific validation (i.e., 96.1%) while a minority of proposals lack any validation (i.e., 2.6%). In turn, just one was applied in a real-world industry case (i.e., 1.3%).

According to this data, we can conclude that there is a lack of studies that are validated in real industries. On the contrary, most studies use datasets, test cases, or small cases in a specific domain.

Table 5.7 Conference primary studies

References	Titles	Years
[22]	Baseline detection on Arabic handwritten documents	2017
[63]	Lightweight multilingual entity extraction and linking	2017
[14]	anyOCR: An open-source OCR system for historical archives	2017
[75]	Assisted transcription of historical documents by keyword spotting: a performance model	2017
[81]	Document image binarization with fully convolutional neural networks	2017
[92]	Checking the statutes in chinese judgment document based on editing distance algorithm	2017
[37]	A novel mixed approach for detecting overlap in document images	2017
[27]	Historical document processing	2017
[28]	Agricultural knowledge extraction from text sources using a distributed mapreduce cluster	2017
[91]	A Handwritten Japanese historical Kana reprint support system: development of a graphical user interface	2018
[86]	Detection of negation in the Serbian language	2018
[61]	dhSegment: A generic deep-learning approach for document segmentation	2018
[56]	Understanding documents with hyperknowledge specifications	2018
[31]	Combination of deep learning and syntactical approaches for the interpretation of interactions between text-lines and tabular structures in handwritten documents	2019
[94]	Federated learning of unsegmented Chinese text recognition model	2019
[29]	Layout and text extraction from document images using neural networks	2019
[83]	Field typing for improved recognition on heterogeneous handwritten forms	2019
[69]	Rethinking table recognition using graph neural networks	2019
[74]	Document domain adaptation with generative adversarial networks	2019
[41]	Deep learning for recognizing the anatomy of tables on datasheets	2019
[39]	Automatic classification and recognition of complex documents based on Faster RCNN	2019
[7]	An approach to estimate skew angle in printed document images	2019
[15]	Model-based integration of unstructured web data sources using graph representation of document contents	2019
[38]	High-precision deep learning-based tabular position detection	2020
[70]	Document processing: methods for semantic text similarity analysis	2020
[71]	A novel deep learning character-level solution to detect language and printing style from a bilingual scanned document	2020
[12]	A comparison of sequential and combined approaches for named entity recognition in a corpus of handwritten medieval charters	2020

(continued)

Table 5.7 (continued)

References	Titles	Years
[66]	Historical document processing: a survey of techniques tools and trends	2020
[26]	Quality evaluation for documental big data	2020
[64]	Fast end-to-end coreference resolution for Korean	2020
[20]	Hierarchical recurrent neural network for handwritten strokes classification	2021
[5]	FinSBD-2021: the 3rd shared task on structure boundary detection in unstructured text in the financial domain	2021
[50]	The FinSim-2 2021 shared task: learning semantic similarities for the financial domain	2021
[53]	ImpactCite: an XLNet-based solution enabling qualitative citation impact analysis utilizing sentiment and intent	2021
[68]	TexRGAN: a deep adversarial framework for text restoration from deformed handwritten documents	2021
[47]	Intelligent document processing method based on robot process automation	2021
[24]	Engineering of an artificial intelligence safety data sheet document processing system for environmental, health, and safety compliance	2021
[88]	Case study of few-shot learning in text recognition models	2021
[67]	Detection and localization of struck-out-strokes in handwritten manuscripts	2021
[30]	HCRNN: a novel architecture for fast online handwritten stroke classification	2021
[93]	End-to-End approach for recognition of historical digit strings	2021
[34]	Inventory and content separation in grammatical descriptions of languages of the world	2021
[76]	LayoutParser: a unified toolkit for deep learning-based document image analysis	2021
[17]	Line segmentation of individual demographic data from Arabic handwritten population registers of Ottoman Empire	2021
[46]	Linecounter: learning handwritten text line segmentation by counting	2022

5.3.3.3 RQ3. What are the Proposals Researched?

This question aims to find out the nature of the proposals. Different types of proposals were observed regarding the main contributions of the study.

As can be seen in Fig. 5.6, *models* is the most frequent type of contribution (i.e., 54.55%) followed by *algorithms* (i.e., 40.26%) and *frameworks* (i.e., 12.99%). The least researched proposals are those concerning the study of the phases of the IDP pipeline *pipeline*, with only one study investigating this branch (i.e., 2.6%), followed by *surveys* (i.e., 3.9%), *theoretical* research (i.e., 9.09%), and *methodological*

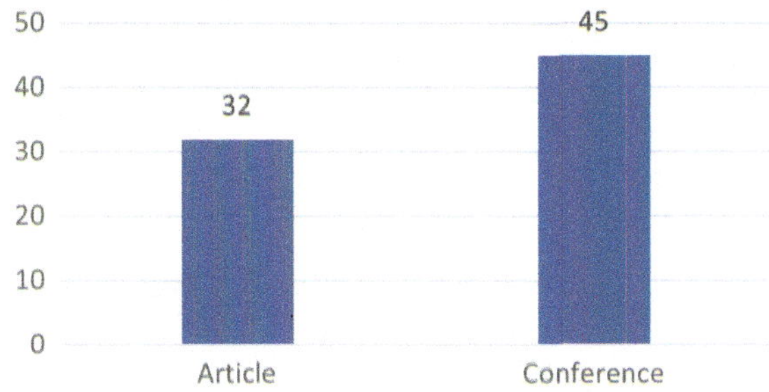

Fig. 5.3 Total number of contributions by type of publication

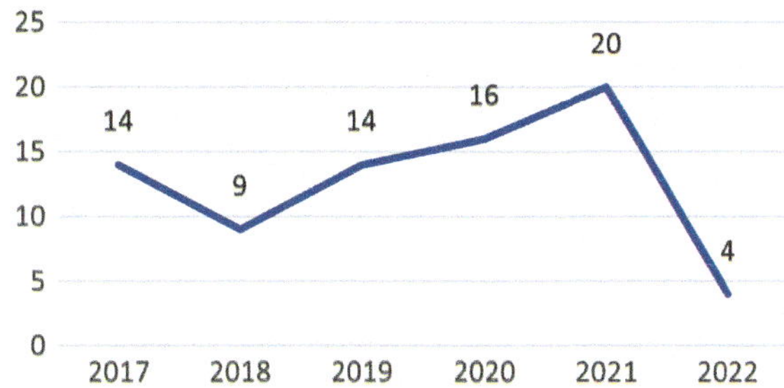

Fig. 5.4 Total number of contributions per year

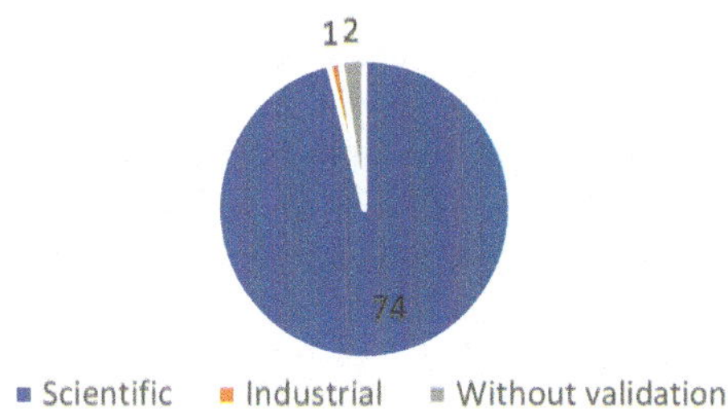

Fig. 5.5 Total number of contributions per validation type

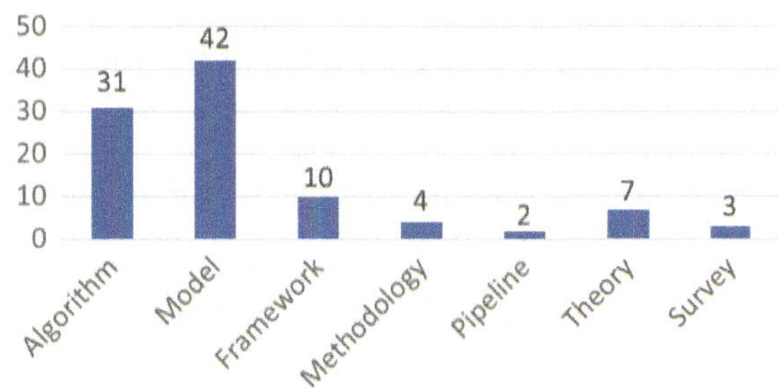

Fig. 5.6 Total number of contributions per types of proposal

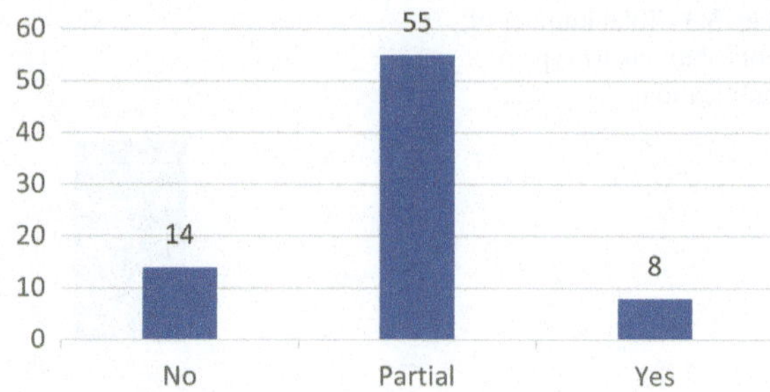

Fig. 5.7 Total number of contributions per level of integration with task automation

research (i.e., 5.19%). Note that one work might refer to more than one type of contribution.

5.3.3.4 RQ4. How Integrated is Task Automation?

Figure 5.7 reflects where the proposals consider any kind of *automation*. The *partial* column label denotes that the proposal can be integrated but does not address automation directly.

Although there are proposals that consider the automation of IDP processes , these are a minority with 8 (i.e., 10.39%). In turn, those that cannot be integrated into any automation initiative account for 14 proposals (i.e., 18.18%). Finally, partial proposals (i.e., 71.43%) do not consider automation but are considered integrable due to the particular nature of the proposal.

Usually, the tools that allow task automation are used in business processes that need IDP techniques, denoting the importance of RPA in the sector [58].

5.3.3.5 RQ5. What are the Types of Documents Addressed?

After analyzing the studies, a wide variety of challenges are encountered when processing different documents. Most of the encountered challenges were related to structured, unstructured, handwritten, or typewritten documents.

Analyzing the results (cf. Fig. 5.8), it can be seen that there are no significant differences when dealing with different types of documents, although structured documents (i.e., 79.22%) stand out above the others. However, unstructured documents (i.e., 55.84%) and handwritten documents (i.e., 58.44%) receive the same attention, while typewritten documents (i.e., 46.75%) receive less interest from the community.

After reading the studies, it can be observed that there are already previous proposals that solve these types of problems. The new solutions only aim to study which ones have better performance or reduce the number of errors produced. It should be noted that this is an area studied for a long time.

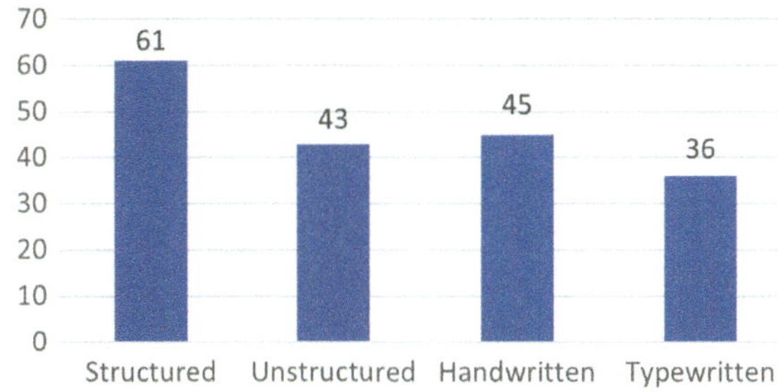

Fig. 5.8 Total number of contributions per type of document addressed

5.3.3.6 RQ6. What are the Main Application Domains?

The studies can be applied to almost any domain that requires document processing, although specific use cases are presented as a basis, e.g., handwritten historical documents. Nonetheless, the proposals could be applied to other domains.

As can be seen in Fig. 5.9, 68.83% of the proposals are of general scope. This is because they are proposals for general use or for improving internal functionalities of the IDP phases regardless of the type of domain to which they apply. Besides general-purpose studies, those based on historical documents are in the second position (i.e., 10.39%) and financial documents in the third position (i.e., 7.79%). Publications that do not belong to any specific domain, technical documentation, spoken language, and authentication documents share the same level of contributions (i.e., 2.6%).

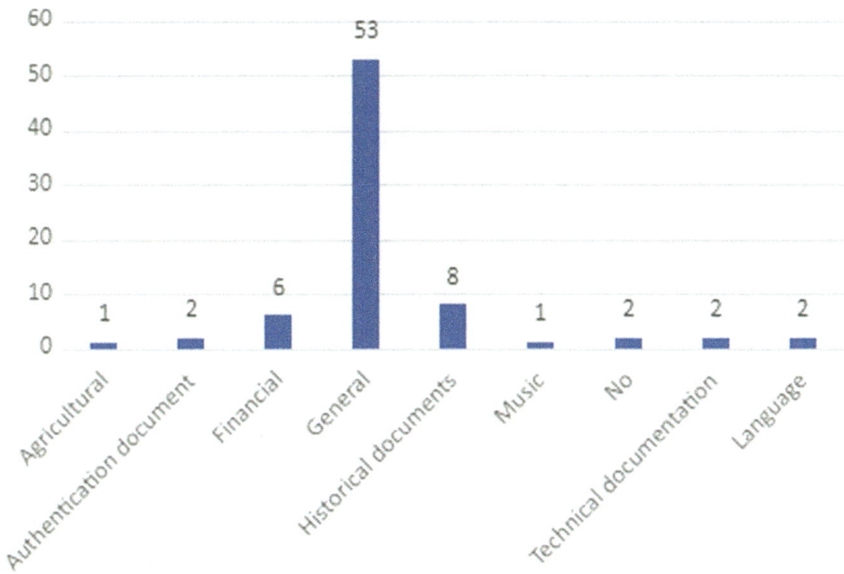

Fig. 5.9 Total number of contributions according to IDP application domains

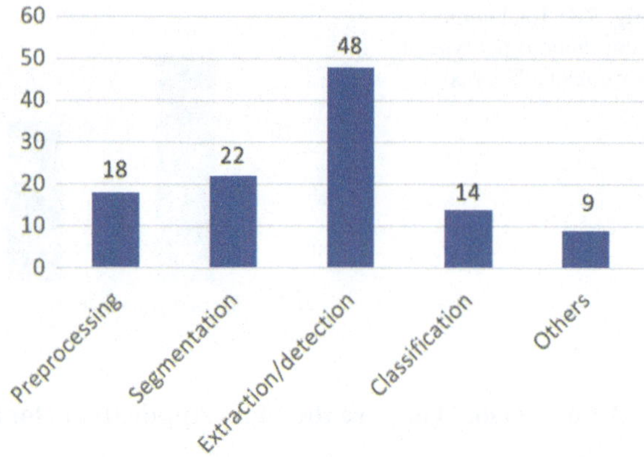

Fig. 5.10 Total number of contributions per document processing phases

Meanwhile, the remaining domains (i.e., music and agriculture) share the lowest level of interest with one publication (i.e., 1.3%).

5.3.3.7 RQ7. What are the Main Phases in Which IDP Proposals are Framed?

This question aims to show which phases receive the most significant attention by the community. It should be borne in mind that the same proposal can deal with multiple document processing phases.

As shown in Fig. 5.10, the field that attracts the most research interest is the improvement of the *extraction/detection* phase (i.e., 62.34%), which encompasses Natural Language Processing, OCR, entity detection, and other functionalities. Document *segmentation* ranks second (i.e., 28.57%). In the tail, *preprocessing* (i.e., 23.38%), *classification* (i.e., 18.18%), and *other* types of proposals not specifically related to a specific phase (e.g., configurations, training, and unitized phases) receive less interest from the community (i.e., 11.69%).

5.4 Industrial Methodology

In this section, IDP-related software or commercial tools are analyzed following the same method illustrated in Fig. 5.1.

Table 5.8 Research questions

RQ	Question	Aim
RQ1	What are the main software tools on the market for intelligent document processing?	Find and identify the main (and the most promising) software tools on the market for IDP
RQ2	What are the main functionalities that can characterize each phase of the IDP pipeline?	Identify the degree of support offered by the main tools available in the market for the functionalities of the IDP pipeline phases
RQ3	How well supported are the main industry tools concerning the general characteristics identified?	Identify the degree of support provided by the main tools on the market for the general characteristics
RQ4	How widespread are AI and automation in the IDP sector?	Determine how and to what extent the analyzed tools use AI and the status of IDP automation

Table 5.9 Keywords

Mandatory words	Related words
Document processing	Intelligent, cognitive, automation, pipeline, software, tool

5.4.1 Planning

In this phase, the need for this review is identified within the industrial environment in terms of tools or software platforms that allow IDP management. For this purpose, a series of RQs are formulated, and the definition of the review protocol.

5.4.1.1 Research Questions

This study tries to answer the following RQs to provide a comprehensive analysis of the state of IDP in RPA (cf. Table 5.8).

5.4.1.2 Search Strategy

The first step in defining the search strategy is to identify the generalist search engines to be used for queries. The search engines selected were Google, Bing, and AOL Search (cf. Table 5.9).[4,5,6]

[4] https://www.google.es/.

[5] https://www.bing.com.

[6] https://search.aol.com/aol/webhome.

Table 5.10 Inclusion and exclusion criteria

Code	Description
C1	Software or service that includes document processing, whether this is a compliment or the primary objective of the tool
C2	Current software with technical support and updates
C3	Read the specifications provided about the tool, including whether it is software that includes IDP that meets the above criteria. If in doubt, read the associated documentation and tutorials
C4	Read the specifications provided about the tool, including whether it is software that includes IDP that meets the above criteria. If in doubt, read the associated documentation and tutorials

5.4.1.3 Inclusion and Exclusion Criteria

The tools which are obtained from the search engines are filtering according to the following inclusion/exclusion criteria (cf. Table 5.10).

The procedure was to try to answer each of the RQs posed and save each of the characteristics that defined it to define a classification criterion for each selected tool. Therefore, a series of defining characteristics will be obtained for each question, cf. Table 5.11.

Those tools for which it has not been possible to obtain transparent information obtain a result of N/A. There could be information about it, but not easily accessible, or it could be that, since they do not have such functionality, no reference to it is explicitly stated in any documentation.

5.4.2 Conducting

The search process described in Sect. 5.4.1 is carried out using different search engines. These searches result in a set of tools shown in Table 5.12.[7]

5.4.3 Reporting over the Research Questions

Once all the industrial tools have been presented together with the knowledge extracted from each one, each research question can be answered.

[7] The relationship between these tools and the characteristics described in Table 5.11 are shown in the sheet with the title "Industrial" in the Excel document available at https://doi.org/10.5281/zenodo.6400519.

Table 5.11 General features

Ref	General features	Description
G1	Automation level	The level of automation available in the IDP tool. If it has some type of built-in automation (Yes), if it does not have any type of automation (No), if it has automation through integration with other platforms (Partial)
G2	Phases of the IDP pipeline supported	Phases of the IDP pipeline are supported by each evaluated tool. Identify what these phases are (Preprocessing, segmenting, extraction/detection, classification, others), indicating whether it is performed (Yes) or not contemplated (No)
G3	Tool provides a web distribution	If the tool is a web version (Yes), desktop (No), both through shared features (Partial)
G4	Free license	Whether the license type is free (Yes), partially free with limitations (Partial), or commercial only (No)
G5	Acceptable learning curve	Level of difficulty of the learning curve, based on a subjective opinion of the author of this report based on the tutorials and documentation found, indicating whether it is relatively easy (Yes), medium difficulty (Partial), or complex (No)
G6	Deployed on own servers with no external visibility	In case it is a web tool if it can be self-deployable: Yes, No
G7	Process structured and unstructured information	Support for structured, unstructured, or both documents
G8	Graphical process editor	The tool provides graphical editors to facilitate the user to build the process execution flow with, for example, drag&drop options (Yes) or not (No)
G9	Visual adaptation and configuration	Allows the adaptability of visual elements (icons, logos, etc.): Yes, No, Partial
G10	Allows process configuration	It allows documenting in detail the characteristics of each process, including objectives, metrics, deliverables, assumptions, team members, scope, stakeholders, customers, input data, output data, customer requirements, other comments, etc. It is classified as Yes or No
G11	Simultaneous management of several processes	The tool allows you to manage multiple document processing processes simultaneously: Yes or No
G12	Reporting and statistics	The tool provides statistics on the documents handled: Yes or No
G13	Robot output flow configuration	The tool allows you to define and configure the exit flow of the process in case of failures in the usual flow: Yes, No, Partial

(continued)

Table 5.11 (continued)

Ref	General features	Description
G14	Debug process mode	The tool has a debugging and step-by-step execution mode and the inspection of variables: Yes, No, in case of having some kind of trace that allows seeing the errors, but it is not a debugging mode, Partial
G15	Export process	The tool allows the export of processes configured in the platform: Yes, No, in case of having some kind of export but it is not an export functionality, Partial
G16	Multiple users	The tool allows access to multiple users at the same time: Yes or No
G17	Supported languages	The tool allows you to process documents in multiple languages. If it allows 5 or more languages, it will be considered multiple languages; if it has only accepted languages, they will be made explicit.

5.4.3.1 RQ1. What are the Main Software Tools on the Market for Intelligent Document Processing?

As previously found, the field of IDP has been studied for decades, so there have been industrial tools for IDP years plaguing the market. An external report (i.e., Everest Group [72]) has been chosen to check the results and compare the tools mentioned in those reports with the tools found in the present study. This report marked as leaders the tools offered by WorkFusion, Automation Anywhere, Kofax, ABBYY, IBM, and Antworks. Then, except for HCL Technologies, the companies and products in the external report match the search results and have not been excluded by the inclusion/exclusion criteria. The conclusion is that these tools are not only the main ones according to the external reports, but they also coincide with the results of this study.

5.4.3.2 RQ2. What are the Main Functionalities that Can Characterize Each Phase of the IDP Pipeline?

The IDP pipeline identified during the scientific domain contains the following phases: preprocessing, segmentation, extraction/detection, classification, and others. During the analysis of the industrial tools, it has been observed that the same pipeline phases are also widely used. Nevertheless, some additional phases were identified, such as document import, data export, and human interaction for data validation or creation of the IDP processes by choosing the phases. These phases are not presented in the scientific study and have not been considered for our IDP pipeline since they

Table 5.12 Industrial primary results

Company	Tool
Automation anywhere	Intelligent document processing
UIPath	Document understanding
Blue Prism	Decipher IDP
Google	Document AI
Amazon AWS	Textract/Comprehend/A2I
IBM	Datacap
ABBYY	FlexiCapture/FineReader engine
Azure	Form recognizer
ROSSUM	DATA CAPTURE
ANTWORKS	Document processing
Hyland	Brainware Intelligent Capture
Foxit	PDFCompresor/Maestro server OCR/Document transformation service
SoftWorksAI	Trapeze
OpenText	Intelligent capture + Magellan
Parascript	FormXtra
Alfresco	Intelligent information capture (via AWS) forms recognition and processing
EPHESOFT	Intelligent document processing
WorkFusion	Document intelligent
Infrrd	Intelligent document processing
Edgeverve	AssistEdge
DMS-solutions	elDoc
Kofax	Kofax transformation
BIS	Grooper
Bizdata	SmartDetect
Hypatos	Hypatos document processing
Kanverse	Kanverse IDP
Parashift	Parashift process documents
HyperScience	Intelligent document processing
Appian	Appian IDP
Datamatics	TruCap+
Docdigitizer	Docdigitizer document processing
Goneutrinos	Intelligent document processing
Acodis	Intelligent document processing
AutomationHero	AutomationHero document processing
Indico	Indico intelligent process automation
OpenBots	OpenBots documents
Nanonets	Nanonets

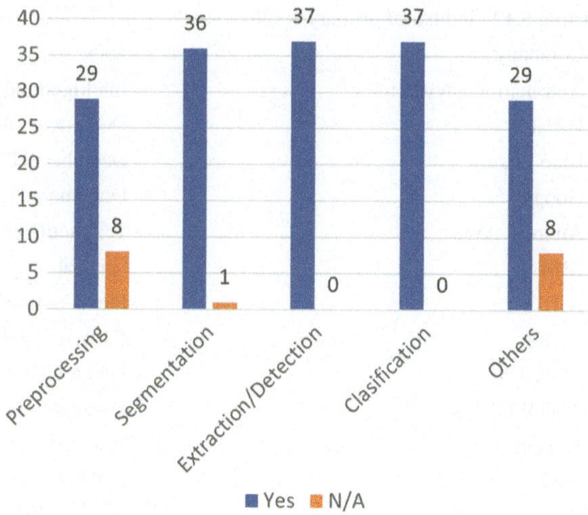

Fig. 5.11 Total number of tools according to their phase in the IDP pipeline

correspond to tasks related but not directly to document processing. However, some activities related to the IDP process, which some companies carry out, were cataloged in the phase *others*. This is because they could not be in a specific phase when performing actions that could be classified in more than one phase or carrying out process preparation work. The most widespread point, and why most of the tools can be included in the *others* phase, is the possibility of training or configuring the intelligent model that would be used in the IDP process.

Regarding the pipeline phases identified during the execution of the industrial report, they coincide with those identified in the analysis of the scientific scope. Figure 5.11 shows how many tools contemplate the fulfillment of the identified pipeline phases.

As can be seen, all the tools perform the phases of *classification* and content *extraction/detection*, thus being the main pipeline functionalities used in the industry. It should be noted that another important phase is *segmentation*, where the different sections that make up the documents are identified. Meanwhile, *preprocessing* and *others* take the last position for some of the tools. In short, 65% of the tools fulfill all the phases of the pipeline since part of the tools do not have information on preprocessing or explicitly report the existence of additional phases (cf. "N/A" label in Fig. 5.11).

5.4.3.3 RQ3. How Well Supported are the Main Industry Tools Concerning the General Characteristics Identified?

Concerning the general characteristics that appear in Sect. 5.4.1, a series of data have been extracted that can be analyzed.

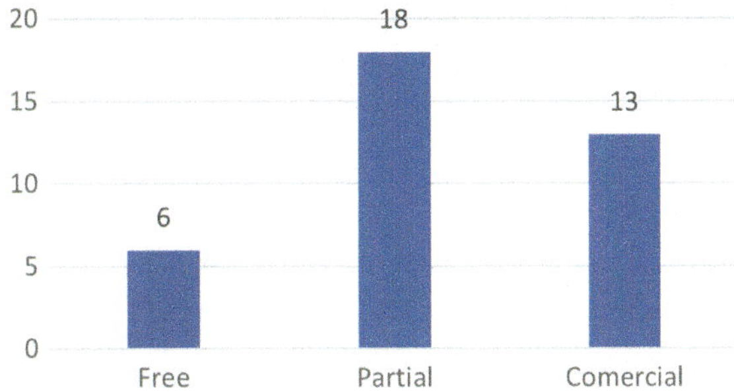

Fig. 5.12 Total number of tools per license type

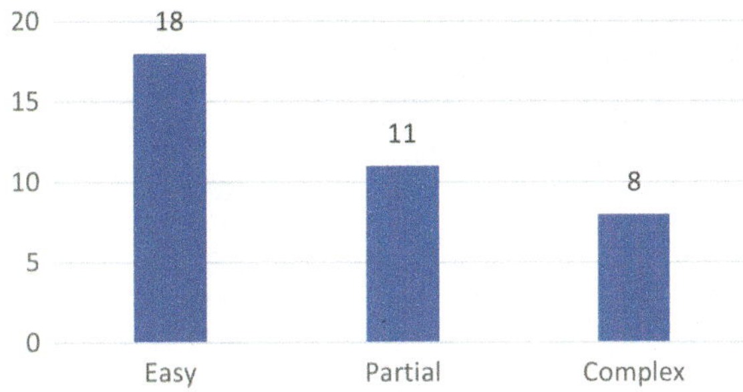

Fig. 5.13 Total number of tools according to their learning curve

To begin with, in Fig. 5.12 can be observed that almost all the tools have a commercial license, but some offer a *community* version with reduced capabilities or a free version with limited time use or with fewer features than the commercial version. Only 2 tools are *open source*, and these are precisely the ones that allow visual adaptation or modification of the software due to the nature of their license type. The graph represents the freeware of the tools, being the *community* and *open source* tools considered free, those with some limited freeware as partial, and those purely commercial as non-free. As can be seen, the industry focuses on commercial tools. Those with limited free, limited *community*, or *open source* versions have fewer features than the paid versions. Only OpenText offers complete open source for free.

Regarding the difficulty of using the tool, it is subjective data, so the perception of the difficulty of use of the tools may vary among users. However, as can be seen in Fig. 5.13, all the tools offer a *user-friendly* interface, increasing complexity concerning the use of advanced functions.

Half of the tools (i.e., 48.65%) in Fig. 5.13 are easy to use, with well-defined steps, and some have slight complexity (i.e., 29.73%).

Also, concerning the deployment of the system, it is extracted from Fig. 5.14 that half of the tools (i.e., 49%) allow self-deploy.

Except for one tool, all of them claim to have features for handling both unstructured and structured documents, either through the tool's functions or through third-party functions that can be included in these tools.

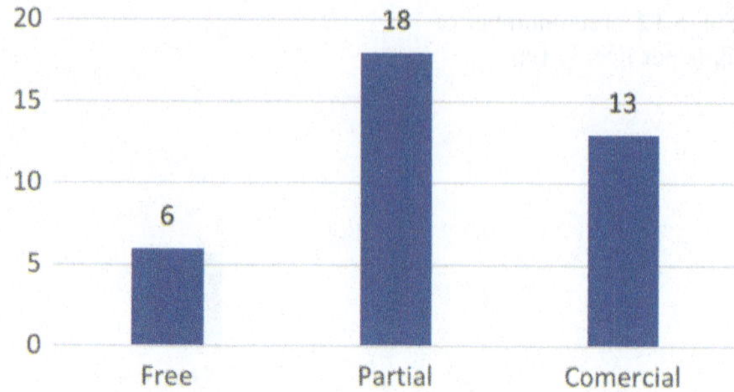

Fig. 5.14 Total number of tools per type of deployment

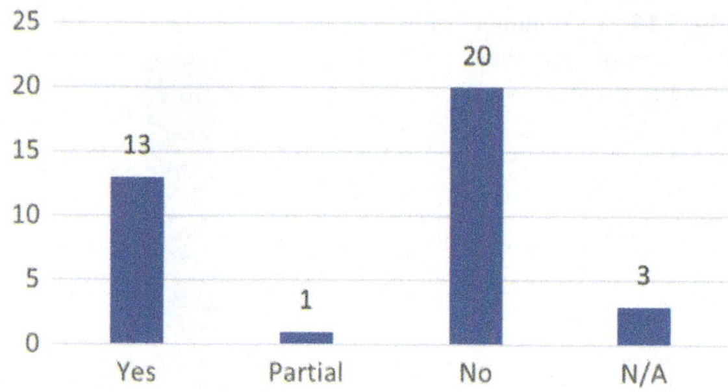

Fig. 5.15 Total number of tools that allows building IDP process

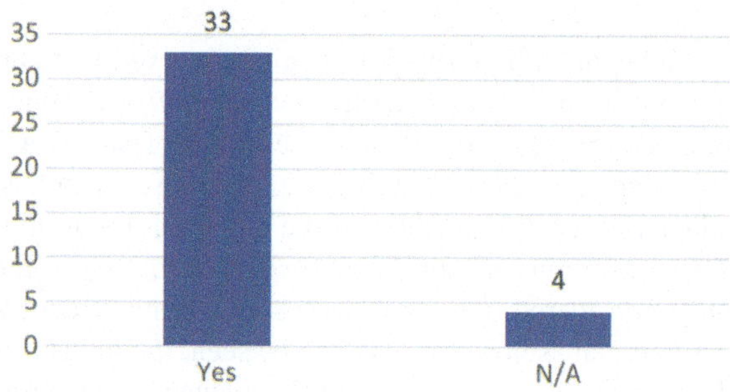

Fig. 5.16 Total number of tools that are able to run multiple processes

Those that allow building an IDP process are scarce, and several of them are included in the elaboration of the RPA creation process. This reduces the IDP process creation level but allows some control over its status.

Only 35% of the tools in Fig. 5.15 contemplate the construction of the IDP process, most of them being a process with a defined pipeline, although they allow parameterizing and configuring the process already defined in the tool.

Most of the tools investigated in Fig. 5.16 allows the management of multiple IDP processes.

Statistical reports of the IDP process are present in most of the tools (i.e., 90%) in Fig. 5.17. However, there are variations in the amount of data shown in the reports, from some exhaustive tools to others that only offer information regarding case counts or process usage.

Fig. 5.17 Percentage of tools that generates statistics

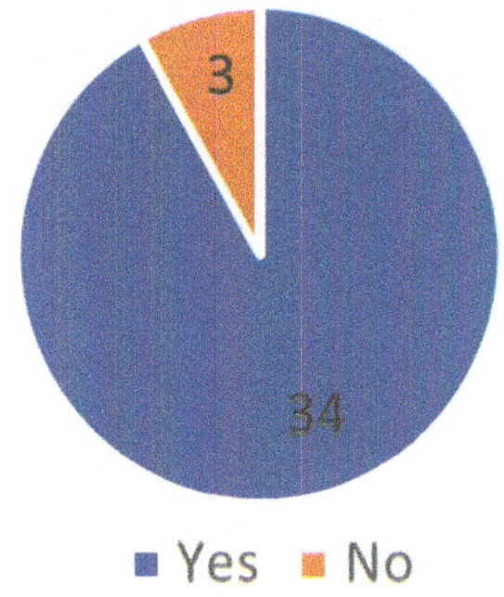

Fig. 5.18 Total number of tools that allows to configure processes

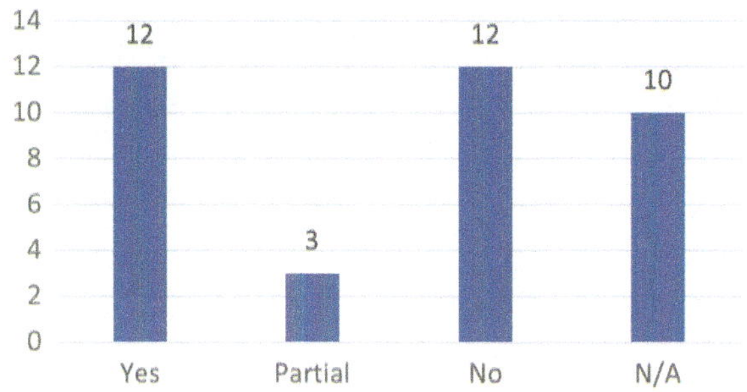

Fig. 5.19 Percentage of tools that allows to debug processes

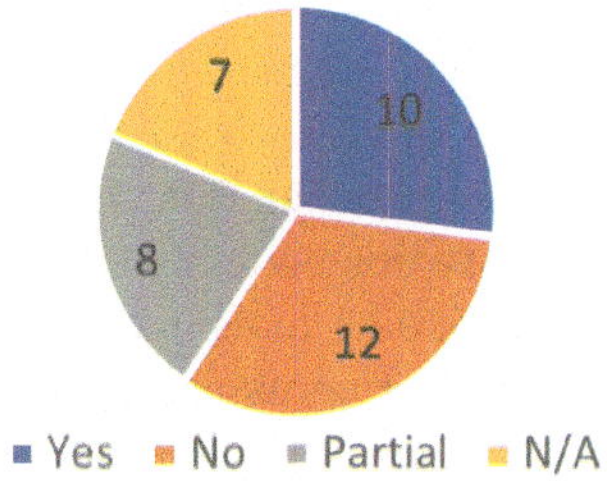

In some tools, the output configuration of the IDP process is automatic, while in others it requires a specific configuration. There is a disparity of options depending on the difficulty of use intended to give to the tool. Although it is important to note that a significant number of tools lack the capability to provide error reporting, requiring manual validation from the user, generating failure reports, or simply failing to process documents that cause errors.

In Fig. 5.18, it can be seen that the same number of tools (i.e., 32.43%) of the tools allow and do not allow configuring the process, while a partial configuration contemplates by 8.11% of the tools.

As for the tools that have error debugging mode, 19 of the tools in Fig. 5.19 have some debugging, either direct or partial. The partial debugging mode corresponds to the use of logs or visuals, indicating the existence of incorrectly extracted data and the need to reconfigure the process.

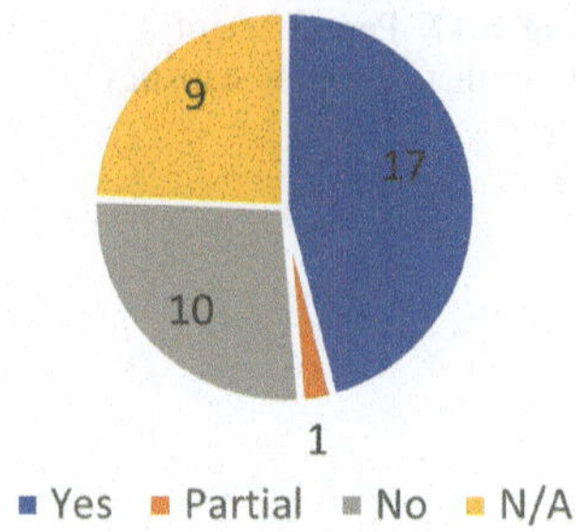

Fig. 5.20 Percentage of tools that allows to export processes

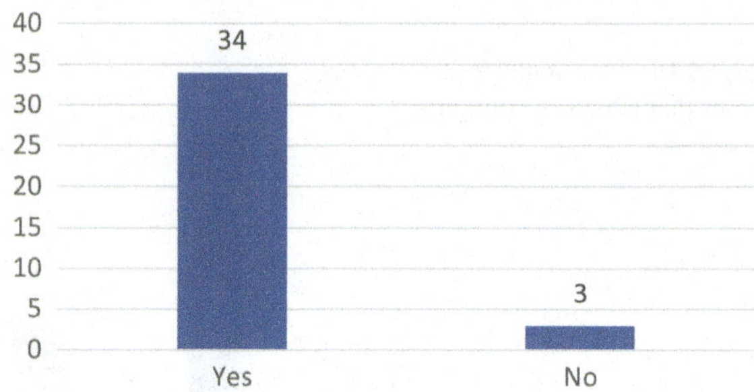

Fig. 5.21 Total number of tools that support multiple users

As can be seen in Fig. 5.19, half of the tools (i.e., 49%) offer some degree of process debugging.

Not all the tools studied in Fig. 5.20 allow the export of the IDP process created, and less than half (i.e., 49%) of the tools offer this possibility or at least suggest that it can be done directly from the application.

Regarding tools that allow multiple users simultaneously, collaborative work is considered necessary in the industry in terms of the number of tools that allow multiple users to work simultaneously.

Most of the tools in Fig. 5.21 (i.e., 91.89%) claim to provide multiple users usage features in a face to tools for a unique user (i.e., 8.11%).

Regarding the tool's scope on a global scale, the tools considered offer more than 5 languages from different countries as tools with a global scope classified as multiple, as they try to reach a more significant number of countries.

As can be seen in Fig. 5.22, most of the tools (i.e., 81.08%) claim to provide multiple-language processing features.

Regarding how the use of the tools is offered to the public, it can be seen in Fig. 5.23 how most of the tools (i.e., 67.57%) are provided in a web-based form or at least partially web-based (i.e., 22%).

Fig. 5.22 Total number of tools that supports multiple languages

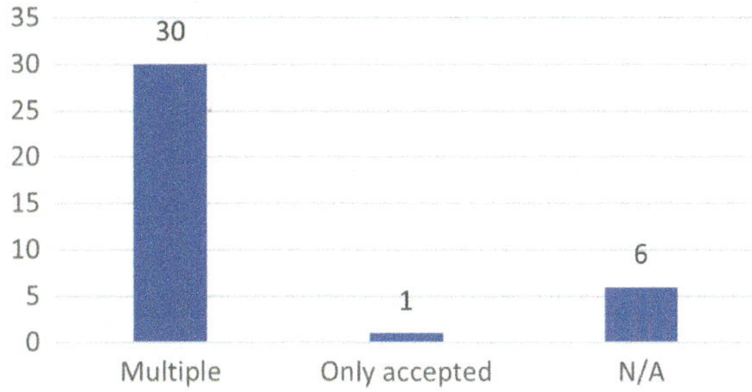

Fig. 5.23 Total number of tools that comprises a web tool

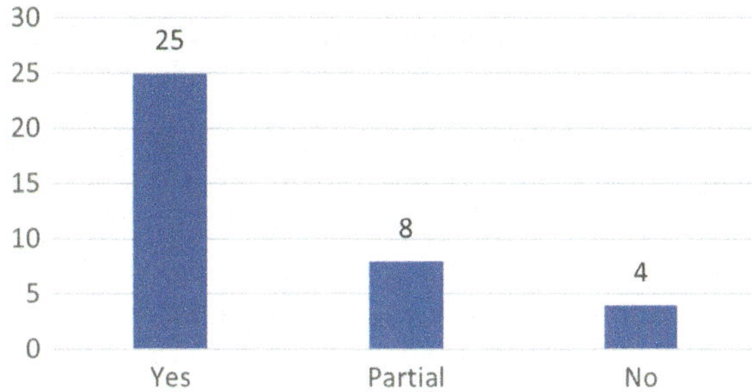

5.4.3.4 RQ4. How Widespread Is the Use of AI and Automation in the IDP Sector?

Although tools and techniques for document processing have existed for a long time, the advance of AI has made its insertion in IDP tools essential for the industry, and all the tools identified have AI-enhanced functionalities to improve the results of document processing. This study has not specified the phases in which they are used, or the type of AI since most private tools do not provide this type of information.

Regarding the integration of automated functions in the IDP industry, all tools are capable of processing batch documents. Some may have scheduled tasks to access the location of the documents and execute the task dictated in the IDP process on existing documents. Others allow the execution of automated tasks more completely to create advanced robots that execute tasks prior to or after processing documents. These fully or partially automated tools will be counted as having automation. Others, faced with the possibility of using task automation instead of developing their system, allow the integration of their IDP system in RPA tools already existing in the market, such as UIPath or Automation Anywhere, classified as having Partial automation. Those tools that do not have the slightest hint of automation will be classified as having No automation.

A total of 33 of the 37 tools (i.e., 89.19%) contemplate some degree of automation, whether complete, partially automated, or through integration with other RPA tools, making it clear that integrating automation into IDP processes is essential for companies and today's industry (cf. Fig. 5.24).

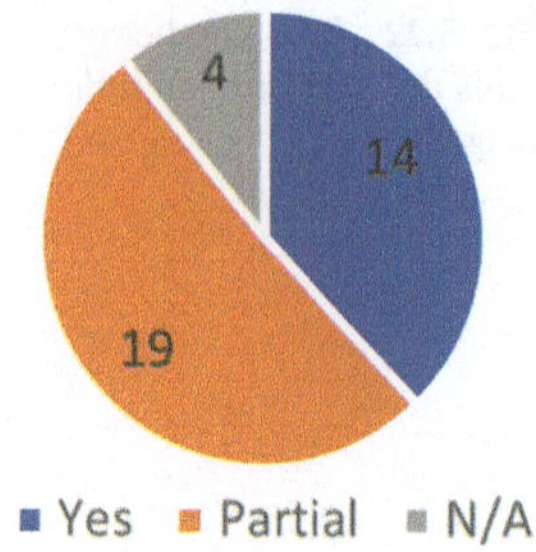

Fig. 5.24 Total number of tools to automate IDP processes

5.5 Discussion and Limitations

This chapter addresses 11 research questions, with seven pertaining to the scientific field (cf. Sect. 5.3.3) and four to the industrial field (cf. Sect. 5.4.3). The answers to these questions lead to the following discussion.

First of all, answering scientific questions, it can be observed in RQ1 in Sect. 5.3.3.1 that the number of publications in IDP increases year by year, with the exception of 2018. A more in-depth study on the factors that have an influence on the trend of publications could be considered interesting. Regarding RQ3 in Sect. 5.3.3.3, a great interest of the scientific community in the application of AI to improve model performance can be observed in recent years. In turn, RQ5 in Sect. 5.3.3.5 points to a greater interest in (1) structured documents, which may be because this type of documents are the most used in industry, and (2) handwritten documents, which may be due to the improvement of techniques associated with the recognition of handwritten texts. Finally, RQ7 in Sect. 5.3.3.7 presents an increased interest in the extraction/detection and segmentation phases of the IDP pipeline, which have been the most complex processes within IDP traditionally. The reasons for this increment seem to be similar to RQ5, since the rise of AI has allowed an improvement in the techniques applied to these specific phases, obtaining better results.

Subsequently, the industrial study was performed and a series of analyzable characteristics were obtained. Answering RQ2 in Sect. 5.4.3.2, the IDP pipeline elements have been considered according to their function because companies call them by different names even though they describe them similarly. What is more, RQ2 considers phases beyond those strictly related to IDP, such as data *import/export* or *human interaction*. Regarding question RQ4 in Sect. 5.4.3.4, although all the tools have AI-enhanced functionalities, most do not indicate what kind of implementation they have or how they work. Therefore, obtaining information about them was considered counterproductive, as it would not provide a clear view of the real state of the industry. In addition, most tools do not include RPA functionalities, but allow the main RPA tools to integrate their solutions as IDP plugins. However, this means transferring part of the functionalities to a RPA tool, such as data operation or process configuration.

Although this study attempts to conduct a comprehensive investigation, some limitations and improvements have been detected. First, the search process considered several words and specific search queries related to the inclusion criteria, which may have resulted in missing articles of interest as a result. Second, although the used digital libraries (i.e., Scopus, Science Direct, IEEE Xplore, and ACM) and search engines (i.e., Google, Bing, and AOL) are the most popular and complete, other systems exist that may contain additional results. And third, related to the scientific exclusion criteria, the filter on the publication year can also be considered a limiting factor but is intentional to get the most recent content and discard the old ones. Also, a filter by journal or conference publication has been specified to select only high-impact studies. This publication filter could have left out the research of possibly interesting and innovative studies that were excluded through this filter.

Finally, after the analysis of the RQs asked at the beginning of the SLR, some points have been found that are prone to further research (cf. Tables 5.13 and 5.14).

Some of the most recurrent challenges identified are technical ones related to version control, inclusion of technological advances, releases, rollback, comparative efficiency of models, etc. These problems are derived from the lack of control over the software, mainly proprietary by the companies, which makes it difficult to carry out these types of actions. Carrying out these actions, either by allowing the use of third-party models, or through Open Source integration solutions as in existing proposals [51], would facilitate and help to solve part of the identified challenges.

Table 5.13 Challenges identified by category (part I)

Methodology	
Name	Description
Challenge 1: Task identification	Create a methodology for identifying tasks that can be automated in the document processing of an organization
Challenge 2: Organizational preparation	Organizations, depending on their business volume or types of documents to be processed, require some adaptation to implement IDP tools. The lack of an organizational readiness framework for IDP implementation has been identified in the scientific literature. A methodological study is considered necessary to assist in the implementation of documentation-related tools in organizations
Roles	
Challenge 3: Roles in document processing	Within the field of document processing, there is no clear differentiation of the roles involved. As in other fields, it is important to clearly differentiate the tasks to be performed by the different roles within an organization. This is to differentiate the roles involved in creating the document tasks to be processed or the automation system, those in charge of their management, to structure the roles involved in a disciplinary way

(continued)

Table 5.13 (continued)

Methodology	
Name	Description
Metrics and indicators	
Challenge 4: Organization evaluation	Similar to organizational preparation, an organization's maturity level for implementing document processing tools needs to be addressed
Challenge 5: Implementation success	There are no clear points in the literature on success or failure factors for implementing or using IDP techniques. A deeper analysis of the success factors may help to better identify and manage the processes involved in IDP
Challenge 6: Profitability assessment	Having identified the tasks in a company that are likely to be transferred to an IDP process, analyze whether their conversion to IDP is profitable. A study on possible metrics that evaluate the ROI of the conversion to an IDP process, with parameters such as volume of documents and number of errors produced
Software engineering	
Challenge 7: Technological advances in document processing	As shown in Fig. 5.6, advances can be integrated through adaptations, but they do not have a clear and defined framework structure to be integrated. A defined structure is considered necessary at the scientific level, which allows the integration of technological advances with each other and with external tools
Challenge 8: Processing techniques update	In both scientific literature and industrial tools, there is a lack of control over the versions of the technology used. When faced with new versions or retraining of cognitive models, there is no clear way to compare results or change the system's version. They are also often seen as separate independent processes, making comparative analysis difficult
Challenge 9: Exceptions control	Both in the scientific literature and industrial tools, there is a lack of consensus on how to act when faced with documents that cannot be processed or documents identified as having been processed erroneously
Challenge 10: Scalability of document processing	In the face of large quantities of documents to be processed, there is no clear statement on the actions to be taken to carry out the scalability of a process. The scientific community does not conduct research in this area, while the industry generally does not indicate how it performs process scalability internally

Table 5.14 Challenges identified by category (part II)

IDP and automations	
Name	Description
Challenge 11: Benefits of IDP implementation with RPA	In the scientific literature, there is no clear exposition of the adoption of RPA with IDP, even though the benefits of RPA implementation in similar processes in the industry are obvious [79]
Challenge 12: Relationship between IDP and RPA	Although the relationship between the two fields is clear, an in-depth study is needed to clearly identify which parts of the processes relate to task automation and which parts relate to document processing
Challenge 13: Methodology for integrating IDP with RPA	Given the strong interrelationship between the two fields, as can be observed at the industrial level, to carry out a methodology for integrating IDP with RPA, this methodology should address the non-intrusive integration of both technologies. In addition, it should guide the design of a process to be automated in which the parts referring to IDP and RPA are identified

5.6 Conclusion and Future Work

This chapter conducts a SLR on IDP in RPA because of their relationship. Therefore, the RQs stated in the current chapter aimed to provide a comprehensive overview of the IDP state of research in the RPA context.

This relation has been previously identified [58, 79], in which document processing is considered to be a complementary technology to RPA. For this purpose, the studies that passed the inclusion/exclusion criteria are analyzed to assess the current state of the art of IDP in the scientific community. A total of 77 studies were analyzed, where it can be seen that IDP is a very studied area with many applications for different contexts. Current literature presents proposals for different types of documents, with different steps in the document processing pipeline. Its aim is to improve existing techniques with new proposals, mainly machine/deep learning to optimize performance, reduce errors, and improve operation. In contrast, it has been observed that task automation is considered necessary in the IDP industry, as can be seen in Fig. 5.24. This fact contradicts what is observed in IDP scientific research (see Fig. 5.7). In summary, the scientific community is significantly active globally, mainly in the USA, China, and Spain. Mainly, general solutions are provided, although some studies are tight to specific domains like the treatment of historical documents or accounting.

Regarding the industrial field, it can be seen in Sect. 5.4.3.2 that the industry intended to maximize the massive processing of documents, allowing the use of the tools by multiple users of the same organization and including the creation of multiple IDP processes simultaneously. It can be observed that document processing is focused on improving the results of the specific stack of documents to be treated, thanks to additional phases of document template configuration and AI retraining. Nevertheless, industrial research shows that the companies are more focused on IDP pipeline phases, such as classification and extraction/detection, which causes some tools to skip some of the phases of the process. All this, combined with the ability to process documents in several languages, aims to reach customers worldwide and process documents on a massive scale to maximize their profitability from implementing the tools.

In addition, the current chapter leads to some future works to be explored. First, integrating IDP processes in business process management or RPA methodology would be beneficial due to (1) the strong relations between them and (2) the research gap acknowledged in the current chapter—overall from an industrial perspective. Second, although there are already proposals to facilitate the use of IDP with developments or Application Programming Interfaces [45], it can be investigated how to best integrate the IDP processes in RPA, how to include new techniques with the advance of research, how to integrate different approaches depending on their needs, and how to train a machine learning models for specific use cases. This point may lead to an IDP/RPA framework to facilitate their use. Finally, going beyond IDP, the performance and suitability of different technologies can be evaluated when paralleling or scaling up tasks with RPA. The critical point here is to identify the challenges in integrating these technologies and the possible limitations of such integration.

Appendix A: List of Abbreviations

RPA	Robotic Process Automation
IDP	Intelligent Document Processing
AI	Artificial Intelligence
OCR	Optical Character Recognition
SLR	Systematic Literature Review
RQs	Research Questions

References

1. Abdallah, A., Berendeyev, A., Nuradin, I., Nurseitov, D.: Tncr: Table net detection and classification dataset. Neurocomputing (2021)
2. Ahmed, R., Gogate, M., Tahir, A., Dashtipour, K., Al-Tamimi, B., Hawalah, A., El-Affendi, M.A., Hussain, A.: Deep neural network-based contextual recognition of Arabic handwritten scripts. Entropy **23**(3), 340 (2021)
3. Alaei, A., Conte, D., Martineau, M., Raveaux, R.: Blind document image quality prediction based on modification of quality aware clustering method integrating a patch selection strategy. Exp. Syst. Appl. **108**, 183–192 (2018)
4. Annabestani, M., Saadatmand-Tarzjan, M.: A new threshold selection method based on fuzzy expert systems for separating text from the background of document images. Iran. J. Sci. Technol. Trans. Electr. Eng. **43**(1), 219–231 (2019)

5. Au, W., Ait-Azzi, A., Kang, J.: Finsbd-2021: the 3rd shared task on structure boundary detection in unstructured text in the financial domain. In: Companion Proceedings of the Web Conference, pp. 276–279 (2021)
6. Baidya, A.: Document analysis and classification: a robotic process automation (RPA) and machine learning approach. In: 2021 4th International Conference on Information and Computer Technologies (ICICT). pp. 33–37. IEEE (2021)
7. Banumathi, K.L., Jagadeesh Chandra, A.P.: An approach to estimate skew angle in printed document images. In: 2019 1st International Conference on Advances in Information Technology (ICAIT), pp. 480–484. IEEE (2019)
8. Baviskar, D., Ahirrao, S., Kotecha, K.: A bibliometric survey on cognitive document processing. Libr. Philos. Pract. 1–31 (2020)
9. Baviskar, D., Ahirrao, S., Potdar, V., Kotecha, K.: Efficient automated processing of the unstructured documents using artificial intelligence: a systematic literature review and future directions. IEEE Access (2021)
10. Bhowmik, S., Sarkar, R., Nasipuri, M., Doermann, D.: Text and non-text separation in offline document images: a survey. Int. J. Doc. Anal. Recogn. (IJDAR) **21**(1), 1–20 (2018)
11. Biswas, S., Riba, P., Lladós, J., Pal, U.: Beyond document object detection: instance-level segmentation of complex layouts. Int. J. Doc. Anal. Recogn. (IJDAR) **24**(3), 269–281 (2021)
12. Boroş, E., Romero, V., Maarand, M., Zenklová, K., Křečková, J., Vidal, E., Stutzmann, D., Kermorvant, C.: A comparison of sequential and combined approaches for named entity recognition in a corpus of handwritten medieval charters. In: 2020 17th International Conference on Frontiers in Handwriting Recognition (ICFHR), pp. 79–84. IEEE (2020)
13. Bourbakis, N., Mertoguno, S.: A holistic approach for automatic deep understanding and protection of technical documents. Int. J. Artif. Intell. Tools **29**(06), 2050007 (2020)
14. Bukhari, S.S., Kadi, A., Jouneh, M.A., Mir, F.M., Dengel, A.: anyocr: An open-source OCR system for historical archives. In: 2017 14th IAPR International Conference on Document Analysis and Recognition (ICDAR), vol. 1, pp. 305–310. IEEE (2017)
15. Burget, R.: Model-based integration of unstructured web data sources using graph representation of document contents. In: WEBIST, pp. 326–333 (2019)
16. Calvo-Zaragoza, J., Castellanos, F.J., Vigliensoni, G., Fujinaga, I.: Deep neural networks for document processing of music score images. Appl. Sci. **8**(5), 654 (2018)
17. Can, Y.S., Kabadayı, M.E.: Line segmentation of individual demographic data from Arabic handwritten population registers of ottoman empire. In: International Conference on Document Analysis and Recognition, pp. 312–321. Springer (2021)
18. Chakraborti, T., Isahagian, V., Khalaf, R., Khazaeni, Y., Muthusamy, V., Rizk, Y., Unuvar, M.: From robotic process automation to intelligent process automation. In: International Conference on Business Process Management, pp. 215–228. Springer (2020)
19. Cristani, M., Bertolaso, A., Scannapieco, S., Tomazzoli, C.: Future paradigms of automated processing of business documents. Int. J. Inform. Manag. **40**, 67–75 (2018)
20. Degtyarenko, I., Deriuga, I., Grygoriev, A., Polotskyi, S., Melnyk, V., Zakharchuk, D., Radyvonenko, O.: Hierarchical recurrent neural network for handwritten strokes classification. In: ICASSP 2021–2021 IEEE International Conference on Acoustics, Speech and Signal Processing (ICASSP), pp. 2865–2869. IEEE (2021)
21. Enríquez, J.G., Jimenez-Ramirez, A., Dominguez-Mayo, F., Garcia-Garcia, J.: Robotic process automation: a scientific and industrial systematic mapping study. IEEE Access **8**, 39113–39129 (2020)
22. Fawzi, A., Pastor, M., Martínez-Hinarejos, C.D.: Baseline detection on Arabic handwritten documents. In: Proceedings of the 2017 ACM Symposium on Document Engineering, pp. 193–196 (2017)
23. Feng, D., Chen, H.: A small samples training framework for deep learning-based automatic information extraction: case study of construction accident news reports analysis. Adv. Eng. Inform. **47**, 101256 (2021)
24. Fenton, K., Simske, S.: Engineering of an artificial intelligence safety data sheet document processing system for environmental, health, and safety compliance. In: Proceedings of the 21st ACM Symposium on Document Engineering, pp. 1–4 (2021)

25. Fernandes, J., Simsek, M., Kantarci, B., Khan, S.: Tabledet: an end-to-end deep learning approach for table detection and table image classification in data sheet images. Neurocomputing **468**, 317–334 (2022)
26. Fugini, M., Finocchi, J.: Quality evaluation for documental big data. In: Proceedings of the 22nd International Conference on Enterprise Information Systems—Volume 1: ICEIS, pp. 132–139. INSTICC, SciTePress (2020). https://doi.org/10.5220/0009394301320139
27. Gatos, B., Louloudis, G., Stamatopoulos, N., Sfikas, G.: Historical document processing. In: Proceedings of the 2017 ACM Symposium on Document Engineering, pp. 1–2 (2017)
28. Gómez-Pérez, P., Phan, T.N., Küeng, J.: Agricultural knowledge extraction from text sources using a distributed mapreduce cluster. In: 2016 27th International Workshop on Database and Expert Systems Applications (DEXA), pp. 29–33. IEEE (2016)
29. Gorai, M., Nene, M.J.: Layout and text extraction from document images using neural networks. In: 2020 5th International Conference on Communication and Electronics Systems (ICCES), pp. 1107–1112. IEEE (2020)
30. Grygoriev, A., Degtyarenko, I., Deriuga, I., Polotskyi, S., Melnyk, V., Zakharchuk, D., Radyvonenko, O.: Hcrnn: a novel architecture for fast online handwritten stroke classification. In: International Conference on Document Analysis and Recognition, pp. 193–208. Springer (2021)
31. Guerry, C., Coüasnon, B., Lemaitre, A.: Combination of deep learning and syntactical approaches for the interpretation of interactions between text-lines and tabular structures in handwritten documents. In: 2019 International Conference on Document Analysis and Recognition (ICDAR), pp. 858–863. IEEE (2019)
32. Ha, H., Horák, A.: Information extraction from scanned invoice images using text analysis and layout features. Signal Proc.: Image Commun., 116601 (2021)
33. Hadjadji, B., Chibani, Y., Nemmour, H.: An efficient open system for offline handwritten signature identification based on curvelet transform and one-class principal component analysis. Neurocomputing **265**, 66–77 (2017)
34. Hammarström, H.: Inventory and content separation in grammatical descriptions of languages of the world. In: International Conference on Theory and Practice of Digital Libraries, pp. 29–40. Springer (2021)
35. Holeček, M.: Learning from similarity and information extraction from structured documents. Int. J. Doc. Anal. Recogn. (IJDAR), 1–17 (2021)
36. Ivančić, L., Suša Vugec, D., Bosilj Vukšić, V.: Robotic process automation: systematic literature review. In: International Conference on Business Process Management, pp. 280–295. Springer (2019)
37. Jalali, F., Ebrahimi, A.: A novel mixed approach for detecting overlap in document images. In: 2017 Iranian Conference on Electrical Engineering (ICEE), pp. 1701–1707. IEEE (2017)
38. Jiang, J., Simsek, M., Kantarci, B., Khan, S.: High precision deep learning-based tabular position detection. In: 2020 IEEE Symposium on Computers and Communications (ISCC), pp. 1–7. IEEE (2020)
39. Jun, C., Suhua, Y., Shaofeng, J.: Automatic classification and recognition of complex documents based on faster RCNN. In: 2019 14th IEEE International Conference on Electronic Measurement & Instruments (ICEMI), pp. 573–577. IEEE (2019)
40. Kajla, N.I., Missen, M.M.S., Luqman, M.M., Coustaty, M., Mehmood, A., Choi, G.S.: Additive angular margin loss in deep graph neural network classifier for learning graph edit distance. IEEE Access **8**, 201752–201761 (2020)
41. Kara, E., Traquair, M., Kantarci, B., Khan, S.: Deep learning for recognizing the anatomy of tables on datasheets. In: 2019 IEEE Symposium on Computers and Communications (ISCC), pp. 1–6. IEEE (2019)
42. Kara, E., Traquair, M., Simsek, M., Kantarci, B., Khan, S.: Holistic design for deep learning-based discovery of tabular structures in datasheet images. Eng. Appl. Artif. Intell. **90**, 103551 (2020)
43. Keeling, R., Chhatwal, R., Huber-Fliflet, N., Zhang, J., Wei, F., Zhao, H., Shi, Y., Qin, H.: Empirical comparisons of CNN with other learning algorithms for text classification in legal document review. In: 2019 IEEE International Conference on Big Data (Big Data), pp. 2038–2042. IEEE (2019)

44. Kitchenham, B., Charters, S.: Guidelines for performing systematic literature reviews in software engineering (2007)
45. Lenc, L., Martínek, J., Král, P., Nicolao, A., Christlein, V.: Hdpa: historical document processing and analysis framework. Evol. Syst. **12**, 177–190 (2021)
46. Li, D., Wu, Y., Zhou, Y.: Linecounter: Learning handwritten text line segmentation by counting (2021). arXiv preprint arXiv:2105.11307
47. Li, J., Lin, C.M., Hu, S.x.: Intelligent document processing method based on robot process automation. In: 2021 Global Reliability and Prognostics and Health Management (PHM-Nanjing), pp. 1–6. IEEE (2021)
48. Ljajić, A., Stanković, M., Marovac, U.: Detection of negation in the Serbian language. In: Proceedings of the 8th International Conference on Web Intelligence, Mining and Semantics, pp. 1–6 (2018)
49. Long, S., He, X., Yao, C.: Scene text detection and recognition: the deep learning era. Int. J. Comput. Vis. **129**(1), 161–184 (2021)
50. Mansar, Y., Kang, J., Maarouf, I.E.: The finsim-2 2021 shared task: learning semantic similarities for the financial domain. In: Companion Proceedings of the Web Conference 2021, pp. 288–292 (2021)
51. Martínez-Rojas, A., Sánchez-Oliva, J., López-Carnicer, J., Jiménez-Ramírez, A.: Airpa: An architecture to support the execution and maintenance of AI-powered RPA robots. In: International Conference on Business Process Management, pp. 38–48. Springer (2021)
52. Memon, J., Sami, M., Khan, R.A., Uddin, M.: Handwritten optical character recognition (OCR): a comprehensive systematic literature review (SLR). IEEE Access **8**, 142642–142668 (2020)
53. Mercier, D., Rizvi, S.T.R., Rajashekar, V., Dengel, A., Ahmed, S.: Impactcite: An xlnet-based solution enabling qualitative citation impact analysis utilizing sentiment and intent. In: ICAART (2), pp. 159–168 (2021)
54. Mijangos, V., Sierra, G., Montes, A.: Sentence level matrix representation for document spectral clustering. Pattern Recogn. Lett. **85**, 29–34 (2017)
55. Mittal, R., Garg, A.: Text extraction using OCR: a systematic review. In: 2020 Second International Conference on Inventive Research in Computing Applications (ICIRCA), pp. 357–362. IEEE (2020)
56. Moreno, M., Schirmer, L., Bayser, M., Brandão, R., Cerqueira, R.: Understanding documents with hyperknowledge specifications. In: Proceedings of the ACM Symposium on Document Engineering 2018, pp. 1–4 (2018)
57. Neji, H., Halima, M.B., Hamdani, T.M., Nogueras-Iso, J., Alimi, A.M.: Blur2sharp: a GAN-based model for document image deblurring. Int. J. Comput. Intell. Syst. **14**(1), 1315–1321 (2021)
58. Ng, K.K., Chen, C.H., Lee, C.K., Jiao, J.R., Yang, Z.X.: A systematic literature review on intelligent automation: aligning concepts from theory, practice, and future perspectives. Adv. Eng. Inform. **47**, 101246 (2021)
59. Nguyen, M.T., Le, D.T., Le, L.: Transformers-based information extraction with limited data for domain-specific business documents. Eng. Appl. Artif. Intell. **97**, 104100 (2021)
60. Obukhov, A., Krasnyanskiy, M.: Application of machine learning for document classification and processing in adaptive information systems. In: Computer Science On-line Conference, pp. 291–300. Springer (2020)
61. Oliveira, S.A., Seguin, B., Kaplan, F.: dhsegment: A generic deep-learning approach for document segmentation. In: 2018 16th International Conference on Frontiers in Handwriting Recognition (ICFHR), pp. 7–12. IEEE (2018)
62. Oral, B., Emekligil, E., Arslan, S., Eryiğit, G.: Information extraction from text intensive and visually rich banking documents. Inform. Proc. Manag. **57**(6), 102361 (2020)
63. Pappu, A., Blanco, R., Mehdad, Y., Stent, A., Thadani, K.: Lightweight multilingual entity extraction and linking. In: Proceedings of the Tenth ACM International Conference on Web Search and Data Mining, pp. 365–374 (2017)
64. Park, C., Shin, J., Park, S., Lim, J., Lee, C.: Fast end-to-end coreference resolution for Korean. In: Proceedings of the 2020 Conference on Empirical Methods in Natural Language Processing: Findings, pp. 2610–2624 (2020)

65. Pastor, M.: Text baseline detection, a single page trained system. Pattern Recogn. **94**, 149–161 (2019)
66. Philips, J., Tabrizi, N.: Historical document processing: A survey of techniques, tools, and trends. In: KDIR, pp. 341–349 (2020)
67. Poddar, A., Chakraborty, A., Mukhopadhyay, J., Biswas, P.K.: Detection and localisation of struck-out-strokes in handwritten manuscripts. In: International Conference on Document Analysis and Recognition, pp. 98–112. Springer (2021)
68. Poddar, A., Chakraborty, A., Mukhopadhyay, J., Biswas, P.K.: Texrgan: a deep adversarial framework for text restoration from deformed handwritten documents. In: Proceedings of the Twelfth Indian Conference on Computer Vision, Graphics and Image Processing, pp. 1–9 (2021)
69. Qasim, S.R., Mahmood, H., Shafait, F.: Rethinking table recognition using graph neural networks. In: 2019 International Conference on Document Analysis and Recognition (ICDAR), pp. 142–147. IEEE (2019)
70. Qurashi, A.W., Holmes, V., Johnson, A.P.: Document processing: methods for semantic text similarity analysis. In: 2020 International Conference on INnovations in Intelligent SysTems and Applications (INISTA), pp. 1–6. IEEE (2020)
71. Rabby, A.S.A., Islam, M.M., Hasan, N., Nahar, J., Rahman, F.: A novel deep learning character-level solution to detect language and printing style from a bilingual scanned document. In: 2020 IEEE International Conference on Big Data (Big Data), pp. 5218–5226. IEEE (2020)
72. Research, E.G.: Everest group peak matrix tm for intelligent document processing (IDP) (2021)
73. Ribeiro, J., Lima, R., Paiva, S.: Document classification in robotic process automation using artificial intelligence—a preliminary literature review. Commun. Intell. Syst., 211–221 (2021)
74. Rusticus, D., Goldmann, L., Reisser, M., Villegas, M.: Document domain adaptation with generative adversarial networks. In: 2019 International Conference on Document Analysis and Recognition (ICDAR), pp. 1432–1437. IEEE (2019)
75. Santoro, A., De Stefano, C., Marcelli, A.: Assisted transcription of historical documents by keyword spotting: a performance model. In: 2017 14th IAPR International Conference on Document Analysis and Recognition (ICDAR), vol. 1, pp. 971–976. IEEE (2017)
76. Shen, Z., Zhang, R., Dell, M., Lee, B.C.G., Carlson, J., Li, W.: Layoutparser: A unified toolkit for deep learning based document image analysis (2021). arXiv preprint arXiv:2103.15348
77. Shidaganti, G., Salil, S., Anand, P., Jadhav, V.: Robotic process automation with AI and OCR to improve business process. In: 2021 Second International Conference on Electronics and Sustainable Communication Systems (ICESC), pp. 1612–1618. IEEE (2021)
78. Sirajudeen, M., Anitha, R.: Forgery document detection in information management system using cognitive techniques. J. Intell. Fuzzy Syst. **39**(6), 8057–8068 (2020)
79. Syed, R., Suriadi, S., Adams, M., Bandara, W., Leemans, S.J., Ouyang, C., ter Hofstede, A.H., van de Weerd, I., Wynn, M.T., Reijers, H.A.: Robotic process automation: contemporary themes and challenges. Comput. Indus. **115**, 103162 (2020)
80. Tafti, A.P., Baghaie, A., Assefi, M., Arabnia, H.R., Yu, Z., Peissig, P.: OCR as a service: an experimental evaluation of Google Docs OCR, Tesseract, ABBYY Finereader, and Transym. In: International Symposium on Visual Computing, pp. 735–746. Springer (2016)
81. Tensmeyer, C., Martinez, T.: Document image binarization with fully convolutional neural networks. In: 2017 14th IAPR International Conference on Document Analysis and Recognition (ICDAR), vol. 1, pp. 99–104. IEEE (2017)
82. Tensmeyer, C., Martinez, T.: Confirm-clustering of noisy form images using robust matching. Pattern Recogn. **87**, 1–16 (2019)
83. Tomoiaga, C., Feng, P., Salzmann, M., Jayet, P.: Field typing for improved recognition on heterogeneous handwritten forms. In: 2019 International Conference on Document Analysis and Recognition (ICDAR), pp. 487–493. IEEE (2019)
84. Toselli, A.H., Romero, V., Vidal, E.: Word graphs size impact on the performance of handwriting document applications. Neural Comput. Appl. **28**(9), 2477–2487 (2017)
85. Tran, M.T., Trieu, L.Q., Tran, H.Q.: Document representation and classification with twitter-based document embedding, adversarial domain-adaptation, and query expansion. J. Heurist., 1–23 (2019)

86. Ubul, K., Tursun, G., Aysa, A., Impedovo, D., Pirlo, G., Yibulayin, T.: Script identification of multi-script documents: a survey. IEEE Access **5**, 6546–6559 (2017)
87. Vinjit, B., Bhojak, M.K., Kumar, S., Chalak, G.: A review on handwritten character recognition methods and techniques. In: 2020 International Conference on Communication and Signal Processing (ICCSP), pp. 1224–1228. IEEE (2020)
88. Wang, J., Si, S., Hong, Z., Qu, X., Zhu, X., Xiao, J.: Case study of few-shot learning in text recognition models. In: International Conference on Web Information Systems Engineering, pp. 394–401. Springer (2021)
89. Würsch, M., Ingold, R., Liwicki, M.: Divaservices-a restful web service for document image analysis methods. Digit. Scholarship Human. **32**(suppl_1), i150–i156 (2017)
90. Xiong, Z., Shen, Q., Wang, Y., Zhu, C.: Paragraph vector representation based on word to vector and CNN learning. Comput. Mater. Continua **55**(2), 213–227 (2018)
91. Yamazaki, A., Sando, K., Suzuki, T., Aiba, A.: A handwritten Japanese historical kana reprint support system: Development of a graphical user interface. In: Proceedings of the ACM Symposium on Document Engineering 2018, pp. 1–4 (2018)
92. Yang, Y., Feng, Y., Ge, J., Zhou, Y., Zeng, J., Li, C., Luo, B.: Checking the statutes in Chinese judgment document based on editing distance algorithm. In: 2017 14th Web Information Systems and Applications Conference (WISA), pp. 197–200. IEEE (2017)
93. Zhao, M., Hochuli, A.G., Cheddad, A.: End-to-end approach for recognition of historical digit strings (2021). arXiv preprint arXiv:2104.13666
94. Zhu, X., Wang, J., Hong, Z., Xia, T., Xiao, J.: Federated learning of unsegmented chinese text recognition model. In: 2019 IEEE 31st International Conference on Tools with Artificial Intelligence (ICTAI), pp. 1341–1345. IEEE (2019)

Chapter 6
Challenges in Banking and Solving Them Using RPA

Ganeshayya Shidaganti, Lakshya Khandelwal, Kushagra Gupta, and Himanshu Vaswani

Abstract Robotic Process Automation (RPA) is a technology used by banks and other financial institutions to automate manual business operations so that they can compete in today's market. Onboarding clients, ongoing services, credit card processing, mortgage processing, loan processing, and information retrieval are just a few of the banking activities that necessitate a huge workforce and take time. The necessity for a big workforce for these procedures is rapidly shrinking with the development of Robotic Process Automation and other cognitive technologies. Through automation, the bank's analysts may devote more time to advanced-value tasks, such as evaluating automated outcomes and assessing complex loans that were previously too complex to automate, which can improve process precision, decrease operating time per loan, and provide the bank with additional analyst capacity for client support. The present chapter addresses the current challenges faced in the banking industry and provides solutions in the form of use cases, thereby highlighting the importance of RPA and AI in the banking industry.

6.1 Introduction

Onboarding services, loan processing, credit card processing, and mortgage processing can be a time-consuming procedure with a lot of paperwork and data checks. The use of RPA (Robotic Process Automation) technology allows banks to automate a number of tasks. We'll look at the extent of banking process automation with RPA in this chapter [1].

G. Shidaganti (✉) · L. Khandelwal · K. Gupta · H. Vaswani
Department of Computer Science and Engineering, M.S. Ramaiah Institute of Technology, (Affiliated to VTU), Bangalore, Karnataka, India
e-mail: ganeshayyashidaganti@msrit.edu

S. Bhattacharyya et al. (eds.), *Confluence of Artificial Intelligence and Robotic Process Automation*, Smart Innovation, Systems and Technologies 335,
https://doi.org/10.1007/978-981-19-8296-5_6

What is Robotic Process Automation (RPA) and how does it work in banking industries?

Robotic process automation [2] is a technology that extracts structured data and processes it to replace human effort for repetitive and redundant operations. Robots are programmed to perform human-like actions based on a set of instructions. They can work with apps faster and with less error than humans. The "bots" can skim through an email or invoice, download files to a specified location, copy-paste, and delve through data to file it. It's designed specifically for process-oriented industries. In the business world, automation has become a popular topic. While practically every instrument used by organisations since the Industrial Revolution falls under the category of process automation, the advancements in business automation over the last 20 years have been astonishing. And the technologies that enable automation are progressing at a rapid pace.

The banking sector is expected to grow by leaps and bounds in the future years as more governments push their projects related to a cashless economy. However, the current industry is beset by its own set of problems. We'll take a closer look at how banking process automation can help in improving customer experience and enhance the growth of RPA and AI in various fields. The use of RPA in banking services is discussed in detail in the further sections of the chapter.

The chapter describes numerous circumstances in which RPA can be employed in the banking sector, including its benefits, methodologies, and use cases. This chapter starts with the challenges faced in processing bank documents and how RPA is used in processing bank documents. The section also provides various solutions to the challenges faced. Further, the chapter covers RPA in loan processing, highlighting some of the advantages and use cases of RPA in loan processing. The next sections of the chapter explain RPA in credit card processing and mortgage banking, followed by the conclusion.

6.2 Literature Survey

Hannah Valgaeren [3] described robotic process automation in the Belgian banking sector in her thesis "Robotic Process Automation in Financial and Accounting Processes in the Banking Sector." Different respondents from banking and non-banking companies were interviewed as part of a qualitative study. According to the findings of the study, deploying RPA delivered significant benefits to banking organisations. Financial and accounting operations, which are primarily manual, repetitive tasks, can be completed quickly while maintaining a high degree of quality. However, the study's breadth was limited because only three banks were examined, and many respondents withheld sensitive information.

In recent writings, some academics have discussed RPA implementation strategies. "The conceptual model for RPA implementation success proposed by Santos et al. [4]" provides an overarching conceptual model of the main RPA topics, including three broad steps for successful RPA implementation, namely strategic goals, process assessment, and tactical evaluation, in their review article "Toward robotic process automation implementation: an end-to-end perspective," aggregating insights from existing RPA implementation studies. The first phase, "strategic goals," suggests that automation targets should be set based on and connected with organisational goals in order to inspire automation and offer a benchmark for evaluating and measuring whether or not goals are met after implementation. According to the model, these objectives must take into account the benefits, limitations, and future challenges of RPA in order to comprehend the potential benefits of automation, avoid setting unrealistic goals that cannot be met, and provide a long-term perspective for actions to address common challenges. After establishing strategic goals, Santos recommends evaluating the process to be automated based on frequent RPA challenges and avoiding procedures that require a lot of system changes, a lack of rules, or other issues identified in the literature. Finally, after selecting the most appropriate processes, the conceptual model suggests conducting a tactical evaluation of how to implement RPA automation based on the same factors identified in the literature (opportunities, difficulties, and future challenges) to identify, for example, integration requirements.

Herm's [5] "Consolidated Framework for Robotic Process Automation Implementation Projects" (2020) recently presented a consolidated and refined model for RPA implementation projects, based on a review of case studies and validation of the findings with RPA experts, in which the common stages of RPA implementation are segregated and general guidelines are suggested for each stage of the implementation process. The model is based on case studies from a wide range of situations, and the conclusions have been generalised to encompass all RPA deployment initiatives.

6.3 RPA in Processing Bank Documents

Maintaining paperwork for a bank's numerous services is a time-consuming operation. Handling and processing documents have gotten a lot easier thanks to automation and cognitive technologies. Commercial banks face considerable problems in document generation as they strive to meet targets, please customers, and stay in good standing with regulators. The overview of this section is shown in Fig. 6.1.

Challenges faced in Document Processing

- **Ongoing and Onboarding Servicing**: Commercial banking customer onboarding can be a time-consuming and costly process. Onboarding a new client entails several procedures. Requesting client data, setting limits, and tracking new data are all possible procedures. Banks should perform legal reviews and inquiries of these materials. For banks, this is frequently a time-consuming, inefficient, and costly practice. Many of these processes are done manually. With the development of RPA, the economic investment in these processes is reduced by 50%.

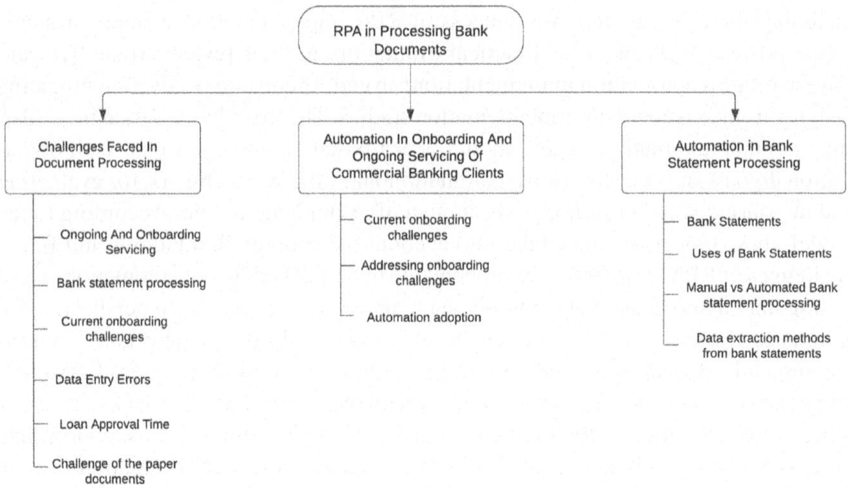

Fig. 6.1 Overview of RPA in processing bank documents

- **Bank Statement Processing**: The client's financial statements need to be digitised to maintain readily available data and solve the problem of data loss. Customers sometimes wish to see a summary of transactions from several accounts, and bank statement processing automation is a straightforward approach to gathering the data and providing a financial picture. Account reconciliations also involve bank statement processing automation, and these statements are used to map out disparities between cash books and passbooks.
- **Data Entry Errors**: Dirty data refers to inconsistent, incomplete, and erroneous data. This frequently happens when data is transferred from paper to computer in commercial banks. Names are misspelt, and figures are entered incorrectly. incorrectly, addresses messed up, and a slew of other problems. Often, these go unnoticed for a long time. When they are uncovered, restoring the data can take a lot of time. To eliminate human error, some commercial banks conduct extensive manual examinations, although this can consume up to half of an analyst's time. A company can not invest that much time, there has to be an efficient way to do it. Automated document generation can capture data accurately, validate dubious information, and alert the user if the information is missing, incorrect, or placed in the wrong field. Data entry becomes easier, and the organisation saves a lot of money. Money and time are both valuable commodities.
- **Loan Approval Time**: Customer management, credit analysis, presentation, approval, and portfolio risk management are all phases of the loan approval process. All of these procedures necessitate the use of paperwork, which takes time. When a commercial bank has to consider competition, impatient clients, and trying to serve as many clients as possible, this procedure will either be rushed, resulting in more errors, or it will be delayed, causing potential clients

to become frustrated and leave for another bank. The fact that Fin Techs can get loans authorised virtually rapidly adds to the strain.

The process doesn't have to be this difficult. Automating the process from the outset will be a smooth and error-free process, with loans being granted in minutes. Document automation removes the need for double-entry of client information, which consumes a significant amount of time and leaves a possibility for errors. Banks can serve customers remotely and customers will be happy as transactions could be completed on mobile devices.

- **Challenge of the Paper Documents**: For any organisation, paper can be a serious issue. First, there's the cost of making sure there's enough paper to print all of the necessary paperwork. The record must be maintained for at least five years. Given the number of clients and transactions a bank handles, this will necessitate a vast and costly storage space that must also be safeguarded. Furthermore, paper documents are much more difficult to track, making it easier for sensitive information to be stolen.

  Except in a few circumstances when a paper document is required, automating document generation is paperless. These papers may simply be stored in the cloud, which saves space and improves security. The software can also be linked to management software, which will delete papers that are no longer needed. Investing in automation software will pay off for a commercial bank.

6.3.1 Automation in Onboarding and Ongoing Servicing of Commercial Banking Clients

6.3.1.1 Current Onboarding Challenge

The process of onboarding commercial banking clients can be broken into eight steps as shown in Fig. 6.2 [6].

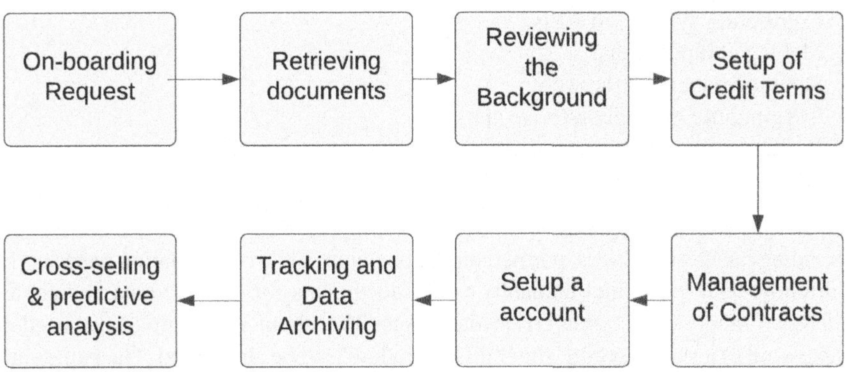

Fig. 6.2 Onboarding life cycle

- **Information Gathering**: To enrol a new commercial banking client, different kinds of data need to be collected. Documentation such as business licences, cooperation agreements, and financial history must be collected from several sources, mandating a considerable investment in staffing, time, and training. Employees must physically look for papers from clients, check them, and put them into bank database systems before any further research can be done. This method can take a long time and cause a delay in the onboarding process.
- **Manual Processing**: Onboarding a client now necessitates a large number of manual activities. Manual processing can be time-consuming thereby increasing the time of the onboarding process. Processing done using automation technology takes lesser time compared to manual processing.
- **Data Validation**: The information gathered during customer onboarding and subsequent actions are utilised by processes such as cross-selling analysis and regulatory compliance. As numerous teams contact the client at various phases of the onboarding process, there is a high risk of human error, which could result in the maintenance of inaccurate client data. The consequences of having inaccurate client data might be severe. False or fraudulent data can cause substantial problems, such as regulatory noncompliance or poor consumer experiences.

6.3.1.2 Addressing Onboarding Challenges

Many of the manual data collection and verification tasks required in the early stages of customer onboarding can be automated with RPA technology. RPA tasks include the following [7]:

- Opening emails and attachments;
- Filling out forms;
- Merging data from numerous sources;
- Copying and pasting data;
- Moving files and directories;
- Extracting structured data from documents;
- Connecting systems to APIs;
- Making computations;
- Scraping data from the web;
- Logging into online/enterprise apps;
- Reading and writing to databases.

Consider an agricultural company that requires a bank loan. For compliance and onboarding checks, the firm would need to supply the bank with a series of papers, including business licences, partnership agreements, and trust formation paperwork. Currently, bank personnel manually verify documents for relevant data and cross-reference them with computerised databases. The client may upload the needed paperwork to a shared secure site if this procedure is to be automated. The papers are then scanned for pertinent information and compared to other databases by a robot. The robot may record that the client's KYC procedure is complete after the data have been validated. As a consequence, a procedure that took weeks to complete may now be done in minutes.

Automation adoption: The ways in which banks can start automating are discussed as follows [8]:

- **Current State Assessment**: Completing a current state evaluation is the first step towards automating client onboarding operations. Banks must have a thorough awareness of the processes involved. They should analyse which processes take more time, which processes are cost-efficient, if there are any obstacles, and if any kind of customer dissatisfaction exists. While some improvements will be common to all banks, each will have its own set of problems that must be addressed and automated. A more targeted design will lead to more efficient solutions to potential risks.
- **Identifying targets that can be Automated**: After a bank has identified its problems, it must pick which processes must be automated first. To identify such processes, they can be grouped based on complexity. The processes must be assigned priorities to each process based on its difficulty, the number of resources it needs to access, the type of action it performs (i.e. does the process just read data, access a system, and upload or download data, or does it need to apply rules and logic to modify data?), the volume of work it generates, and the time it takes to complete it. Once each process is assigned its priority, banks should begin with processes involving low complexity and progress to higher complexity progresses.
- **Choose a Technology Partner**: A bank must locate the right technology provider to assist make those goals a reality after doing a current state evaluation and identifying candidates for automation. There are a lot of companies in the automation market, so picking the right one is crucial. While selecting a vendor, banks should consider not just the pricing, RPA, and analytical and cognitive functionality, but also continuous vendor support, vendor experience, and development. Integration with the existing architecture must also be considered.
- **Complete a Pilot Programme**: As a technology provider is chosen, its main role is to start deferentially. A pilot programme may be used by banks to assess what an automated system might look like and what sort of system access is needed to automate the indicated operations. The bank and the technology partner may cooperate on a process design document and a solution design document that describes the automated process flows as part of a pilot programme. Because conducting a pilot programme also involves unit and functional testing of in-scope operations, banks can assess how successful bots will be in a real production situation. Banks can better understand the consequences they can expect to see by adopting automation technology in a controlled setting for a limited period of time, allowing them to plan the next part of the project correctly.
- **Development and Automation GoLive**: After a pilot programme has been completed successfully, banks can begin automating entire procedures. This implementation can be done in stages, with the technology partner preparing thorough "as-is" and "to-be" process documents that the bank must approve.

Following the development, testing, and modification of code by the technology partner, bots can be deployed.

- **Scale-up Automation Activities**: The previous phase's bots' work may now be assessed, and new bots can be deployed in other parts of the bank. To obtain better efficiencies and savings, more complex technologies may be stacked in. The automated staff of a bank can evolve as technology advances.

The following case study [9] demonstrates the capability of RPA:

A bank employed its creatively skilled analysts to gather important information, to begin with, the KYC procedure, which takes two hours to complete, before deploying RPA technology. When RPA was applied, this data collection was automated, reducing the processing time from two hours to two minutes. RPA methods saved time while also allowing employees to devote more time to the essential analyses for which they were trained, resulting in more informed and faster decision-making.

6.3.2 Automation in Bank Statement Processing

The way firms manage and process financial statements are changing thanks to automation. The banking sector is fiercely competitive, and internet banking is increasingly replacing paper-based statements to improve consumer satisfaction and enable real-time access to premium services. Electronic statements help to avoid identity theft and fraudulent activities by allowing you to see a complete list of transactions in an easy-to-read format. Bank statement processing solutions use robotic process automation (RPA) and optical character recognition (OCR) technologies [10].

According to statistics, introducing automation technologies into bank statement processing workflows results in a 2 to 5% boost in revenue for 35% of financial institutions. Companies cannot ignore the deployment of automation in bank statement processing for customer onboarding and underwriting any longer. In reality, the BFSI industry's automation adoption rate is rapidly expanding in the post-covid age.

The purpose of this book is to teach the reader the fundamentals of bank statement processing and automation. The guide also highlights the importance and advantages of automated data extraction versus manual data extraction from bank statements, as well as assisting the reader in selecting the best bank statement automation system.

6.3.2.1 Introduction to Bank Statements

Bank statements [11] are an official record of all financial transactions that take place over a month. These account statements show the inflows and outflows of funds into and out of your account. Depending on the bank from whence the statement is generated, each statement has its format. However, there are a few things that all bank statements have in common:

- **Opening balance**—This describes the balance of an account before any transactions take place.
- **Withdrawals**—Withdrawals are debits and indicate that you've withdrawn money from your account. Any time you make a withdrawal, your total amount contained in that account gets reduced.
- **Credits** (or deposits)—Credits are referred to as deposits. The term is used to refer to transactions flowing into.
- **Date**—Dates of transactions are recorded in bank statements.
- **Transaction IDs**—Transaction IDs are reference numbers of your transactions that took place. These are used to map transactions coming from vendor accounts and verify income or credit sources.
- **Account holder name**—The account holder is the beneficiary or the person who has an account in the bank.
- **Account number**—The account number is generally between 8 to 12 digits in length. Individuals have to furnish their account numbers to borrowers, lenders, and vendors before being able to make a transaction (Fig. 6.3).

6.3.2.2 Uses of Bank Statements

In most cases, customers use bank statements to keep track of their transactions; however, lenders and insurers use these documents for identity and address verification. The following are the most common uses of bank statements:

- **Verify Income Sources**: Bank statements contain financial information that proves employment, annual turnover rates, and individual income sources. Before accepting loans or mortgages, credit lending firms frequently request them.
- **Loan and Mortgage Approvals**: Banks and NBFCs use bank statement processing to perform a credit profile analysis of borrowers. Cash flow analysis of statements gives an idea about the financial standing of borrowers which helps in determining loan or mortgage amounts.
- **For Documenting Finances**: Processing your bank statements can help you realise how much money you make, how much interest you earn, how much you owe in fines, and how much you owe in other settlements. It is considerably easier to manage your finances if you receive these statistics every month.

6.3.2.3 Manual Versus Automated Bank Statement Processing

Reading data from PDF documents or physical bank statements and manually inputting it is known as manual bank statement processing. Account reconciliation involves reading transactions from bank statements and recording them in a general bank ledger. Automated bank statement processing eliminates the need to physically scan documents and manually input data from tables printed on pages, saving countless hours of data entry time (Table 6.1) [11].

Account Name	
Address	TULASI BUILDING, 3RD CROSS GANDHINAGAR, BELLARY, 583103
Date	29 Mar 2022
Account Number	
Account Description	Savings
Branch	BASAVESWARANAGAR BELLARY
Drawing Power	0.00
Interest Rate(%p.a.)	9.7000
CIF No.	74095027001
IFS Code	
MICR Code	583002122
Nomination Registered	Yes
Balance as on	28 MAR 2022 INR 25918.45
Search for	25 NOV 2021 to 28 MAR 2022

Date	Details	Ref No./Cheque No	Debit	Credit	Balance
28 MAR 2022	TRANSFER TO 5097623162090 - UPI/DR/208785171780/AJIO/PYTM/paytm-5394/OidPSIMp		977.88	-	25918.45
28 MAR 2022	TRANSFER TO 5097542162090 - UPI/DR/208721835077/KWALITY /YESB/q257008914/UPI		94.00	-	26896.33
26 MAR 2022	TRANSFER FROM 3199424044303 - NEFT*UTIB0003395*AXIC220657411911*MAHAVEER TEA COM		-	25000.00	26990.33

Fig. 6.3 A typical bank statement

6.3.2.4 Data Extraction Methods from Bank Statements

The following are the various methods to extract data from bank statements:

- **Template-based Optical Character Recognition Method (OCR)**: For scanning and reading bank statements, OCR is a pattern-based recognition tool. It works for specific formats because it is a template-based approach.

  OCR isn't a flexible option because it can't read documents of the same type in multiple layouts or formats. This is because when it comes to extracting data from bank statements and detecting them, OCR frameworks must be designed with rules and meet specified criteria. Using multiple templates for each bank

Table 6.1 Comparison between manual and automated bank statement processing

Manual bank statement processing	Automated bank statement processing
A time-intensive process that involves entering data, making corrections, and validating by hand	Limited human intervention is needed and data is automatically scanned, read, and input into systems using intelligent OCR solutions
Limited to a few hours of the day	Automated bank statement processing solutions can scan and document statements in real time, 24X7, without being prone to fatigue or breaks
Re-corrections are involved if data entries are wrong	No need for re-corrections since data is validated automatically during entry
The cost of hiring employees is high in the long term	Initial costs involved in paying for software are high, but the long-term costs of running it are low
Transactions can be delayed due to slow processing	Transactions are fast due to high processing speeds
There are many chances of document fraud	There are almost no chances of document fraud
There is a limit to the number of statements that can be processed in a day	Unlimited bank statements can be processed by using automated solutions
Error rate can be as high as 30%	No margin for human error and accuracy rate is 99%
Cannot convert bank statements to different formats	You can convert bank statements to Excel, JSON, and a variety of formats

statement gets challenging as the number of vendors grows, and it becomes a time-consuming process.

- **AI-based or Cognitive Data Extraction**: For successful decision-making, AI-based or cognitive data extraction is a self-learning system that mimics human behavioural patterns. Bank statement data extraction software can automatically learn from large volumes of data to interpret bank statements in multiple formats. Different document layouts, changes in header/column/table placements, and small variations in the arrangement of data displayed in bank statements can all be recognised by AI. What a manual operator could take hours to learn and adapt to numerous formats, AI can do in seconds, and its accuracy improves as you feed more data into models. As AI technology is capable of analysing larger datasets, it will eventually assist users in automatically digesting bank statements. You also won't need to bother about building fresh templates for scanning new bank statements because AI eliminates the necessity for them.
- **Combining Machine Learning and Template-based Approach**: The bank statement processing paradigm uses standard templates for reading existing bank statements, combining the best of both worlds. When a new bank statement format is introduced, the machine learning [12] system adapts and no longer uses the default templates. It then generates a template for the new statement without the requirement for human participation. The system becomes wiser as more bank

statements are processed, and its database expands to handle a wider range of financial records. After a short amount of time, the accuracy of the technology improves as well.

6.4 RPA in Loan Processing

RPA in loan processing can speed up procedures, improve reliability, and eliminate labour-intensive activities by transferring responsibility to robots and allowing employees to focus on more complex tasks. RPA in loan processing can strengthen current banking systems while also assisting the sector in scaling its growth. According to Mckinsey [13] research, a bank improved its corporate credit evaluation through automation, resulting in an 80% increase in production (Fig. 6.4).

What are the obstacles that automated loan processing can overcome? [14]

Customers' expectations have changed over time when it comes to loan processing. Loan processes are sometimes bureaucratic and lengthy in today's fast-paced environment, where everyone seems to be battling to find enough time. The financing process is complicated, and it can lead to client unhappiness. It is becoming increasingly important for the sector to meet the expectations of customers and to apply cutting-edge technology to improve procedures. Quick and easy access to loans will make the process less stressful for customers, who will benefit from streamlined products and services. Automated loan processing can personalise the loan process for borrowers, servicing them at a time, place, and media that is convenient for them. By overcoming these obstacles, lending and borrowing will become more profitable, resulting in increased client loyalty.

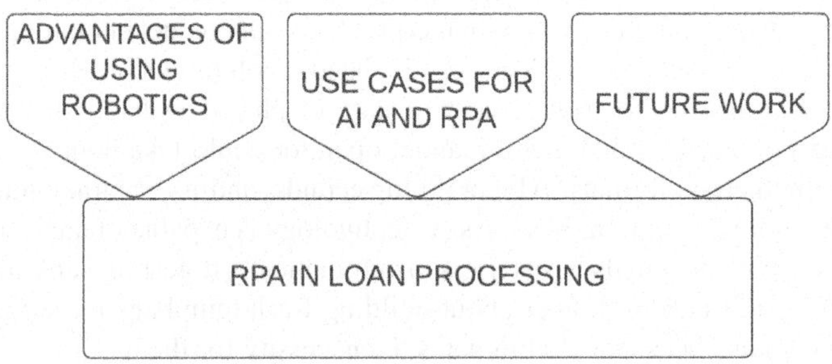

Fig. 6.4 RPA in loan processing

6.4.1 RPA's Advantages in Loan Processing

Loan processing automation can help with everything from loan origination to post-closing. Workflows will be smoother and more productive as a result. RPA in loan processing can improve the industry's operations in the following ways [15]:

- **Growth in Productivity**: Data input, document routing, task allocations, and email sorting are just a few of the tasks that can be automated. Automation may dramatically enhance productivity by removing large chunks of time spent on repetitive chores. Lenders can use the time saved to complete other duties, such as finalising and processing more applications in less time and increasing earnings [16].
- **Fraud Detection**: There has been a rise in fraud within the banking sector, which has resulted in massive monetary losses. RPA can help reduce losses by detecting fraud early on with the use of LOS (Loss Origination Systems), which uses powerful predictive analytics to determine how hazardous it is to lend money to a certain borrower. Companies can tailor the kind of loans that are more susceptible to fraud and employ RPA applications to do so.
- **Improved Customer Service**: Processing loan applications can be a time-consuming and inconvenient process with a negative reputation. By improving the client experience [17], RPA in loan processing can save the company's reputation. Along with the individualised experience that RPA can provide, the reduced time taken may be a big draw for clients. When compared to manual operations, there are fewer risks of errors [18].
- **Easier Audits**: When RPA is used in loan processing, papers are frequently categorised and processes are often quick, making compliance rules easier to apply. The auditor's job is made easier by having quick access to files and operations. With the help of RPA, following regulations becomes a breeze.
- **Revenue Prediction**: RPA can predict revenue, making it a very profitable technology. Automation can forecast how much money a lead will make on a loan throughout the course of the loan cycle. The expected revenue assists in dealing with market volatility and regulatory requirements (Fig. 6.5) [19].

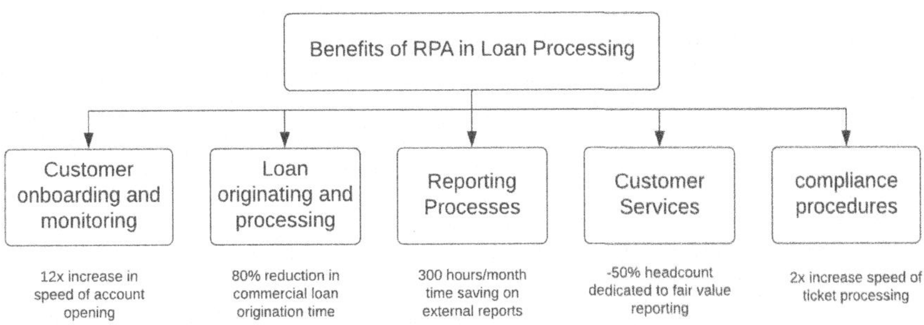

Fig. 6.5 Benefits of RPA in loan processing

6.4.2 Use Cases of RPA in Loan Processing

With the help of over-the-desktop automation programmes, RPA is mostly used in commercial banking activities. You can automate the transcription of banking data and eliminate the need for human interaction. Here are a few examples of how RPA can help in loan processing [20]:

- **Reports are generated automatically**: RPA can generate Suspicious Activity Reports and develop compliance reports that flag fraudulent transactions (SAR). In comparison to humans, RPA can quickly read compliance papers and type data.
- **Customers' Onboarding**: Customer onboarding is made more difficult by manual verification, which takes a long time and is inconvenient. RPA can extract information from documents and compare it to information provided by consumers.
- **Lending for Mortgages**: Loan start, document processing, and financial comparisons are all aspects of mortgage lending that will be automated, resulting in a speedier loan approval procedure. Customer satisfaction rises as a result of this.
- **Report on Loans Due**: RPA bots can identify the loans which are due. It can collect records from different branches and areas and combine them into a single file. The bot can send an automated email to the customers before the scheduled date.
- **Balance Register**: To ensure the customer has paid all of the costs required for loan processing [21], the balance register process is automated. The credit and debit details are auto-updated to balance the information according to predefined standards.
- **Loan Closure NOC**: Automate the no objection certificate procedure for loan closing (NOC). The bot checks the primary banking system for any further pending amounts after receiving the accompanying paperwork and the customer's cheque for the outstanding loan amount. The bot stops the process if a request is pending and modifies the system's pertinent comment. If the process of creating the NOC isn't already underway, it begins it and automatically emails it to the pertinent stakeholders for review and dispatch.
- **Modifications to Retail Asset Details**: To automatically verify documents obtained through loan applications, RPA bots can be employed. If the information varies from the data already present in the primary banking system, the system starts a replacement operation to update a system equivalent, which is then forwarded to a professional authorised person within the Client Asset Modification team. With the assistance of a professional authorizer, the bot alters the loan application system if the customer only has a loan account and no prior savings account.
- **Booking a Loan Quickly**: To book express loans, use RPA. To disburse the loan, the bot prepares a loan or uses an approved template, followed by a loan voucher. Through several touch points within the core financial system, the bot refreshes the acquired data.

- **Repayment of a Loan**: Use the RPA bot to find the account that has an overdue loan balance. Check to see if the customer has enough money in his or her account to pay back the loan. Calculate the right value of the data on which repayment is made.
- **Loan Exposure Report**: To compute the final loan amount due report, combine all records from various files into one file. Send the file to all stakeholders automatically.
- **Posting of Loan Requests**: In the core loan management system, enter loan requests received through omnichannel processing and the customer service desk.
- **Processing of Loans**: Check credentials, process, and disburse loans, including new loans, PO financing, and partial distribution.
- **Detection of Fraud**: Use RPA bots and AI algorithms to collect data from a variety of sources, including data received via a loan application. To check for suspected fraudulent loan applications, run the loan application via an anti-money laundering (AML) system and third-party organisations [22].
- **Migration of Loan Documents**: Transfer loan applications, scanned documents, and support from several sources to a central document management system (DMS). The loan processing department can speed up the procedure of converting loan applications into digital assets by including intelligent data collection in the process [23].
- **How to Apply for a Loan**: Using RPA bots on a digital platform, speed up the processing of high-value loans. To reduce the technique from weeks to one or two days, quickly gather client data, analyse the value of the mortgaged property, and obtain valuation reports from many sources.
- **Processing of Loan Payments**: Obtain missing information and additional data from third-party websites as needed for the procedure. Promptly, apply payment procedures to borrowers' accounts. Fulfil the requirements for audits by statutory agencies evaluating the timeliness with which you respond to borrower requests.
- **Control of loan documentation quality**: To extract individual files from loan management systems and merge them in a PDF, use RPA bots. RPA can be used to automate the usual control procedure and make loan documents more accessible for quality assurance and auditing [24].
- **Home Loan/Top-up CERSAI**: Through the CERSAI portal, you can check for fraud in lending against equitable mortgages as well as multiple loans taken on an equivalent asset by different institutions. This is a technique that is typically unique to India.
- **Maintenance of a Foreclosure on a Loan**: Validate loan foreclosure requests received from the banking industry automatically.
- **Consolidation of Debt**: Collect large-scale records from numerous channels containing information such as the most recent closing balance, current transactions, and so on. Reconcile the data through a bot. Manually process the exceptions.

6.4.3 The Future of RPA in Loan Processing

Every sector is waking up to RPA's potency and potential for revolution. The banking industry is no exception. To boost efficiency, the banking industry is turning to automation to provide better, more secure, and more dependable services. The pandemic has also shown vulnerabilities in the credit industry that can now be filled with automation. The challenge of performing everything virtually and digitally is solved with RPA. RPA helps businesses manage competition by incorporating low-cost, high-quality banking processes that benefit customers while staying current with technology. According to a study done by McKinsey, one-fourth of banking processes would be automated in the next few years to improve services. This demonstrates RPA's utility in the banking business.

When it comes to implementation, RPA can be utilised for loan processing in a variety of ways. After doing a thorough examination of its operations, the organisation can sit down and pick which use cases it wants to focus on. Repetitive and redundant tasks can readily be automated, resulting in increased efficiency. The next stage is to create an execution strategy that meets the needs of the organisation. You can speak with an Accubits specialist at any time and they will assist you in making the best selection for your company. Then it's only a matter of reaping the rewards of robotic process automation [25].

With increased client demands, regulatory requirements, and competing technology, the lending picture has become more complicated. Customers want access to loan products and services that are tailored to their unique needs and available at their preferred time, place, and channel. Lenders, for their part, are employing a variety of technologies and processes to boost productivity, expedite approvals, and reward consumer loyalty. As a result, the majority of clients fail to understand the actual terms and conditions of the process, resulting in unnecessary confusion. Robotic Process Automation (RPA), which can be employed in the loan process in the form of lending management automation, can address these issues. RPA, or Robotic Process Automation, meets the needs of clients while also assisting lenders in improving the loan origination process and providing them with new chances for growth [26]. RPA refers to the use of software robots to assist in the automation of business processes such as document filing and routing, email reminders, notifications, data synchronisation, and so on. It may be used to create workflows and automate operations from beginning to end using Enterprise Automation platforms. Automation enables the integration of heterogeneous systems, the delivery of consistent and trustworthy data at any level of the loan origination process, and the acceleration of overall operations while providing significant audit and control benefits. An automated loan processing system can take data from paper documents electronically, index them, and route them to the proper system throughout the loan process, from origination to underwriting and post-closing. Workflows can be built up for specific tasks to ensure repeatable, uniform processing and speedier turnaround times. It is possible to automate manual repetitive processes that would otherwise be conducted by loan operations staff.

- Lenders can also benefit from an automated loan processing system [19];
- Following a slew of ever-changing rules;
- Maintaining accountability within processes and subprocesses;
- Finding vital information within organisational silos without making it a difficult, time-consuming chore.

Maintaining compliance in order to reduce risk

It's critical to remember that automation isn't meant to take the role of existing loan origination or quality-control procedures. The front-end integration necessary for RPA deployment does not necessitate any changes to the IT architecture. The most effective tools always operate in tandem with existing technology to increase data quality and speed.

6.5 RPA in Credit Card Processing

The credit card business is increasing, and with it comes a large amount of data. The average consumer uses more than one card at a time, and online payments are growing four times faster than retail payments. At this rate, it is important to automate the processing of credit cards. This section discusses the current challenges and ways to implement RPA in credit card processing (Fig. 6.6).

6.5.1 The Current Challenges in Credit Card Processing

The push by most governments to implement a cashless economy has made the use of credit cards commonplace. When compared to flat currency, customers are increasingly preferring to carry plastic money. The average customer has more than one credit card, and internet payments have surpassed retail payments by four times. The ever-growing credit business, on the other hand, has overburdened financial

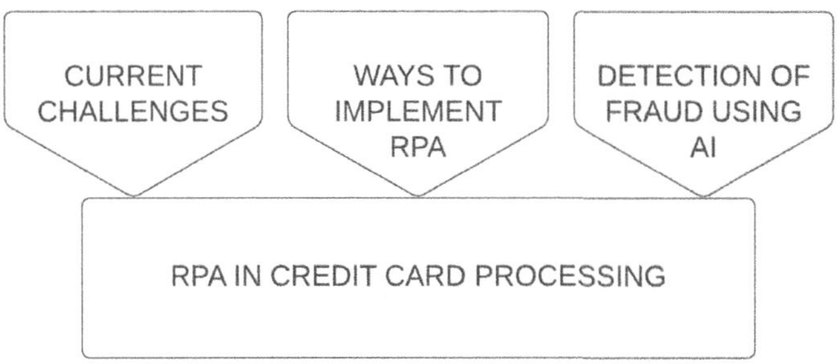

Fig. 6.6 RPA in credit card processing

institutions that still rely on manual operations. Traditional credit card processing takes weeks since a variety of processes and checks must be followed from confirming consumer information to accepting the credit card [27]. As a result, the entire system has become inefficient, and the likelihood of errors and fraud has increased. This is understandable, given that banking institutions process up to 500,000 credit card applications per day. This is When Robotic Process Automation Can Help.

The challenge is undeniably genuine. But that is no longer the case. The introduction of robotic process automation (RPA), often known as RPA, has proven to be a huge help to banking and financial institutions. For starters, an RPA might be a good mix of artificial intelligence (AI) [28] and machine learning capabilities that makes managing high-volume, recurring activities simple and error-free. When it comes to credit card processing, an RPA can assist by speeding up the entire process and decreasing excessive wait times. Long wait times not only cost financial institutions a lot of money, but they also make customers unhappy. RPA allows banks to issue credit cards to customers in as little as 24 h. An RPA can instantly interface with many systems, validate the needed information, run background checks, and approve or disapprove the application based on the principles.

All this will translate to following [29]:

- Reduced costs;
- Streamlined operations;
- Happy staff;
- Digitised data;
- Reduced business response time;
- Better customer service.

6.5.2 *Implementing RPA in Credit Card Processing*

The following are the ways in which RPA can be implemented to simplify credit card processing with accuracy in optimal time.

- **Processing of Credit Card Application**: The KYC (Know Your Customer) process can be automated using RPA. The bot is capable of identifying anti-money laundering compliances and other checks by assigning a score to the applicant after accessing the risks and collecting data. It also handles the paperwork involved in the process using automated interfaces.
- **Personalization**: With straight-through processing, the card is personalised, sent, and activated automatically after approval. Referrals are handled using an autodialer for customer and issuer interactions, as well as permission.
- **Transaction Clearing and Settlement**: The numerous processes optimised by RPA include exception handling, verification of the bank's charges, reviewing, updating, and re-process validation of unsuccessful data.
- **Monitoring**: RPA finds past-due accounts, blocks cards when there is questionable activity, alerts customer support staff, keeps track of collaterals, and

secures the transaction. Additionally, it manages and supervises the data-gathering organisations.

- **Customer Service**: RPA can register and handle a customer's complaint, as well as provide transaction alerts, offers, and rewards, as well as other event-based periodic messages, such as a personalised New Year's or birthday greeting. It can also cross-sell and analyse client behaviour through automated telemarketing.
- **Management of Merchants and Partners**: RPA has a wide range of applications in credit card processing. Merchants and partners are subjected to PoS testing, certification, activation-deactivation, settlement, and reconciliation.

6.5.3 Detection of Credit Card Fraud

According to the European Central Bank's fifth report on card fraud, the total value of fraudulent transactions using cards issued inside the Single Euro Payments Area (SEPA) and acquired globally totalled €1.8 billion in 2016. Furthermore, between 2011 and 2016, there was a 66% increase in card-not-present (CNP) fraud, leading to a 35% increase in overall fraud.

Worse, Europe isn't the only player on the field. Economies appear to be cast in the same mould all across the world! In 2018, $574.3 million of the $788.6 billion in transaction value on Australian cards was fraudulent. According to the Reserve Bank of Australia, there is a current fraudulent rate of 72.8% per $1000, demonstrating the growing trend of online card fraud and cybercrime. With such figures, the banking industry has never had a greater need or reason for early fraud detection and prevention methods.

The use of traditional methods to combat fraud has many drawbacks. They are frequently inflexible when it comes to integrating into numerous systems, lack access to external data and real-time behavioural profiling and analysis, and provide different fraud management solutions across the value chain.

Using cognitive thinking, several machine learning algorithms are put in place that constantly execute a series of checks and balances with the primary objective of discovering anomalies, identifying patterns, and making decisions about fraudulent activity. When fraudulent behaviour is detected, an automated call or SMS notice is delivered to the user to verify their credentials and key information. The user gets flagged and reported if they are unresponsive.

Simply said, RPA gathers and captures all of the data from the user in an application layer, including voice recognition features, and delivers it to be stored and profiled later. Additional real-time analysis of user spending trends and behaviours is performed using Cognitive RPA and Cognitive Virtual Reality. It may also retrieve and validate patterns by SMS or phone call.

6.6 RPA in Mortgage Banking

Banks Benefit from Mortgage Process [30] Automation in several ways. Banking and mortgage lending entail numerous complicated processes, necessitating ongoing compliance. One of the most significant advantages of RPA in mortgage processing is that it aids in the systematisation, standardisation, and strict definition of all procedures within a lender's operations.

Anyone who has ever purchased a house understands how much paperwork is involved: credit checks, employment verifications, insurance statements, inspection reports, and so on. Even the tiniest mistake, such as a missing date on a paper, can derail the process and complicate things, causing stress for both the buyer and the seller.

RPA solutions, which automate routine, rule-based operations without requiring human participation, can greatly improve the mortgage lending experience. While banks have made significant progress in automating core business processes, workflows, and mortgage-related operations, RPA can improve efficiency by complementing existing workflows.

Furthermore, finding and resolving faults and exceptions, which now need a high level of human interaction, is a big opportunity for RPA. Process exceptions in applications like healthcare claims may present complex problems concerning policies and coverage guidelines that require human reasoning, judgement, and knowledge to answer. However, in the mortgage lending industry, a large percentage of errors and process exceptions are common and may be handled by using properly executed rules-based algorithms. As a result, RPA is well-suited to automating the resolution of errors and exceptions, as well as removing bottlenecks that frequently stymie mortgage origination.

Banks and other financial service providers have a unique potential to use smart tools [31] to highlight the commercial value in addition to enhancing the operational efficiency of lending procedures. Consider the mortgage customer: he or she may be an excellent candidate for a variety of extra services offered by the bank, such as insurance, savings programmes, and retirement funds. The difficulty for a bank is to connect the dots between a specific client—and what is known about that customer—and the bank's other offers that will be of interest to that customer.

In many contexts, that process is currently labour-intensive and slow. Combining cognitive applications that utilise pattern recognition and logic to analyse data and form conclusions with RPA as a processing engine presents an opportunity. So, for example, from John and Mary's mortgage documentation, a cognitive tool may deduce that they have two school-aged children and identify them as prospects for the bank's college savings programme. RPA solutions could make it easier to move John and Mary into the appropriate product line's marketing funnel.

The smart tools would complement a broader customer-focused strategy based on data analytics, social media, and other technologies while exploiting RPA's basic strengths. Furthermore, the tools can greatly improve data collecting and analytics, allowing for that vital customer insight. Robots may also be readily extended to other

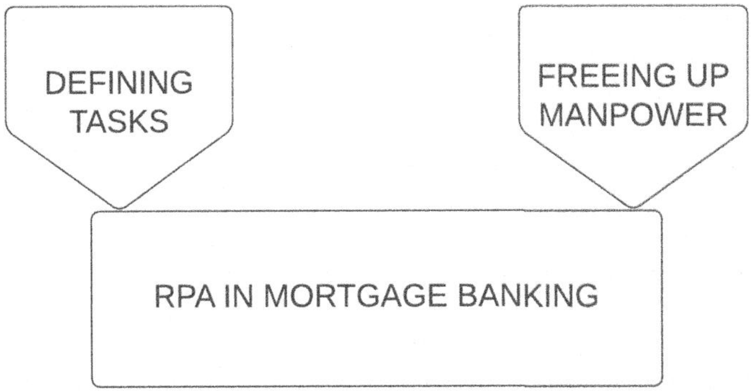

Fig. 6.7 RPA in mortgage banking

bank services, such as client self-service, by following up on ATM interactions with recommendations for improving the safety and security of a customer's savings or other assets, for example.

Today, customer retention may be a significant requirement for banks. Millennials, unlike previous generations, would switch banks at the drop of a hat; therefore, an opportunity to provide a pleasant customer experience must be taken advantage of. Smart technologies that improve operational efficiency and provide insight into client preferences are frequently a significant component of a winning competitive recipe for increasing customer loyalty and improving results (Fig. 6.7).

6.6.1 Defining Tasks and Freeing up Manpower

By clearly and precisely outlining each stage in the process, a mortgage lender stands to benefit from increased efficiency and productivity if a process is conducted repeatedly over some time. Many banks have processes that are so inadequately defined that if ten employees were asked to recite the steps for performing a task, the questioner would likely get ten different responses.

Interdepartmental and even employee-to-employee cooperation can be difficult with this type of haphazard approach to business-critical tasks, and it could be a primary cause of errors. An intrinsic component of implementing an RPA system is always going to be precisely identifying those jobs and then fully automating them, eliminating both inefficiency and error while freeing up humans to work on things that require more brain power.

RPA technologies can take over drudge work from humans while causing minimal impact to existing systems. RPA agents can interact with existing systems and software in a way that is practically indistinguishable from humans, whether it be a simple spreadsheet application or sophisticated bespoke corporate software. As a result, existing systems and processes do not need to be disturbed when outdated platforms are replaced with new ones.

You'll use RPA in a variety of ways as you branch out into other areas of automation. All you have to do now is figure out which option is best for you. If you're not sure what your best option is, start by asking yourself these questions about how you handle claims. Do you have to direct them to the carrier's website? Is it possible to handle the claims yourself, or do you need to find someone who can help you in an emergency? Individuals who add house or vehicle insurance are in the same boat, because your customers need to know that they can go to your website and get help right away.

RPA can interpret structured or semi-structured data and determine where it should be entered into a different programme. Document routing, data input, customer notifications, and even risk management are all activities that it can handle.

Most importantly, RPA solutions can be implemented immediately on top of existing systems, requiring minimal changes to the company's current operations.

6.7 Conclusion and Future Work

Commercial banks spend a lot of time and money on bringing new clients on board, and new and amended rules will force them to spend even more time and money on these processes. RPA and cognizable technologies have the potential to help banks save time and money while also reducing errors by automating activities. Furthermore, automating credit card processing, loan processing, and mortgage processing has allowed banks to avoid the time-consuming processes of document review and clear client requests in the shortest time possible, allowing them to serve several customers at the same time. Businesses that use RPA have witnessed a 50% decrease in time and expense, as well as a significant improvement in the accuracy of their work. The benefits of RPA can help customers have a better experience, potentially allowing banks to form long-term relationships with them. Future work will incorporate more sophisticated automation tools to address existing challenges in the current framework.

We are looking into ways to automate every sector of the banking industry with the help of robotics. With the rapid growth of RPA technologies in day-to-day life soon most if not all banking tasks will be performed using RPA tools. RPA, or Robotic Process Automation, services, when properly implemented, can be truly transformative for the banking sector by automating manual, repetitive, and time-consuming tasks. The result of automating such mundane tasks would be increased productivity, a significant reduction in error rate, and an impressive turnaround time.

References

1. Pokharkar, A.P.: Robotic process automation: concept, benefits, challenges in banking industry. IIBM'S J. Manag. Res., 17–25 (2019). http://iibmjournalofmanagementstudies.in/index.php/iibm/article/view/143
2. What is Robotic Process Automation? Uipath, https://www.uipath.com/blog/rpa/what-is-robotic-process-automation
3. Valgaeren, H.: Robotic Process Automation in Financial and Accounting Processes in the Banking Sector (2019)
4. Santos, F., Pereira, R., Vasconcelos, J.B.: Toward robotic process automation implementation: an end-to-end perspective. Bus. Process Manag. J. **26**(2), 405–420 (2019)
5. Herm, L.-V. et al.: A consolidated framework for implementing robotic process automation projects. International Conference on Business Process Management. Springer, Cham (2020)
6. Gupta, A., Ranjan, S.: Automation in On-Boarding and Ongoing Servicing of Commercial Banking Clients (n.d.). Deloitte. Retrieved March 29, 2022, https://www2.deloitte.com/content/dam/Deloitte/us/Documents/financial-services/us-cons-automation-in-on-boarding-and-ongoing-servicing-of-commercial-banking-clients.pdf
7. Noelle, C.: How Do Banks Benefit from Robotic Process Automation (RPA)? ProcessMaker (2019), https://www.processmaker.com/blog/how-do-banks-benefit-from-robotic-process-automation-rpa/#:~:text=RPA%20enables%20banks%20to%20complete,without%20overhauling%20existing%20operating%20systems
8. Seven Proven Automation Strategies for Banking and Financial Services. Sutherland (2017). https://www.sutherlandglobal.com/ourthinking/seven-provenautomation-strategiesfor-banking-and-financial-services
9. Gruzauskas, V., Ragavan, D.: Robotic process automation for document processing: a case study of a logistics service provider. J. Manag. **36**, 119–126 (2020). https://www.ltvk.lt/file/zurnalai/Vadyba_2020_2_36_2.pdf#page=119
10. Ling, X., Gao, M., Wang, D.: Intelligent document processing based on RPA and machine learning. 2020 Chinese Automation Congress (CAC). IEEE (2020). https://ieeexplore.ieee.org/abstract/document/9326579/
11. Docsumo: Bank Statement Processing and Automation. Complete Guide to Bank Statement Processing and Automation (n.d.). Retrieved March 29, 2022, from https://assets.websitefiles.com/5f689f82910c6b4f1ffb855b/611ca5e8e2c55ee88e986bde_BankStatementAutomationGuide.pdf
12. Mitchell, T.M.: Machine Learning. McGraw-Hill, New York (1997). ISBN: 978-0-07-042807-2
13. Business Process Automation in Banking & Finance. Accelirate, https://www.accelirate.com/industries/business-process-automation-in-banking-finance/
14. How is Robotic Process Automation (RPA) used in Banking Industry? FITA, https://www.fita.in/robotic-process-automation-rpa-used-banking-industry/
15. Nelito: 10 ways Robotic Process Automation can improve the Loan Origination Process (2019). https://www.nelito.com/blog/10-ways-Robotic-Process-Automation-can-improve-the-Loan-Origination-Process.html
16. Top 10 Benefits of Robotic Process Automation (RPA). 10xDS (2019), https://10xds.com/blog/insights/advantages-of-robotic-process-automation
17. Ghosh, M.: HDFC Becomes 1st Bank to Deploy Robots for Customer Service; Fires 4,581 Employees due to 'Improved Efficiency'. Trak.in (2017). https://trak.in/tags/business/2017/01/30/hdfc-bank-automation-humanoid-ira/
18. Siklos, P.L.: The Changing Face of Central Banking: Epilogue. The Changing Face of Central Banking, Cambridge University Press, 2002, pp. 300–308. Syed. R. et al. "Robotic Process Automation: Contemporary Themes and Challenges." Computers in Industry, vol. 115, 2020. "Top-5 Benefits of Robotics Process Automation (RPA) Adoption for Your Company." Prov (2018), https://www.provintl.com/blog/top-5-benefits-of-robotics-process-automation-rpa-software

19. V, M.P.: How to Get Started with Loan Process Automation with RPA. Accubits Blog (2021). Retrieved March 29, 2022, from https://blog.accubits.com/how-to-get-started-with-loan-process-automation-with-rpa/
20. Wilds, C.: Robotics in Banking with 4 RPA Use Case Examples. The Lab (2018), https://thelabconsulting.com/robotics-in-banking-with-4-rpa-use-case-examples/
21. Datamatics: RPA Use Cases in Loan Management Processes. Datamatics (n.d.). Retrieved March 29, 2022, from https://www.datamatics.com/intelligent-automation/rpa-trubot/use-cases/banking-financial-services-loan-management
22. Thekkethil, M.S. et al.: Robotic process automation in banking and finance sector for loan processing and fraud detection. 2021 9th International Conference on Reliability, Infocom Technologies and Optimization (Trends and Future Directions)(ICRITO). IEEE (2021), https://ieeexplore.ieee.org/abstract/document/9596076/
23. Aguirre, S., Rodriguez, A.: Automation of a Business Process Using Robotic Process Automation (RPA): A Case Study, 65–71 (2017). https://doi.org/10.1007/978-3-319-66963-2_7. Available from: https://www.researchgate.net/publication/319343356_Automation_of_a_Business_Process_Using_Robotic_Process_Automation_RPA_A_Case_Study
24. Moffitt, K.C., Rozario, A.M., Vasarhelyi, M.A.: Robotic process automation for auditing. J. Emerg. Technol. Account. **15**(1), 1–10 (2018)
25. Neetika: Future of Robotic Process Automation in the Banking Sector. Convedo (2019), https://info.convedo.com/future-of-robotic-process-automation-in-the-banking-sector
26. Robotic Process Automation Market—Growth, Trends and Forecast (2019–24). Mordor Intelligence, https://www.mordorintelligence.com/industry-reports/robotic-process-automationmarket#:~:text=The%20robotic%20process%20automation%20market%20was%20valued%20at%20USD%202094.3,the%20forecast%20period%202021%20%2D%202026
27. Bryer, T.: MasterCard using AI to turn staff into top sellers. CNBC (2016). https://www.cnbc.com/2016/03/03/mastercard-using-artificial-intelligence-to-turn-staff-into-top-sellers.html
28. Williams, D., Allen, I.: Using artificial intelligence to optimize the value of robotic process automation (2017). Available from: https://www.ibm.com/downloads/cas/KDKAAK29
29. Nallicheri, N.: Automating Bank Operations? Keep Eyes Wide Open. Accenture (2018), https://bankingblog.accenture.com/automating-bank-operations-keep-eyes-wide-open
30. Aljuhani, N. et al.: Robotic process automation and reengineering using Bizagi and UiPath: case study on mortgage request process. Int. J. Simul. Process Model. **17**(2–3), 166–177 (2021)
31. What is RPA? How does it work? Top RPA tools of 2020. AI Multiple (2020), https://research.aimultiple.com/rpa/

Chapter 7
Robotic Process Automation in Healthcare

Jagjit Singh Dhatterwal, Kuldeep Singh Kaswan, and Naresh Kumar

Abstract Robotic process automation (RPA) is a contemporary breakthrough that automates repetitive, routine, and rule-based human activities to benefit businesses that choose to adopt such software. RPA is a commercially available technology, although there is little scientific research on the issue. As a result, the goal of this chapter is to examine how the academic community defines RPA and how much of its status, trends, and applications have been studied in the literature. The differences between RPA and business process management are also covered. The chapter offers the results of a systematic literature review (SLR) on RPA, including an overview of the terms and uses of RPA in real-world settings as well as the advantages of using RPA in different sectors of the economy.

Abbreviations

RPA	Robotic process automation
CRM	Customer Relationship Management
IT	Information Technology
ITPA	Information Technology Professionals Association
UI/UX	User Interface/User Extensions
BPO	Business Process Outsourcing
BPM	Business process management
BPMS	Business Process Management Systems
EHR	Electronic Health Record

J. S. Dhatterwal
Department of Artificial Intelligence and Data Science, K L Educational Foundation, Vaddeswaram, Andhra Pradesh, India

K. S. Kaswan (✉) · N. Kumar
School of Computing Science and Engineering, Galgotias University, Greater Noida, India
e-mail: kaswankuldeep@gmail.com; kuldeep.kaswan@galgotiasuniversity.edu.in

N. Kumar
G. L. Bajaj Institute of Technology and Management, Greater Noida, UP, India

S. Bhattacharyya et al. (eds.), *Confluence of Artificial Intelligence and Robotic Process Automation*, Smart Innovation, Systems and Technologies 335,
https://doi.org/10.1007/978-981-19-8296-5_7

EMPI	Enterprise-Wide Master Patient Index
ERP	Enterprise Resource Planning
ECM	Enterprise content management
SLRP	Systematic Literature Research Protocol
HR	Human Resource
CEO	Chief Executive Officer
KPMG	Klynveld Peat Marwick Goerdeler
HFS	Hierarchical File System

7.1 Introduction

Many functions spend a significant amount of time communicating with various apps for commercial batch processing. Some examples of these applications are capacity management, Customer Relationship Management (CRM), and Payroll system. The majority of the positions that may benefit from automation and artificial intelligence outsourcing are office-based jobs that need the ability to do a variety of activities in sequential order [1]. Normal factory automation, on the other hand, focuses on selecting a portion of a process or even just one action and developing robots that are specialized in doing that particular task. Office work typically requires the same kinds of job and task repeats, but a physical robot is not necessary since data is being processed across a variety of platforms and devices [2]. Instead, robots capable of launching and landing are deployed and programmed to run extra software. RPA acts as a computing device for persons, automating the boring and typical tasks that occupy a part of the day for every worker in a business. Because of its accessibility and efficacy, robotic process automation (RPA) may become increasingly desirable to a large number of enterprises, particularly those that still use older technologies and networks. Because it is designed to work with the vast majority of legacy services and devices, automation, and artificial intelligence, automating is easier to implement than other corporate automated solutions [3].

Consider the possibility that certain hierarchically organized, repetitive, and difficult activities may be performed by a computer, thereby freeing up the time of skilled people to focus on other important duties. This is the promise offered by Robotic Process Automation (RPA), which has arisen in the last five years as a collection of software applications and technology [4]. Technologies capable of automating tasks inside a process that is based on rules within an organization. We don't necessarily mean machines when we talk about Robots in Advanced Automation operations, systems, or tools; rather, we mean programming blocks or programs that can assume specific duties or jobs previously performed by individuals and carry them out in a manner that is both quicker and more effective [5].

7.1.1 *The Center of Attention*

Both RPA and conventional IT-based automation target various parts of the company and have distinct goals in mind. Automated methods improve operational efficiency. By decreasing costs and increasing delivery times, IT-based process automation is mainly entrusted with increasing profitability. Rather than being limited to the IT department, robotic process automation extends to almost every aspect of the organization, from accounting to marketing. The end users' tasks are being made simpler and more efficient via RPA. On the IT infrastructure side, ITPA looks at the demands and weaknesses of the company's complicated IT system.

Quality of experience (UI/UX)

Robotic Process Automation (RPA) technologies and apps are easy to use and have superior user interfaces (UI) (UX). Creating user interfaces that are easy to grasp and can be controlled by anybody with even rudimentary knowledge of computer technology is the ultimate objective.

7.1.2 *Examining the Use of Robotics in Health Care*

As a matter of fact, when it comes to UI/UX, conventional IT-based process automation tools and apps are less appealing to end users. Traditional IT-based process automation systems, on the other hand, are more complicated and need a higher level of IT expertise to implement. Even though these ITPA apps may be used without any programming knowledge, they are often developed using code.

7.1.3 *Non-disruptiveness*

It is common for traditional IT-based process automation advancements to focus on improving the efficiency of existing less-than-optimal processes and systems. As a result, the present IT infrastructure is sure to change. RPA tools, on the other hand, are more IT-light and, for example, do not disrupt underlying computer systems. The robots access end user computer systems through the user interface and existing access control mechanisms, therefore, no underlying system modification is necessary.

Using both of these ways of automation, a business will have a distinct advantage over the competition. Two different schools of thinking exist regarding the application of RPA in various industries. The backend or back-office duties done by Business Process Outsourcing (BPO) organizations would be particularly vulnerable to RPA's job-killing potential, according to one portion, since software robots will be able to accomplish these activities without becoming weary or making mistakes. Violino says that the technology will give chances since firms require knowledgeable individuals to implement, manage, and maintain the systems. Still, there is also another school of thinking. Lower and middle management will need new skill sets, such as the ability to work with RPA platforms and an understanding of how to manage them, as a result of this development. Some of the displaced people may be relocated to more exciting and demanding positions in IT or other parts of the organization.

7.1.4 Healthcare RPA: An Overview

The healthcare business is a major source of income and jobs in almost every nation. It includes, for example, medical gadgets, clinical trials, health insurance, and other medical supplies. One of the biggest challenges in any healthcare system is keeping track of and processing data from a variety of different internal and external sources such as clinical apps, lab information systems, third-party portals, and insurance portals. Because of the difficulty of integrating various technologies, healthcare companies must depend on people to conduct manual labor-intensive operations to process information.

Health care is made up of patients, physicians, insurers, and other stakeholders. An urgent need exists for a more efficient and accurate back-office procedure to keep up with the rising number of patients, documentation required for follow-up and insurance claims, and so on. By using Robotic Process Automation (RPA), a cutting-edge automation solution, medical facilities can boost productivity while also lowering operating expenses and reducing the risk of human error when processing data such as physician credentialing and enrollment as well as clinical documentation and billing for Medicare as well as secondary claim management in the future.

In addition, pharmaceutical businesses confront difficulties in bringing novel treatments to market because of the requirement to preserve product quality while also increasing productivity and profits. In the healthcare business, process automation technologies may be used to overcome regulatory and reporting difficulties. These tools help healthcare organizations enhance patient safety and bring successful pharmaceuticals to market.

7.1.5 RPA Might Have a Positive Impact on the Healthcare Sector

- Implementing Robotic Process Automation (RPA) in the healthcare sector today will help in the following ways:
- Mitigate the healthcare industry's complexity of processes, the volume of patients, and the number of hospitals.

- Autonomously collecting and combining data from several sources through the integration of dissimilar systems via software robotics.
- Healthcare apps, laboratory systems, third-party portals, insurance portals, and radiology systems.
- Allow workers to put their talents to use in situations that require a human touch.
- Minimize healthcare budget and human resource expenditures while enhancing healthcare speed, intelligence, effectiveness, and effectiveness processes.
- In order to effectively communicate with providers and patients, automate processes such as eligibility requests.
- Improve revenue cycle management by automating claims status queries and conducting claim reviews.
- The goal is to create a digital workforce that will operate alongside the human workforce in order to increase productivity.

7.1.6 Methodology of RPA

The methodology is based on software and algorithms to perform routine human labor [6]. Fundamental rules and application logic largely control it, and it communicates with many information managements using current user interfaces. Its capabilities include automating repeated and rule-based processes by employing an ou pas programming robot known as a bot [7].

7.2 Literature Survey

According to the exploratory literature review, RPA uses specialized technologies. RPA definition has recently been expanded to include artificial intelligence (AI), advanced analytics, business process management, and business intelligence [8]. With modern digital technology's advent, RPA may be reallocated from executing monotonous and error-prone procedures in corporate processes to more complicated experience and understanding and valuation jobs [9]. Forrester evaluated 12 RPA companies delivering entrepreneurship and comprehensive systems that can satisfy the needs of a "public–private partnership" or organization of RPA utilities to determine the current state of the RPA industry [10]. Even though certain RPA providers provide industry-specific options, "the overall notion of RPA is sector-neutral." RPA vendors' collaboration with going to lead artificially intelligent suppliers, on the other hand, enabled the expansions of conventional RPA features and functionality with innovative technological innovations such as personality from requirements management, training robotic systems, AI-screen acknowledgment, unsupervised learning, and automatic production supporting documents production [11]."

The bulk of the 400 organizations questioned by Deloitte has begun their RPA journey, with nearly a fourth more planning to do so within the next two years. They also note that payback timeframes are averaged a little over a year and that their goals for cost savings, accuracy, timeliness, flexibility, and increased conformance have been met or surpassed. According to Forrester, over 4 million robotics will execute recurring activities by 2021, but the emphasis will shift to Ai and robotics and RPA analytic enhancements [12]. Similarly, Everest Group notes that, while many consumers are pleased with the Analytical framework, analysis, and mental function must be improved [13].

Despite the many benefits of RPA, just 5% of organizations in the Deloitte analysis have far more than 50 robotics in their processes. Organizational competence and a business manager's goals for RPA adoption are critical for RPA project performance. Key hurdles for automating the process have been noted as a lack of awareness of RPA and how it can be used, a lack of leadership commitment, and a workforce absence of job termination [14]. A change in management strategy, a movement in corporate structure, and a shift in thinking might assist in bridging the gap between RPA as an IT technology and its business applications. Everest Group people surveyed, on the other hand, identified excellent customer assistance, training, instructional materials, RPA scheduled maintenance, and a solid RPA vendor environment for complementing innovations as critical determinants of RPA implementation. Furthermore, the advent of new Technology raises concerns regarding robot administration, centralized power, and accountability [15]".

7.2.1 Integration of Robotic Process Automation in Organization

It is critical to analyze the commonalities and dissimilarities, as well as the potential synergies, between RPA and associated Technology. Because RPA and BPM are complementary disciplines, the BPM science community should examine business process management systems (BPMSs) and RPA integration [16].

BPM is a multifaceted method for improving corporate performance through continuous improvements, Optimization, and technology trends. One of the inescapable aspects of the BPM endeavor is BPMS as a complete development platform that spans various activities such as methodology, analytics, and" management [17]. RPA, however, deals with discrete and repetitive movements, and executes operations the same way a human would, while BPMS coordinates end-to-end analysis and improves human, robot, and network interactions. RPA is in charge of repeated sequences of activities that may be assigned entirely to software robots [18].

Though both techniques are frequently used independently, business strategy writers highly recommend integrating them to generate even more significant commercial benefits. Without sufficient "resources and time to fully deploy BPMS, RPA can be a beneficial and"reasonably affordable technique for addressing or supplementing some of the unmet aims [19].

RPA enables you to build a technology infrastructure that collaborates with your people to increase productivity [20]. RPA automates nearly every traditional manual activity or job. Cognitive computing robotics with sophisticated and psychological changes manage operations that people would otherwise execute by replicating what a user would do in an application and supplementing automated update technologies [21]. The Kofax Kapow innovative manufacturing platform is the quickest and most effective method to create intelligent machines that handle learning and memory from nearly any program or database schema, including webpages, gateways, office applications, and business applications scripting. Kapow enables you to rapidly create, deploy, and operate automated computer robots interacting with bidirectional communication between internal company systems and external entities [22]. APIs and intricate connection code as a consequence, healthcare businesses may automate nearly any front and backend procedure by installing autonomous robots that replicate the user's movements while also implementing stored procedures and regulations along the way. Whether it's customer registration, complaints refusal investigations, or any other repetitive industry sector, implementing RPA to rules-based, unskilled labor, and error-prone operations in health coverage may speed up operations, put more time, and free up internal employees to focus on measured process deviations [23].

7.2.1.1 Features of Kofax Kapow

Create complex autonomous robots that perform data-driven processes, or create unique systems that include many robots. Collected information and translation from clinical apps, lab information security, third-party portals, healthcare portals, radiography systems engineering, scheduler programs, ERPs, and HR implementations should be automated [24]. Publish robots autonomously with a standard Java,.NET, SOAP, and RESTful interface that may be used to command robotic procedures from application software and remote systems. Use sophisticated rules-based robotics to communicate with corporate applications such as EHR, EMPI, ERP, and ECM systems to automate processes in Fig. 7.1. It follows standards and offers role-based security management [25].

7.2.2 Systematic Literature Research Protocol

An SLR technique was used to meet the aims of this work and answer the investigation problems [26]. SLR technique began in biomedical research, but it has gained popularity in electronic records management systems field studies over the last twenty years since it systematizes information from a preceding body of research and assures the integrity, thoroughness, and validity of conclusions in Table 7.1. Our literature extraction was carried out in three steps, as per standard SLR guidelines:

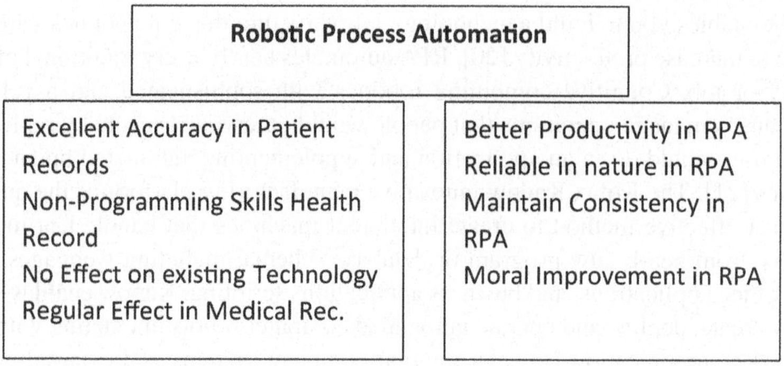

Fig. 7.1 RPA framework

Table 7.1 RPA protocol

RPA protocol	Searching of RPA
Digital environment	Scopus and WoS collection
Item search	Related to terms of RPA
Searching technique	Research publication
In-search schedule	Searching different automation scheme
Out-search schedule	Search robotic publications

(1) SLR protocol development and material search and screening;
(2) Quality evaluation and extraction of relevant articles; and
(3) Qualitatively analyzing and evaluating the approved news items.

7.3 The Industry Benefits of RPA

Automating, as a subscriber and outlay solution, offers various benefits that attract consideration from businesses in a wide range of sectors [27]:

- High Precision—Robots are incredibly exact and regular. They make far better progress or inaccuracies than a human workforce. When people are duplicating data or supplying another computer, higher rates of errors and typos have been found.
- No Technical Knowledge—We don't need to be computer experts to construct a technology robot. Because this is a password innovation, any clear and understandable staff member may utilize a frictional pressure processes design to build up a needed bot or even record their instructions to automate a procedure using a processes recorders tool. After the bot is launched, it would carry out the same actions it does [28].
- Regulatory Adherence—Bots only carry out the commands that have been programmed into them and offer independent auditor information for each step.

Various areas of conformity oversight activities can be improved by using robots. Screening and validation are two of the most potential automated candidates. The capacity of a robot to retrieve and consolidate data from several sources may also increase the effectiveness of regulation, non-financial, and hazard management by eliminating or reducing the day when operations of gathering, assembling, cleaning, and summarizing vast volumes of data. Due to its regulated nature, bot work is well suited to fulfilling even the most stringent quality requirements. There will be no influence on the current system. RPA does not cause any disturbance to the procedures in addition or Technology. Tasks can be automated across the display layer of existing apps in the same way humans do. Robotics are beneficial for old systems when APIs may not be immediately available or businesses lack the trained personnel to establish a deep interface with preexisting legacy applications [29].

- Enhanced Productivity—Compared to manual data processing techniques involving humans, robots' process cycle times are defined and more economical, and they may be done faster.
- Dependability—Operations may be carried out 24 h a day, seven days a week, since these bots can function indefinitely and independently without needing employees to activate bots constantly. If a human is required to participate, it is used to make a choice or correct an error [30].
- Reliability—RPA also offers the convenience of understanding correctly. Based on a predetermined recorded flow, these bots may conduct regular activities, in the same way, each time.
- Increased Morale—RPA can perform some of the most important jobs a company must accomplish daily. Workers can use bots to outsource manual chores such as filing documents, entering data, and seeking content on the web. This will boost morale in your company's HR and operational departments, which are critical to its growth. Employees would have more time to devote to more compelling and exciting projects.

7.4 Effective RPA in Healthcare

A changing population trend has evolved, as shown in a study by the United States Census Bureau in 2016. People over 65 will dominate youngsters under five by 2020. By 2050, the proportion of persons aged 65 and up will be more than double that of early childhood in Table 7.2 [31]. This segment of the population earthquakes is projected to impact the worldwide labor market [32].

This indicates, among other things, that perhaps the number of individuals needing hospital treatment is increasing. The pharmaceutical company must find the money to respond to this changing population trend. Automation and artificial intelligence management in hospitals should thus be fully utilized to give acceptable medical treatment to more people [33].

Table 7.2 Rapid age

Country	Percentage aged below 65	Percentage aged above 65
Japan	36.5	23.0
South Korea	34.9	11.1
Spain	34.5	17.1
Italy	33.0	20.3
Germany	32.7	20.8
France	25.5	16.8
UK	24.7	16.6
China	23.9	8.3
Brazil	22.5	6.9
US	21.4	13.1

The obvious advantages in the medical business go beyond just one element. Furthermore, it may significantly contribute to cost reduction and improved customer service. It is worth noting that, as shown in 2013 Health Leaders research, these are the same medical goals listed by three-quarters of hospital CEOs. The benefits of employing robotic process automation to simplify operations are minimal, and the link between simplified procedures, effectiveness, and fuel savings is self-evident. When you consider the predicted growth in health care expenses of 6.5 percent, "according to UiPath, and KPMG research reveals that" digitalization may enhance savings by up to 50 percent, you should examine the ability of automated to bolster reduction in costs [34].

Another method RPA may help to reductions is by more precisely detecting government healthcare applications "that do not fulfill the standards (approximately 30–40%, according to the study as mentioned earlier), and thereby reclaiming a significant amount of money that would otherwise have gone"overpaid in Table 7.3.

Furthermore, suppose back-office regular duties are delegated to automated processes. In that case, doctors will be able to employ their talents to better serve their consumers' requirements rather than constantly typing in names, periods, and addresses. Customer relations in healthcare insurance might be aided by automated processes [36].

So, according to Accenture Quarterly figures, 36% of nursing jobs—primarily management and back-office—are accessible to automated. So, let us now outline the sudden power and pertaining "that robotic process automation might offer to medicine."

Table 7.3 Health issues in different countries [35]

Country	Percentage health care problem
Hungary	72
Poland	62
Brazil	46
United Kingdom	42
China	38
United States	38
Russia	27
Australia	25
Japan	17
Peru	17
France	14
India	14
Germany	13
South Korea	6
Turkey	3

7.5 Innovation of RPA in Healthcare

The advantages of automated processes in medicine include improvements in efficiency, production, and cost-effectiveness. In truth, there is much more to matter than this deductive methodology; we must not overlook the qualitative component of RPA in medicine. Peter B. Nichol, for example, points out that "the discourse has broadened beyond reducing the cost to performance, involvement, and inventiveness." But first, let's go through the fundamentals using data from the HFS Architecture Study [37].

The applicability of automating for structured information with predefined model parameters, the KPMG research we referenced earlier advises a broad scope of RPA application for the whole financial statements of a"hospital. Administrative duties and tasks required for the billing process are included—account management for patients, from preregistration through bill payment [38].

Revenue cycle activities such as medical professional consultation applications, passenger "pre-arrival and presence, and claim rejections are especially well suited for mechanization. Consider how expensive mistakes such as erroneous data entry may be! So, why not go with an error-free collaborator, as RPA promises? Furthermore, RPA can help with"conformance with a plethora of safety and quality in healthcare, which is critical when working with claimed rejection.

Bots can accomplish tasks involving many solutions, such as patient records, insurance adjudication, management and implementation, or patient information, in an error-free and timely manner. As a result, the expenditures of re-processing may be eliminated, lowering the total healthcare spending. According to Peter B.

Nichol, process automation can also aid members, practitioners, and hospital administrators. This entails delegating operations like registration accounting, practitioner accreditation, or care collaboration and consultancy services to bots, which may be "time-consuming and error-prone if performed"internally [39].

UiPath cites another outstanding achievement to support the argument for automated processes in medicine. The hospital in question is a Berlin hospital with almost 2,000 staff, 70,000 emergency department visits, and 300,000 outpatient visits every year. The primary motivation for using RPA was to acquire better control over the company, which was predicted to considerably add to the happiness of their large number of patients. RPA software adoption was a huge victory for both providers and patients."Because of the 80 percent mechanization, more efficient supply operations (such as complaints management or paying) resulted in a cost decrease per claim. Before RPA connectivity, over three-quarters of consumers preferred to use cloud capabilities such as arranging appointments, availability of medical history, or payment information rather than phoning the medical provider to seek such knowledge [40].

By delegating mundane work to intelligent automation, hospital personnel may devote more time to acute management, which helps patients, healthcare administration, and the administration as a whole; Table 7.4. Last, but not least, we want to emphasize that RPA should not be viewed as a formidable rival to be dreaded in the employment market. This reels the previous idea from a new angle [41].

It is more evident in medicine "than in other businesses that mechanization is not intended as an "alternative for manpower," but rather as a facilitator of effective international division of labor. In other words, if machines assume over the mundane administrative work, human" people may focus on their distinctively human abilities, such as expressive expressiveness and diagnosing thinking [43].

7.6 RPA in Health Insurance

In the health insurance market, it is automating employs computerized software products to expedite procedures and minimize the manual work required to handle health coverage documents, such as complaints. When used with robotic process automation (RPA) solutions, applications such as Blue Prism and UiPath may drastically reduce the effort and skill necessary in health coverage operations. Previously, health coverage mechanization happened at the software code stage; however, with medical insurance, RPA may currently operate at the keyboard level without even any traditional programming. When designed correctly, RPA connects mainframe data, cloud-based technology variables, and desktop apps like PowerPoint into a single, simplified, standardized, and autonomous procedure. Although it acts similarly to an Office macro, RPA in the insurance market differs significantly. Unlike Excel, it can cut and paste data between numerous programs and technologies, ranging from outdated infrastructure to cutting-edge virtualized apps at the desktops level.

Table 7.4 Effect of automation work [42]

Benefits and business risks	Percentage business leader	Percentage of employee
Total benefits	99	81
A reduction in manual errors	52	44
Better quality work product	45	30
An increase in speed	43	41
Higher productivity in experience level	43	24
Increase availability of employee	41	34
Increased utilization	38	21
Decreased labor costs	31	25
Demerits	2	19
Complete business risks	92	89
Lack of personal touch	47	52
Potential job loss	40	48
Employee inability	36	32
Poor customer service	32	37
Less Industry experience	23	24
Inaccuracies	2	2
No business risks	8	10

In the context of clinical insurance payments, RPA does not allude to the concept of robots offering extraordinary determinations. On the opposite. RPA healthcare applications entail automating the simplest, most repetitious secretarial and administrative tasks at the keyboard level. Consider the thoughtless, redundant chores of cutting "and pasting data from one program to another. By automation these easy, repetitive" operations, your health insurance claims team will be able to focus on more entertaining, fascinating, and vital higher-level activities. The computers do what it does best—faster and more accurately.

Some refer to what RPA performs in healthcare insurance providers as the "final mile" of automation technology because it mediated the relationship between various applications after massive system implementations. Configurations such as Oracle Insurance Claims Dispute resolution,"Plexis, and Pega fall short of the target of"horizontal handling in most health insurance claims departments while being promoted as being capable of doing so. The focus of health insurance RPA "is on the micro-task or keystroke level, combining operations that large systems could not fully handle, or were never able to"achieve. To fully receive the benefits of RPA in the insurance sector, it is necessary to recognize that real growth opportunities from robotics can only be obtained since there are many standardized, repetitious activities on a big scale to automate. Outside of existing systems, most healthcare insurance carriers have relatively few defined processes, make when it relates to manual workarounds.

7.7 Advantages of RPA in Health Insurance

- "RPA is now regarded as the ultimate frontier and final piece in the automation of insurance coverage processes. Insurance coverage firms are riddled with repeated minute-level administration chores and mechanical procedures outside of existing systems; thus, if those jobs and activities can be standardized and computerized, personnel costs are lowered, and performance increases dramatically. Job roles can be enhanced, relieving individuals from monotonous duties and reducing the possibility of mistake."
- Rewards of using RPA in healthcare utilization include:
- Robots help full-time professionals by eliminating time spent on repetitive work and accurately synthesizing data from various sources across distributed platforms.
- Increases claim to handle time by up to 70% by automating and validating client data and entering workers' compensation submittal data across many systems
- RPA insurance procedures may operate 24 h a day, seven days a week. Robots do not require overtime pay, medical coverage, or being fired.
- A 60% reduction in mistake rates in medical billing processing. Healthcare coverage RPA can automatically detect particular product or service anomalies and share personal and financial facts through databases to ease claims processing.
- Reduced policy processing period. In less than 2 s, health coverage robots manage data input for new client boarding, acquire and transport consumer data across many systems, and produce premiums.
- Healthcare coverage RPA achieves 100 percent compliance with regulatory requirements. Robots keep a running history of their actions and do automatic periodic reviews.

7.8 Health Coverage in RPA

"A real-world RPA in insurance companies' administration examples will assist in simplifying what appears to be a complex procedure. Assume you wish to decrease the amount of data transcription and verification labor necessary to handle new inbound claims for treatment from a doctor or hospital."

"The present pre-RPA insurance" coverage claim procedure is as follows:

- Electronically connect the allegation type to the allegation management system.
- Choose search and "see if the individual seeking a claim is a consumer in the claim management system."
- Then individually copy–paste the claim form data into conventional disciplines in the allegation management system."
- The keyboard shortcut informs customers via Outlook electronic mail that their allegation is being handled."That's a lot of menial work for a firm that has already been"marketed fundamental medical billing technologies with clean, thorough processes four times. This is the current situation for all significant health insurance firms. But don't worry, RPA is here to fill in the blanks.

  Let's look at the claims procedure once RPA has been applied:

- RPA launches the email attachment and"the pdf file containing the claims, form data."
- "RPA immediately copies the pdf files from the application and copy–paste them into the web-based medical billing data management system, while also connecting the actual pdf form."
- "RPA scans the claim processing platform for the claim submitter's name to ensure that they are a client. the invoice and searching for the returns management system."
- "RPA uploads the bill to the complaints management system."

"Finally, the RPA submits the complaint to the back director for approval while" also sending the consumer an automated Outlook email informing them that the claims are being handled.

The following are the essential stages "that the health coverage RPA solution executed autonomously in this use case: UiPath logged into the program, automatically navigating the appropriate screens to identify the required fields holding the information, and then copy–pasted the relevant data into the document in the DMS and complaints control system. Insurance RPA also operated the programs at the mouse or keyboard levels, directing the pointer to the relevant areas and inputting data and keyboard permutations as needed. It sounds magical, doesn't it? It almost is, but it requires measurements come and process uniformity to make it work."

7.9 Steps of RPA in Health Insurance Claims

Widen the priority of the medical insurance RPA initiative to discover possible use scenarios. While not all activity processes are treated equally, some are unquestionably better choices for your first medical insurance RPA assignment than others. Is the task you're considering automating reliant on a lot of social assessment? Then it's most likely not a strong choice for RPA. With its repetitious task, medical billing administration is an exciting choice for RPA. The first stage is to scope the project, which requires you to select a short, reasonable list of operations that would benefit from automation. These should ideally be monotonous, secretarial administration operations that demand little social contact and occur regularly. But what if you don't have any established mechanisms in place for physical labor? I'll go into more detail about that later.

Evaluate the average advantage received from RPA deployment by determining the benchmark medical billing operation cost. After you've defined the breadth of your pilot program, it's time to minimize the total cost of your medical billing organization. You'll need to have this knowledge to calculate the full advantages of RPA in the billing process—you're sometimes during, if you will. How many financial practices do you save? Evaluating your actual operating expenses and comparing them to your lowered operational cost is critical to ensuring the project's success. You want to demonstrate the benefits of organizational processes and RPA implementation.

"How do you establish a starting point for an RPA reinsurance project? Take the scopes you've established above and head over to the Human resources department to collaborate on determining how much each person in the division costs, fully stocked with perks and everything. If they refuse to provide you with these statistics, you may discover pretty accurate compensation statistics for various insurance claim roles on websites such as Indeed and Glassdoor."

Analyze existing government health insurance claims procedures to identify the potential for intelligent automation—all down to the mouseover level takes time to absorb—the story of a mouse clicks. It feels like a bunch of effort. Once you've gathered your benchmark financial information,"it's time to diagram the processes—this is how you'll find RPA potential in medical billing processes. You can use process mapping tools such as Microsoft Visio to illustrate the operations in focus. It is critical to employ a stencil, such as the BPMN 2.0 stencil (Business Intelligence Model and Nomenclature), the benchmark used by all IT experts. You'll save time down the line when you pass over the procedures to the person implementing the automated process."

The only thing to make this function is to monitor a sizeable medical representative "at their workstation for a few days with a notepad or computer in hand, or on their lap. You must use tremendous caution. Comprehensive records of every Bluetooth keyboard click in all the programs you want to manage using RPA Organizations such as The Lab will do this sort of "day in the life" analysis for you, using computer tools such as GoToMeeting or WebEx, or even confront if you like. Many businesses choose the former"option because distant observation through screen-sharing significantly reduces project expenses.

Regardless of how you conduct your inspections and take "notes, it's critical to watch numerous health coverage claim employees for a few days at a time—enough to establish a good sample group. That way, you'll have solid data to create an accurate RPA enhancement economic case. A decent rule of thumb is to keep an eye out. Scale up to 50–500 staff."

Standardize injury companies' procedures before, but not after, implementing health coverage RPA. Why is it critical to standardize processes and procedures before or during the deployment of medical insurance RPA? Let us pause for a moment to consider this. Because the system was standardized before the robot's deployment, a robot on a manufacturing floor can weld metallic components together rapidly and with few mistakes. The robot welders place "A before spot B and spot B before spot C, allowing it to work swiftly and accurately. Indeed, all the robots on the automated assembly flooring doing the same task are designed to weld areas A."

7.10 Conclusion

This Chapter discussed UiPath, the Institute of Healthcare Technology Revolution, automation replaces health informatics realistic, scalable, and sustainable. This includes all of the advantages of automated processes in healthcare that are addressed in this essay. The application of next-generation technologies in the industry appears to be the ideal example of putting technical advancements to good use in the service of humanity, allowing them to thrive in the age of Technology. In addition, as noted in the KPMG above mentioned paper, RPA may only be the beginning it may be combined with machine learning and cognitive technologies in the (not-too-distant) future, expanding its functionality. According to the KPMG analysis, even the most difficult tasks, such as diagnosis, will eventually be receptive to this type of technology.

References

1. Kuffner, J. et al.: Online footstep planning for humanoid robots. In: Proceedings IEEE International Conference on Robotics and Automation, Vol. 1, The Grand Hotel, Taipei, Taiwan, 14–19 September 2003, pp. 932–937. IEEE Service Center. https://doi.org/10.1109/ROBOT.2003.1241712
2. Hashimoto, S.: Humanoid robots in Waseda University Hadaly-2 and WABIAN. IARP First International Workshop on Humanoid and Human Friendly Robotics, pp. 1–10 (1998). https://doi.org/10.1023/A:1013202723953
3. Pratt, J., Pratt, G.: intuitive control of a planar bipedal walking robot. In: Proceedings of the IEEE International Conference on Robotics and Automation (ICRA '98), Leuven, Belgium, 16–20 May 1998, pp. 1–12. IEEE Service Center, ISBN 0–7803–4300-X
4. Zeglin, G.J., Uniroo, A.: One-Legged Dynamic Hopping Robot. Master's Thesis, Massachusetts Institute of Technology (1991)
5. Chestnutt, J. et al.: Footstep planning for The Honda ASIMO humanoid. In: Proceedings of the IEEE International Conference on Robotics and Automation, pp. 629–634 (2005). https://doi.org/10.1109/ROBOT.2005.1570188
6. Yanase, T., Iba, H.: Evolutionary motion design for humanoid robots. In: GECCO '06 Proceedings of the 8th Annual Conference on Genetic and Evolutionary Computation, Seattle, Washington, USA, 08–12 July 2006, pp. 1825–1832. ACM, New York, NY, USA. https://doi.org/10.1145/1143997.1144291
7. Feng, S. et al.: Optimization based controller design and implementation for the atlas robot in the DARPA robotics challenge finals. In: IEEE-RAS 15th International Conference on Humanoid Robots (Humanoids), Seoul, Korea, 3–5 November, 2015, pp. 1028–1035. IEEE. https://doi.org/10.1109/HUMANOIDS.2015.7363480

8. Feng, S. et al.: Optimization based full body control for the atlas robot. In: IEEE-RAS International Conference on Humanoid Robots, Madrid, Spain, 18–20 November 2014, pp. 120–127. IEEE. https://doi.org/10.1109/HUMANOIDS.2014.7041347
9. Du, C., Lee, K.H., Newman, W.: Manipulation planning for the atlas humanoid robot. In: IEEE International Conference on Robotics and Biomimetics (ROBIO 2014), Bali, Indonesia, 5–10 December 2014, pp. 1118–1123. IEEE. https://doi.org/10.1109/ROBIO.2014.7090482
10. Long, X. et al.: Task-oriented planning algorithm for humanoid robots based on a foot repositionable inverse kinematics engine. In: IEEE-RAS 16th International Conference on Humanoid Robots (Humanoids), Cancun, Mexico, 15–17 November, 2016, pp. 1114–1120. IEEE. https://doi.org/10.1109/HUMANOIDS.2016.7803410
11. Yang, Y. et al.: iDRM: humanoid motion planning with real time end-pose selection in complex environments. In: IEEE-RAS 16th International Conference on Humanoid Robots (Humanoids), Cancun, Mexico, IEEE, 15–17 November, 2016, pp. 271–278. https://doi.org/10.1109/HUMANOIDS.2016.7803288
12. Chestnutt, J. et al.: Locomotion among dynamic obstacles for the honda ASIMO. In: IEEE/RSJ International Conference on Intelligent Robots and Systems, San Diego, CA, USA, 29 October–2 November 2007, pp. 2572–2573. IEEE. https://doi.org/10.1109/IROS.2007.4399431
13. Claudio, G., Spindler, F., Chaumette, F.: Vision-based manipulation with the humanoid robot romeo. In: IEEERAS 16th International Conference on Humanoid Robots (Humanoids), Cancun, Mexico, 15–17 November, 2016, pp. 286–293. IEEE. https://doi.org/10.1109/HUMANOIDS.2016
14. Mao, X., He, H., Li, W.: Path finding for a NAO humanoid robot by fusing visual and proximity sensors. In: 12th World Congress on Intelligent Control and Automation (WCICA), Guilin, China, 12–15 June, 2016, pp. 2574–2579. IEEE. https://doi.org/10.1109/WCICA.2016.7578365
15. Heaston, J. et al.: STriDER: self-excited tripedal dynamic experimental robo. In: IEEE International Conference on Robotics and Automation, Roma, Italy, 10–14 April 2007, pp. 2776–2777. IEEE. https://doi.org/10.1109/ROBOT.2007.363891
16. Arikawa, K., Hirose, S.: Development of quadruped walking robot TITAN-VIII. In: Intelligent Robots and Systems '96, IROS 96, Proceedings of the 1996 IEEE/RSJ International Conference on, Vol. 1, Osaka, Japan, 8–8 November 1996, pp. 208–214. IEEE. https://doi.org/10.1109/IROS.1996.570670
17. Kurazume, R. et al.: Experimental study on energy efficiency for quadruped walking vehicles. Proceedings IEEE/RSJ International Conference on Intelligent Robots and Systems, Vol. 1, pp. 613–618 (2003).https://doi.org/10.1109/IROS.2003.1250697
18. Wooden, D. et al.: Autonomous navigation for BigDog. In: IEEE International Conference on Robotics and Automation, Anchorage, AK, USA, 3–7 May 2010, pp. 4736–4741, IEEE. https://doi.org/10.1109/ROBOT.2010.5509226
19. Raibert, M. et al.: Bigdog, therough-terrain quadruped robot. Proceedings of the 17th World Congress the International Federation of Automatic Control, Seoul, Korea, 6–11 July 2008. https://doi.org/10.3182/20080706-5-KR-1001.01833
20. Kang, T. et al.: Design of quadruped walking and climbing robot. Proceedings of the 2003 IEE/RSJ International Conference on Intelligent Robots and Systems, Las Vegas, NV, USA, 27–31 October 2003, pp. 619–624. IEEE. https://doi.org/10.1109/IROS.2003.1250698
21. Hirose, S., et al.: Quadruped walking robots at tokyo institute of technology. IEEE Robot. Autom. Mag. **2**, 104–114 (2009). https://doi.org/10.1109/MRA.2009.932524
22. Quinlan, M.J., Chalup, S.K., Middleton, R.H.: Techniques for improving vision and locomotion on the sony AIBO robot. In: Proceedings of the 2003 Australasian Conference on Robotics and Automation, Brisbane, Australia, 1–3 December 2003
23. Besari, A.R.A. et al.: The study on optimal gait for five-legged robot with reinforcement learning. Lect Notes Artif Int. Vol. 5928, pp. 1170–1175 Springer, Berlin, Heidelberg (2009). https://doi.org/10.1007/978-3-64210817-4_114
24. Dhatterwal, J.S., Kaswan, K.S., Jaglan, V., Vij, A.: Machine learning and deep learning algorithms for IoD. In: Book Entitled "Internet of Drones: Opportunities and Challenges" in "Apple Academic Press (AAP), Canada, Published, Hard (2022). ISBN: 9781774639856

25. Delgado, D. et al.: Statically stable hexapod robot body construction. IEEE 4th Workshop on Advances in Information, Electronic and Electrical Engineering (AIEEE), 1998, Vilnius, Lithuania, 10–12 November 2016, pp. 1–4.https://doi.org/10.1109/AIEEE.2016.7821819
26. Skaburskyte, A. et al.: Hexapod_robot gait stability investigation. IEEE 4th Workshop on Advances in Information, Electronic and Electrical Engineering (AIEEE), Vilnius, Lithuania, 10–12 November 2016. https://doi.org/10.1109/AIEEE.2016.7821803
27. Stoica, A. et al.: Cassino hexapod: experiences and new leg design. IEEE International Conference on Automation, Quality and Testing, Robotics (AQTR), pp. 1–6 (2010). https://doi.org/10.1109/AQTR.2010.5520756
28. Preety, K.S.K., Dhatterwal, J.S.: Fog or edge-based multimedia data computing and storage policies. In: Book entitled "Recent advances in Multimedia Computing System and Virtual Reality", Published by Routledge Taylor & Francis Group, ISBN No. 9781032048,239 (2021)
29. Schiele, A. et al.: Nano Khod exploration rover—a rugged rover suited for small, low-cost, planetary lander mission. IEEE Robot. Autom. Mag. **15**(2), 96–107 (2008).https://doi.org/10.1109/MRA.2008.917888
30. Kaswan, K.S., Dhatterwal, J.S., Gaur, N.K.: Blockchain of iot based earthquake alarming system in smart cities, book entitled "Integration and Implementation of the Internet of Things Through Cloud Computing", Published in IGI Global, ISBN 13: 9781799869818, ISBN10: 1799869814, EISBN14: 9781799869832 (2021)
31. https://www.census.gov/acs/www/data/data-tables-and-tools/data-profiles/2016/
32. Dhatterwal, J.S., Kaswan, K.S., Preety: Intelligent agent based case base reasoning systems build knowledge representation in Covid-19 analysis of recovery infectious patients. in book entitled "Application of AI in COVID 19" published in Springer series: Medical Virology: From Pathogenesis to Disease Control, July 2020, ISBN No. 978–981–15–7317–0 (e-Book), 978–981–15–7316–3 (Hard Book). https://doi.org/10.1007/978-981-15-7317-0
33. Tadakuma, K. et al.: Mechanical design of the wheel-leg hybrid mobile robot to realize a large wheel diameter. In: Proceedings of the IEEE/RSJ International Conference on Intelligent Robots and Systems, IEEE Service Center, Taipei, Taiwan, 18–22 October, 2010, pp. 3358–3365. https://doi.org/10.1109/IROS.2010.5651912
34. Vargas, G.A. et al.: Simulation of a wheel leg hybrid robot in webots. In: IEEE Colombian Conference on Robotics and Automation (CCRA), IEEE, Bogota, Colombia, 29– 30 September 2016, pp. 1–5. https://doi.org/10.1109/CCRA.2016.7811403
35. https://www.who.int/data/collections
36. Yokota, S. et al.: Control law for rough terrain robot with leg-type crawler. Proceedings of the IEEE International conference on mechatronics and automation, IEEE, Luoyang, Henan, China, 25–28 June 2006, pp. 417–422
37. Kim, J. et al.: A wheel &track hybrid robot platform for optimal navigation in an urban environment. In: Proceedings of the SICE annual conference, IEEE, 2010, Taipei, Taiwan, 18–21 August 2010, pp. 881–884
38. Michaud, F., et al.: Multi-modal locomotion robotic platform using leg-track-wheel articulations. Auton. Robot. **18**, 137–156 (2005)
39. Paul, J.: Military robots and drones: a reference handbook, Springer: ABC-CLIO, 2013, ISBN: 978-1-59884-732-1
40. Wang, Y., et al.: A new redundancy resolution for underwater vehicle-manipulator system considering payload. Int. J. Adv. Robot. Syst. **14**(5), 1–10 (2017)
41. Kamamichi, N. et al.: A snake-like swimming robot using IPMC actuator/sensor. In: Proceedings 2006 IEEE International Conference on Robotics and Automation, IEEE, ICRA, Orlando, FL, USA, 15–19 May 2006, pp. 1812–1817. https://doi.org/10.1109/ROBOT.2006.1641969
42. https://data.world/datasets/automation
43. Klaassen, B., Paap, K.L.: GMD-SNAKE2: a snake-like robot driven by wheels and a method for motion control. In: Proceedings. 1999 IEEE International Conference on Robotics and Automation, Vol. 4, Detroit, MI, USA, 10–15 May 1999, pp. 3014–3301. https://doi.org/10.1109/ROBOT.1999.774055

Chapter 8
Intellectual Property Management in Healthcare Using Robotic Process Automation During COVID-19

Aranya Nath and Usha Saha

Abstract In the current scenario, when the whole world is under stress due to the upcoming variant of the Novel Corona Virus, the Healthcare system, earlier practiced in the traditional mode, has been replaced in Digital mode with various scientific innovations like robotics which shows new thinking of perception among the people how it works in reality. Robotic Process Automation referred to as RPA, plays a significant role in today's healthcare system. The authors, through this chapter, seek to discuss the advancement of Robotic Process Automation and its link with Intellectual Property Management elaborately. The chapter begins with the authors tracing recent developments of Robotics in Healthcare through automation during COVID-19 as it has significant inception. Secondly, the authors also elaborate on how the management of IP assets created out of such a digital revolution due to accelerated and ramped-up innovation comes into the picture quite significantly. It shall enable the readers to gradually grasp the connection between the two fields.

8.1 Introduction

The current medical services are improvised with lots of amendments/changes. Intellectual property (IP) law turns into a significant player in the scene to comprehend those changes.

Initially, people have to maintain a long queue in the hospital for billing purposes/fulfilling the details for admitting the patient, non-availability of Doctor's appointment; as a result, the fundamental right of the people sign it as patient as the

A. Nath (✉) · U. Saha
Phd Scholar, Damodaram Sanjivayya National Law University, Visakhapatnam 531035, India
e-mail: subhamitanath002@gmail.com

U. Saha
e-mail: ushasaha991@gmail.com

U. Saha
LLM IPR & Cyber Law, GITAM School of Law GITAM University, Visakhapatnam 530045, India

S. Bhattacharyya et al. (eds.), *Confluence of Artificial Intelligence and Robotic Process Automation*, Smart Innovation, Systems and Technologies 335,
https://doi.org/10.1007/978-981-19-8296-5_8

right to get proper and fair treatment under article 21 of our Indian Constitution gets violated [1].

Furthermore, with advancements in technology in every sector, healthcare prioritized upgrading the possible techniques to provide better treatment facilities that need the utmost quality during a Pandemic. RPA, better known as Robotic Process Automation, has taken an initial step in utilizing healthcare resources to sort out technical issues like scheduling doctors' appointments for patient requirements in virtual and physical modes. Further, it helps in the billing of medical services rendered by patients. It reduces the workload of human beings by performing all the repetitive work through its technological measures.

As we're in the fifth generation of the Healthcare System, it's required to get acquainted with some terminologies like Electronic Healthcare records, online paying systems, etc. It establishes the connection between IP and Robotic Process Automation. It is essential to examine what could protect types of intellectual property. Intellectual property rights are legal rights that govern the use of inventiveness and technological advances. Essentially, intellectual property necessitates a different and more challenging set of regulations for larceny and possession since these lines are often much more indistinguishable and difficult to verify than the actual property [2].

When actual possessions take, the victim experiences a lack of property or riches while the cheat profits. In any instance, improvements and inventive work might suffer primarily due to the invasion. When licensed innovation thief sufferers possess rights to a melody or patent for an invention, they do not need to lose such rights to face harm. All else being equal, the existence of a replica or a competitor might reduce the value of the first maker's effort and result in a loss of benefits.

In India, there are four main types of intellectual property law: trademarks, patents, copyrights, and licenses. We'll look at how they are related to Robotics Process Automation further down.

8.1.1 Objective of the Study

The research is purely analytical based, it has been carried out so that both technological and IP Law will be merged together which will help researchers from various disciplines to understand the Robotic Process Automation and the booming concept of IP Law which trigger the privacy protection of enforcement of Robotic Process Automation in healthcare during COVID-19. Henceforth to understand the lacuna, research has been carried out.

8.1.2 Need for the Research

Jain and Bhatnagar [3], in their research, contributed to the overview of Robotic Process Automation (RPA) in healthcare. In this research, In healthcare, why is robotic process automation used? Could the system work automatically to balance expanding the number of patients and reducing paperwork and the insurance process? Authors have a variety of tasks that are handled automatically by the software.

Sriram [4] in his research, contributed to the informatics of healthcare over here new concepts like electronic health records, medical devices, and robotics.

Yan Chow in his research, contributed about impact of Automation process in healthcare using robotics, Artificial Intelligence helps in healthcare system during COVID-19 pandemic as it has a significant inception during pandemic [5].

In their research, Basant and Srinivasan [6] contributed to IPR Law protection in the implication of healthcare innovations as RPA is one of the prime techniques in the healthcare system. TRIPS Agreement looks at how the new IP law impacts affordable healthcare, whereas others look at how IP affects medical innovations. Surprisingly, the two threads do not meet. Furthermore, several research studies see IP-driven innovations as barriers to accessibility since they are likely to be controlled entirely by the IP owner. We contend that seeing healthcare access and innovation as complementary processes has value.

In this above the first part of the paper has been discussed about the Evolution of smart healthcare using RPA technique, utilization of resources like Artificial Intelligence, IoT during COVID-19 pandemic. In the second part of the chapter, the authors further demonstrate how, as a function of expedited and ramped-up innovation, the management of IP assets generated by such Robotics Process Automation enters the picture considerably. The main question that will be answered over here is Why does IPR need to be managed in healthcare data during COVID-19 using RPA?

8.1.3 Research Methodology

As we're writing a research paper on the current healthcare system using RPA during COVID-19 and also how IP Laws play an important role in safeguarding the rights of the actual inventor along with legal infringements and mesne profits of the infringer, we have conducted an interview among the healthcare practitioners and patent attorneys to know the actual challenges behind it. How the hospital staffs are getting freaked out in scheduling appointments with the doctor manually and the billings and the challenges which can be met up by IP Laws. After collecting all the research data through the interview, the research has been conducted purely qualitatively in nature. Further, the research has been conducted in an analytical method to analyze the prominent issues of RPA in the healthcare sector.

Primary sources of Data include various interview surveys conducted on healthcare practitioners and Patent attorneys.

Secondary Sources of Data include various medical journals, e-books, IPR-research journals, WIPO journals, etc.

8.2 RPA in Healthcare—A Review

A coming technology revolution called robotic process automation (RPA) aims to relieve people's everyday responsibilities of repetitive and monotonous chores. The community of researchers is presented with a brand-new field of study, and several research studies are being conducted in this area. It is not robotics; instead, it is an entirely different technology. RPA is a new and rapidly expanding sub-domain of robotics. The researchers of such research outline the significant characteristics of RPA and examine its use in critical healthcare.

Robotic process automation (RPA) is a new type of commercial procedure automation technology based on the idea of software robots or artificial intelligence (AI) workers [7]. RPA is a software-based automation approach that automates business processes by examining previous procedures and practices. In essence, it is software that simulates a virtual human workforce and performs repetitive jobs and activities, eliminating the participation of people in the process. Today's corporate environment is incredibly competitive, and everyone wants to be one step ahead of their competitors to obtain an edge. As a result, when certain recurring operations are automated, RPA technology effectively enhances the portability and efficiency of company houses.

Such a form of automation means allowing employees to focus on more vital activities, be more inventive, and dedicate time to improving their subject knowledge and abilities. The three critical phases of every RPS project are planning, implementation, and assessment and monitoring. RPA is regarded as a significant technological evolution of this technique in that new software platforms are emerging that are sufficiently advanced, resilient, scalable, and reliable to make this approach viable for large enterprises [8]. Researchers believe that RPA will usher in a new wave of efficiency and productivity improvements in the current workforce once the innovation becomes more solidified and stable in the coming decade [9].

RPA is the future technology to provide a protracted solution that saves costs and delivery time while enhancing the quality, speed, and production performance of a business process. According to Global Industry Insights, Robotic Processing Automation (RPA) research will likely reach $5 billion by 2024. According to the research, RPA is gaining traction in professional and training services. Robotic Process Automation is thus a Disruptive Technology, and its applicability is limited to fields with task repetition.

RPA delivers cost savings and operational efficiencies for firms that use it. It aids in minimizing human participation in repetitive operations and aids in boosting accuracy by reducing mistakes. It operates continuously, without interruption throughout the year, reducing risks and providing a measured output with tangible effects for businesses (Fig. 8.1).

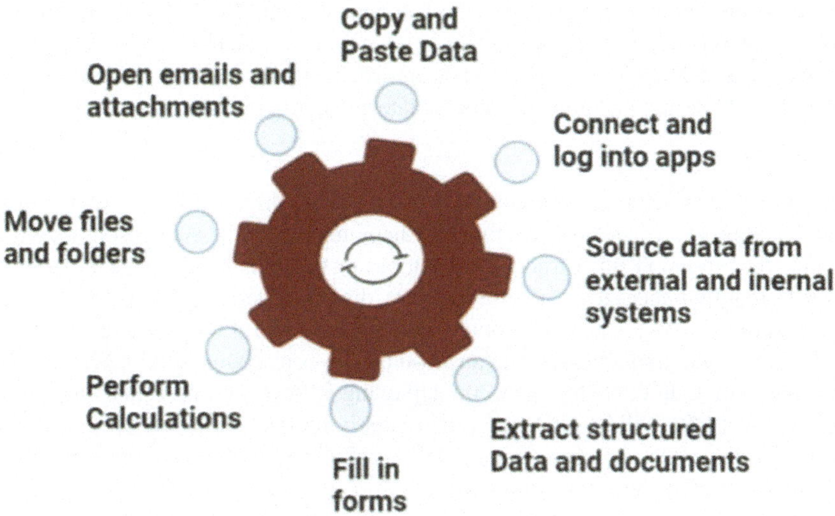

Fig. 8.1 Robotic process automation workflow diagram

8.2.1 RPA Versus Traditional Information Technology-Based Program

The distinction between RPA-based and IT-based business process automation remains marginal. Both strategies may appear identical to the layperson, but they are opposite. These preceding are some of the characteristics that distinguish RPA from traditional IT-based process automation.

(a) Its focus's nature

RPA and standard IT-based robotics both focus on various aspects of the company and are developed with various objectives. Automation leads to the improvement of operational procedures. IT-based process management is primarily tasked with enhancing predictability by minimizing costs and increasing end-user delivery time. Robotic process automation encompasses a greater range of basic operations, extending far beyond the IT sector and into nearly every aspect of the organization, from accounting to marketing. RPA is being used to make end-user jobs more straightforward and efficient. ITPA is primarily concerned with the demands and weaknesses of the company's complicated IT infrastructure.

(b) User experience (UI/UX)

One of the benefits of RPA is that the tools and apps are more user-friendly, with improved user interface design (UI) and user experiences (UX). The objective is to design straightforward user interfaces tailored to individual users' needs and can be used by anybody with a basic grasp of IT. On the other hand, traditional IT-based process automation tools and apps offer less appeal to end users in terms of UI/UX. As a result, typical IT-based process automation systems are more sophisticated and need advanced IT expertise. These apps are frequently produced using coding, and even these ITPA applications may necessitate some programming knowledge.

(c) Non-disruptiveness

Traditional IT-based process automation innovations aim to modify or alter current inefficient processes and systems to make them more efficient. As a result, they will disrupt the present IT infrastructure. RPA tools, on the other hand, are geared toward light IT requirements and do not, for example, disrupt underlying computer systems.

Both of the above ways of automation are crucial, and if a firm can utilize them in tandem, it will undoubtedly gain an advantage over its competitors. There are two schools of thought opposing views on deploying RPA in various fields. One group believes that RPA will be a job killer, particularly for backend or back-office duties done by Business Process Outsourcing (BPO) organizations, because software robots would accomplish those operations without tiring or making mistakes. Indeed, there will be a demand for new skill sets in lower and middle management, including employees who can work with or even maintain RPA systems [10].

8.2.2 RPA in Healthcare Overview

In terms of revenue and employment for large-scale workers, Healthcare Centers are one of the prominent sources [11]. It consists of medical equipment, Clinical trials, Mediclaim, and medical instruments. Although it's healthcare 5.0 generation yet certain challenging tasks are still there like appointment schedule of patients in accordance to the doctor, managing all the internal and external sources through Clinical applications, Mediclaim, etc. It's becoming complex as manual laborers cannot handle all the tasks single-handedly; therefore, in this regard, RPA has been introduced in Healthcare Centers to increase the efficiency of the employees/staff in the various administrative department [12]. It also reduces the chances of errors caused by employees/staff due to the lack of complex procedures like enrollment of patient details, appointment schedule of patients, Mediclaim, documentation, and preparation of invoices. Further, RPA also sorts out the problem of the viability of drugs in the market for safety precautionary measures [13].

8.2.3 Benefits of RPA in Healthcare Sectors

i. Capability to make better use of data [14]

Being digital has its own set of advantages. They are using RPA in conjunction with Optical Character Recognition (OCR) aids in digitizing all paperwork, indexing it, and storing it for future reference. RPA assists in processing health claims, medical diagnosis reports, and updating the same set of information parallel to various healthcare enterprise systems, eliminating swivel chair operations. The initial digitized data is now available for slicing, dicing, and re-purposing in more creative and efficient ways to provide innovative healthcare.

ii. Enhanced knowledge repository

RPA slowly and securely builds an information base while processing transitional work such as health claims. It aids in defining minimum and maximum thresholds in the system for accepting and rejecting claims. These thresholds will be applied upfront in all future claim processing, reducing turnaround time.

iii. Improved customer service

RPA assists in automating routine tasks, synchronizing all digitized information, keeping track of appointments, and organizing all records in an up-to-date manner. Employees can focus more on their core competencies and customer service due to the extra hours a well-coordinated effort to keep all customer information in one place aids in providing innovative service to patients.

iv. Enhanced compliance

The healthcare industry is highly regulated. Organizations must adhere to many statutory compliance and regulations in almost every process. RPA not only aids in the generation of audit trails and automated reports, but it also aids in the fulfillment of observations involving third-party systems by posting the required data and supporting documents to them regularly via role-based access. It ensures high accuracy, improved performance and compliance with regulatory bodies, and improved customer service.

8.2.4 International Standard of Utilization of RPA in Smart Healthcare During COVID-19

RPA is a revolutionary innovation [15] that optimizes internal operations by merging technology, artificial intelligence, and machine learning skills and strategies to imitate human worker activity and automate manual processes inside circulation and procedure applications and functions. Essentially, this technology enables people to delegate tedious, rule-based tasks to software robots. RPA technology allows increased competitiveness, cost reduction, and improved financial performance. Above all, such solutions enable increased process efficiency by introducing new, sustainable practices, such as digitization and automating specific operational operations and whole corporate processes. This problem may be especially critical in light of the present epidemic and potential unforeseeable catastrophes. Because RPA is thought to be a valuable technology, it should examine the characteristics of such a solution and the predictors of and hurdles to RPA adoption during the pandemic. RPA technology allows companies to become more competitive by lowering expenses and boosting financial performance. Above all, these technologies enable increased process efficiency by establishing new, sustainable practices, such as digitization and automation of specific operational operations and whole corporate processes. This problem may be especially critical in light of the present epidemic and potential unforeseeable catastrophes. RPA can support company continuity when the globe is witnessing a possible disruption in the available workforce. Organizations are embracing automation to assist remote employees and service delivery to ensure long-term business continuity in the face of the pandemic.

Due to the sheer COVID-19 epidemic, even the most routine business operations have become vital and time-consuming. Amid this pandemic, enterprises must modernize and implement electronic document management. In the COVID-19 pandemic, several businesses have opted to use contemporary IT solutions and new technologies to support the long-term preservation of company processes and assure employee safety. Automated machines have emerged as critical systems for entities to integrate this modifying workflow, intended to help remote working patterns, deliver services, increase operational productivity, and, most notably for this paper, manage business organization and continuity. After the SARS-CoV outbreak, technology has advanced rapidly to prevent disease spread. The new UiPath Health Screening Bot, introduced in various APAC nations, is one tool that the globe now has at its disposal. We designed the Health Screening robot to make workplace health inspections less onerous using our significant knowledge in applying Robotic Process Automation (RPA) to tackle tricky business and social challenges. The robot distributes and collects surveys via software and may be used or adjusted by human resources (HR) professionals to meet the demands of their firm. There are ready-made templates that may quickly adapt for people who do not require customization. It benefits geographically distributed enterprises with limited resources by allowing for faster deployment.

8.2.5 National Standard of Utilization of RPA in Smart Healthcare During COVID-19

During COVID-19, however, one tool that will reshape the automation environment in the healthcare industry will be robotic process automation (RPA) [16].

RPA is a business process automated system that reduces interpersonal interaction by leveraging software robots or artificial intelligence (AI), often known as digital workers or software robotics. RPA uses a user interface and data capture to modify applications to interpret, interact, and prompt replies with other systems to do repeated activities. RPA robots may do tasks like editing and inserting data, shifting files, filing documents, accessing information, and so on. It helps companies reduce the tedious chores their workers complete, improves productivity and accuracy, aids in instant potential savings, enhances compliance, and boosts flexibility.

8.2.6 Role of RPA in Social Distance During COVID-19 [17]

RPA (Robotic Process Automation) is a slashing technology for automating structured business procedures. It works like any other employee, interfacing with current application user interfaces and automating operations. RPA can help businesses and institutions make better use of their human resources. To avoid congestion in these times of social separation, several industries, such as healthcare, which has significant expenditures in human resources, must adopt alternative operating models, shifts, and staggered employee attendance. They indeed have an impact on working hours and workload. RPA, on the other hand, may be trusted to do a variety of monotonous jobs, allowing the workforce to spend their human energy where it is most required. Allowing staff to automate operations with less physical involvement helps even the most experienced personnel tackle the changing difficulties of COVID-19.

8.2.7 Automation and Patient Care During COVID-19 [18]

Healthcare organizations have historically emphasized patients by lowering costs and enhancing service quality. The market is built on human-centered services, and a tailored approach substantially influences patient experience and outcomes. Most hospitals' major issue is keeping track of information, document processing, patient information processing, billing, lengthy lines, complaint management, patient registration, reporting, and so on. Integrating and analyzing this data becomes a tedious operation, where automation comes in and makes a big difference. On the other hand, hospitals have faced a substantial lack of human resources due to the epidemic. By letting medical personnel focus on patient care, RPA focused on hospital patient management systems can help hospitals expedite their digital transformation.

8.2.8 Automation in RPA for Health Insurance During COVID-19

RPA in healthcare is a method of streamlining procedures using automated robotic software. RPA helps reduce human Labor in processing health insurance documents such as claims. Hospitals use blue Prism to reduce the workforce and skills required in Medicare operations. RPA in health insurance can work without conventional coding by integrating desktop programs like Excel, cloud-based software fields, and mainframe data into a standardized and automated process. RPA isn't judgmental, but it makes things easier by automating tedious and repetitive administrative tasks and type-level paperwork. By automating these simple, repetitive tasks, your health insurance staff are free to focus on higher-level tasks that are more interesting, exciting, and important, letting the computer do what it does best: speed and precision. For example, natural language processing improves the technique of documentation. It can improve the efficiency and accuracy with which complaints are handled.

Now the next chapter will mainly be focused on Intellectual Property Asset management in smart healthcare industries, followed by Big Data regarding the challenges to data privacy will be addressed by IPR.

8.3 IP Management in Healthcare System Utilizing RPA

8.3.1 IPR Protections for Digital Health-Tech

In a country like India where the medical sector is still under the surge of innovation, Intellectual Property protections play a vital role. For example, Copyrights find a significant place in company-used algorithms and software code for websites used by a digital health device which is later used to collect data from the consumer. Similarly, Trademarks determine consumer preference which becomes crucial for a company to understand the requirement of such protection [19]. There are three criteria that have to be taken into consideration which are:

(a) The knowledge isn't, as a body or within the precise configuration and assembly of its components, generally known among or readily accessible to persons that normally affect the kind of information in question;
(b) The information has actual or potential commercial value because it's secret;
(c) The person lawfully on top of things of the knowledge has taken reasonable steps under the circumstances to stay it a secret.

Financial records, customer lists, customer details along with strategies, collected data, algorithms, and policies of the companies may also be considered trade secrets. However, in India, details of customers were held not to be trade secret property. In ***American Express Bank Ltd. v. Ms. Priya Puri***, it was held that the objective of secrecy is utility, so a trade secret has to be utilitarian in nature.

8.3.2 Maneuvering the Capital Base of IP Assets

Companies may record particular categories of intellectual property as intangible assets on financial statements because they consider them to be capital assets. Intellectual property has a broad sense. It can take many distinct shapes like Patents, copyrights, trade secrets, and other forms of intellectual property are examples. Though they may be shown on a company's balance sheet as financial assets, determining the actual valuation of the same is nearly impossible. A capital asset is a substantial property, such as a car or a house, or an investment in stocks. It also has collectibles. These are easy to interpret since they are relevant. Human capital, industry expertise, and expertise are regarded as intellectual property when considered a capital asset. Because such assets are intangible, evaluating them becomes problematic. Thus, in the medical profession, such intellectual property must be considered a part of a company's capital asset, which safeguards IP and gives maximum value to the company, striving to create in the long term.

8.3.3 IP Asset Management and Smart Healthcare

The authors would like to briefly overview IP assets before proceeding to the next chapter and why they are essential to managing them. Intellectual Asset Management (IAM) is a management method that focuses on exploiting patents, trademarks, trade secrets, copyrights, know-how, and other intellectual assets to support and improve overall business performance. IP Asset is a collection of intellectual property creations such as trademarks, patents, copyright, and trade secrets that entrepreneurs select based on their company needs [20]. For example, a publishing business will need to handle copyright and trademark to obtain economic worth since it increases financial value in the market. By adopting the word "assets" [21] business managers and legislators understand that intellectual property (IP) is more than just a legal right; it also provides an economic benefit to all owners. Intellectual property is part of a broader economic environment where human capital defines a productive and competent workforce or a generation of academics and scientists. Human capital has low economic worth in the absence of intellectual property since, by definition, it is non-owned human ingenuity that cannot be held and has no legal standing. Intellectual property will not be generated, safeguarded, or developed without human capital. Intellectual property has been essential for monetary progress. With Present economic knowledge, the confluence of IP rights and human resources represents a substantial financial power.

The intellectual property created is an asset having a theoretical economic worth. This value, however, cannot be realized in practice unless the IP has employed precise, tangible, and practical ways to generate money or other economic advantages. Strategic IP asset planning and development are required preconditions for the dynamic use of IP for micro and macroeconomic development. The best innovation in the world will not generate cash if it is not successfully promoted and exploited. IP asset management is all about making the best use of your human resources.

8.3.4 Why IP Assets Are Required to Manage?

Intellectual property refers to speculative assets legally protected and controlled by a specific firm, implying that others may not utilize them outside the company. The most significant benefit of intellectual property is that it gives businesses a competitive advantage. Speculative assets had protected to the same extent as physical objects. The significance is: (1) Competitors will be unable to infringe on your efforts, which is necessary for web-based or mobile-based businesses. (2) It adds value to the company since it includes the goods and services offered to customers. The organization may enable external parties to utilize the property, but royalty rights or other legal constraints protect this privilege. Various approaches are employed to identify, protect, and enforce intellectual property rights. Multilateral treaty frameworks and international organizational structures are examples of this. As previously said, intellectual property is essential for the economic growth of companies. Many accounting practices in the United States and other nations put pressure on businesses to categorize all intangible assets. After all, preserving assets is essential for today's organization. Ben Bernanke, a well-known US economist who recently spoke at an economic growth conference, understood this; the value of intangible capital, in particular, has been a driving factor for many US firms.

Intangible assets, human capital, and intellectual property have increasingly been recognized as essential aspects of global development in the financial market. Consequently, lawmakers in governments, universities, and research institutions seek to develop constructive IP policies that encourage the production, accumulation, and use of intellectual property assets as a critical tool in monetary strategy. There are methods for producing a company's intellectual property portfolio, and there is growing acknowledgment that current proactive policies may increase a government's productive capacity and IP asset allocation.

Knowledge is unlimited, and those who have supported and promoted the sharing of ideas and information have been at the heart of contemporary economic and social growth, stated former Romanian President Ion Iliescu, a member of the WIPO Policy Advisory Commission (PAC). Intellectual property is at the Center of business strategies, as seen by its growing share of fixed assets in company value.

According to the National Knowledge and Intellectual Property Task Force, which is located in the United States, a company's value in the knowledge era is mainly defined by its capacity to turn individual and organizational knowledge into net worth in time to grab new market segments. As product cycles shorten and rivals lower the time to market, a competitive corporation's methods for developing and commercializing new ideas must be continually validated and improved. The administration of intellectual property (IP) lies at the Center of this transition. It's a method for dealing with intangible asset growth and its influence on a company's strategic market position and shareholder value.

Trademarks, international patents, copyright allocation and utilization, trade secrets, geographical indication, domain names, registered designs, plant breeder rights, and technology are all examples of intellectual property assets that must manage to generate value, special privileges, profits, and consumer goodwill and loyalty. An IP asset assists in the income generation of products through licensing or franchising; it also helps promote money for research and development, hence improving the end outcome. The product's value also rises, which aids in transfer pricing negotiation.

All readers now understand why IP Assets are necessary to manage; the authors would like to conclude how to manage IP Assets in the healthcare industry. Many companies rely on intellectual property assets ("IP"), including commercial healthcare transactions. This article provides an overview of how IP rights may be effectively secured, transferred, and kept throughout transactions. The writers' next thought is, "Who is eligible to hold IPR in healthcare?" Healthcare intellectual property is a broad field. It also encompasses the intellectual property of influential organizations, such as health centers performing clinical trials at universities and medical research firms for biotechnology and pharmaceutical firms. The authors would now want to explain the notion of IPR in the healthcare industry. Patents, trademarks, copyright, and even trade secrets are all examples of intellectual property rights. A medical research board owns a method or system patent for a unique approach, whereas a pharmaceutical business has a medicinal or new drug patent. Health services and organizations can also provide trademark rights for books, rules, regulations, and processes.

8.3.5 Scope of Intellectual Property Rights in Robotic Processing Automation

As we have come across the concept of Smart health in the fifth generation of healthcare, Robotic Process Automation or RPA has a booming effect. RPA or software botnets help solve the medical field's miscellaneous works in the hospital. The primary purpose of using RPA is to solve human errors or mistakes that they usually make while enrolling the name of the patients in their databases and scheduling the appointment according to the necessity. An important question arises about the confidentiality of medical records and patient details carried out by RPA or software botnets. IPR thus came into the picture as we know that under the umbrella of IPR, we have Copyrights, Trademarks, and Patents. Therefore, Copyright protects the data carried out by RPA or software botnets to enroll the patients' names in their databases and schedule the appointment according to the necessity and Medicare billing all are encrypted. Patents protect the novelty of RPA in innovative healthcare to protect from any infringement [22].

Healthcare is one of the most inefficient businesses; eliminating inefficiencies would result in better healthcare delivery, beneficial to the industry and the general population. Every company has inefficiencies, but few confront the healthcare industry's issues, stringent laws around patient data, and a lack of resources to cope with them. Financial services are subject to comparable high levels of regulation, although banks have easier access to money and have historically invested more in technology. As a result, healthcare has more inefficiencies and manual procedures than nearly any other business. The capital for IT and healthcare services comes entirely from healthcare providers' earnings, so RPA enables healthcare providers to avoid expensive, long-running digital transformation implementation projects and gain quick results, allowing them to contribute to patient care significantly.

Along with that, RPA using in smart healthcare for patient scheduling, Claim Management, Regulatory Compliance, Data entry, Migration, Extraction, etc. These benefits the healthcare industry by reducing costs, increasing appointment turnout, eliminating human error, better patient experience, and better employee satisfaction.

In a larger sense of the technology sphere, Artificial intelligence and robotics have worked wonders in resolving the health sector's grave demands throughout the pandemic situation. Diagnostics, Hazard identification, monitoring, mobile health, supply and distribution network, service automation, sterilization, faster research, and pharmaceutical development are available services. They have all benefited greatly from robotics and AI services during the pandemic.

8.3.6 RPA and IP Assets Management

As it has already been mentioned, the intellectual property rights are legal rights that govern the use of inventiveness and technological advances. It helps in safeguarding the rights of the inventor's inventiveness, therefore, the question came to the forefront that Why does IPR need to be managed in healthcare data during COVID-19 using RPA? When we talk about Robotic Process Automation, the first thing that comes to our mind is software bots. With the advancement of information science, Robotic Process Automation has gone far beyond our imagination and marked in different forms in various sectors. Robotic Process Automation is the science of developing software technology used to carry out repetitive work and minimize human errors intertwined with complicated business. From a healthcare perspective, Robotic Process Automation is used in different sectors to help people easily accomplish complex tasks, keep track of any data, schedule appointments of patients and client services, and simulate models and predictions that took years to complete. RPA helps streamline the front office support that is essential to provide better customer support. But every technique has both pros and cons. The main cons of this technique are the lack of proper investment in the initial development of RPA. As a result, IPR plays a significant role in the R&D investments of its competitors [23].

IPR Law protects inventions, creative activities, and ideas, usually a large bag of intangible markets. It ensures they reach the right people and are put to proper use in saving lives. It is necessary to make patent innovations and inventions in the health sector. Therefore, it is clear that copyrights and patents are needed to manage as an IP assets in robotic process automation. Copyright helps protect the database records of every patient, and patents help cover the process of using the machines and the aesthetic aspect of the product. Since it bases on computer software, R&D of healthcare industries in valuation & economic benefit as valuation is an art that helps assess the value of the product through due diligence report. Due diligence is one of the essential aspects which is needed to be done by the team as it is helpful for the mathematical valuation of Intellectual Property (Fig. 8.2).

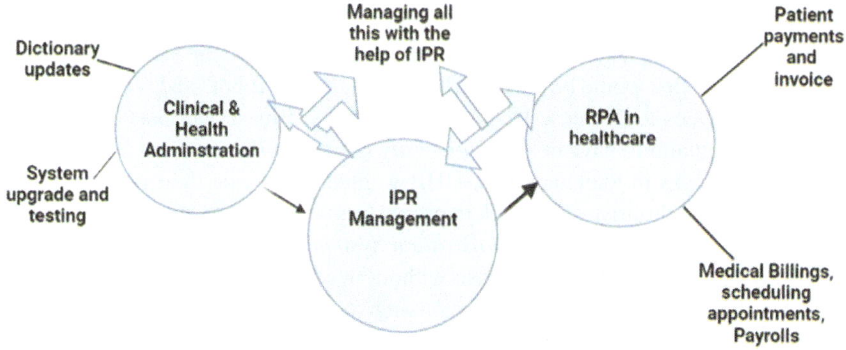

Fig. 8.2 Diagrammatic structure of intellectual property management in robotics

8.3.7 How IPR Management Taken Place Through RPA (Examples)

Given the technical advancement of the previous several decades and the growth of information systems in society, the great majority of services supplied by businesses and organizations are now digital services. Healthcare 5.0 is the fifth industrial revolution, in which technology and automation are making significant changes. When it comes to automating organizational and commercial procedures, robotic process automation (RPA) offers various advantages. In addition to improvements, the complementing application of Artificial Intelligence (AI) methods and methodologies provides improved accuracy and execution of RPA processes in information extraction, recognition, classification, forecasting, and process optimization.

Robotics, the discipline of technology that drives robot development, has played a role in automobile industries, building sites, schools, hospitals, and private residences for decades. However, emerging research disciplines, such as artificial intelligence and sensing, have recently joined with robotics to build powerful autonomous robots with many more potential uses.

Now the question arises what a robot is?

Generally, a robot can comprehend its surroundings and change its actions to achieve a goal. The earliest modern-day robots were created to speed up industrial production processes using automated robotics. However, robots have evolved into autonomous systems capable of operating and making "decisions" without human intervention.

In today's competitive edge, it has been visualized that many players are involved in the RPA process as well as AI-related inventions, therefore, RPA can be used in IP Management as the authors already discussed IP Management before; now the authors would like to give some examples that how robotics can be automated using Intellectual Property Management.

Like Blue Prism, which is a software example of an RPA vendor selling self-service and self-check-in solutions to hospitals, they claim that their RPA platform is focused on filling the gaps in jobs that people can do, but they would be better suited to a job that takes more judgment and critical thought. Their technology automates tasks that would be purportedly better left to a "robot" and is made to serve human employees to decrease friction in their tasks. Blue Prism lists a case study in which they claim to have helped *University Hospitals Birmingham (UHB)* install self-service kiosks in their facilities. UHB wanted to provide their patients with a user-friendly and intuitive self-check in process, as well as integrate their system into the *National program Patient Administration System (PAS).* This would purportedly allow them to access a useful database without needing to update cybersecurity. As we know that Blue prism is a software having such a high rate of utility, therefore, the protection requires much effectiveness in order to prevent infringement, thus IP Management, which is a compilation of all IPs together where copyright, Patent, Trademark, and Trade secrets are, therefore, to foster the growth of technological protection.

Henceforth, it is clear that Blue prism as an RPA is used in IPR for the protection of innovations. As the objective of technology being used to handle business logic, structured inputs are to automate business operations. RPA integrates user interface and workflow execution. It combines human movements like mouse clicks and keystrokes with workflow and business rules to provide a meaningful result.

Operating costs remain a considerable expenditure in hospitals and healthcare institutes. Due to the pandemic, most healthcare players have experienced a substantial loss in income due to a decline in volume. These firms must discover opportunities and methods to increase overall efficiency and minimize costs.

RPA, in conjunction with artificial intelligence and machine learning, may manage some of the time-consuming and labor-intensive commercial processes. IPR is used as a scientific-legal discipline to protect all kinds of innovations that have taken place through RPA. As RPA is used to minimize cost, therefore, IP Audit and IP Data mining is used to safeguard all kinds of procedures that have taken place through RPA. As IP Audit is a tool for identifying your potential IP assets. Ideally, an audit should be carried out by professional IP auditors, but often a preliminary audit can be done in-house, within your company.

Henceforth, it is clear and evident that IPR Management is a pandora's box! which consists of all kinds of Intellectual property which are required for scientific innovation protection as RPA has lots of scientific features and various functionalities attached to it so it requires a solidified protection which can take place through Intellectual Property only

8.3.8 Challenges of AI and RPA in Smart Healthcare

The efficacy of artificial intelligence [24] is of grave concern in the healthcare sector because it lacks clarity, is too bias in nature, and some issues on privacy, safety are of great concern, so some ethical laws implemented to safeguard the challenges.

(a) **Bias in data**

The training of the AI model necessitates a large amount of input in the form of health data or other data. When insufficient or partial data utilize for training AI models, there may be unrepresentative data due to societal prejudice [25].

(b) **Personal**

Health service data are the most sensitive information an individual may have about another person [26]. It provides respect for independent privacy is an essential ethical concept in healthcare. Confidentiality relates to patient autonomy, personal identity, and well-being.

(c) **Ethical issues in research and biomedical medicine**

Like other new scientific approaches, biomedical ethical norms must be followed by AI in healthcare applications [27]. They are autonomy, advantage, non-crime, and justice. They manifest as permission, privacy, safety, voluntary involvement, independent decision-making, and so on, all of which should be considered.

(d) **In RPA, particular challenges are there, which are listed below**

Many of the jobs we undertake in healthcare are laborious and repetitive, resulting in hours upon hours of data input, with personnel frequently re-entering data that already exists and can be obtained elsewhere in the system.

It is frequently why the time lag between submitting a claim to a payer and getting reimbursement from them is so long. Healthcare workers also waste a lot of time gathering information from medical databases and clinical documentation for public health reporting. We are losing money and competitive advantage by significantly reducing the productivity of our human resources on boring jobs. RPA can solve a wide range of process difficulties in healthcare, encompassing invoicing and compliance, electronic health records, clinical documentation, banking institutions, outpatient appointments, and various internal and external customer contact areas.

8.4 Role of IPR in Safeguarding the Challenge of AI and RPA in Smart Healthcare

8.4.1 AI-Robotics and Intellectual Property [28]

Intellectual property protection is critical in all R&D-intensive sectors, including robots.

Robotics companies sometimes invest years of extensive (and costly) research before selling their products and achieving commercial success. The lengthy and expensive process of delivering lucrative goods underscores the importance of intellectual property rights, which are required to recuperate up-front costs and ward off competitors attempting to capitalize on their rival's R&D spending. We will now acquire and manage intellectual property rights in smart healthcare.

The ideal approach to the work of computer-related databases is to assess the compatibility between "data security" and "intellectual property regulation". The "practice, authority, and decision" parts of an individual's intellectual property rights are founded on the "practice, authority, and preference" aspects. The owner's work which involves literature, fiction, poetry, art, and film, needs protection as there's a chance of infringement. The Copyright Act makes it difficult to distinguish between data protection and security. Data protection aims to preserve people's privacy, whereas database protection protects the creativity and cost of gathering, validating, and displaying databases in novel ways. All partnerships follow the essential legal principles of entrance, anonymity, and confidentiality.

8.4.2 Copyright

What is an infringement in one country that may not be so in another? When seeking to describe the complex cloud world in terms of copyright, the courts ought to be careful. In the cloud arena, the extent of copyright is in doubt. In this industry, the software is critical, with robots unable to function without underlying programming—robots without the software would effectively be unable to execute their intended jobs. While conventional robot functions include path-finding, control, locating, and exchanging data, some programming code tries to provide robots with the potential to generate artistic, literary, and musical creations. As a result, relying on copyright to safeguard such software is critical for the robotics sector. There is no specific liability for the copyright-protected content provided by intermediaries. Some countries encourage people and close friends and family circles to make copies of songs and movie files for private use.

8.4.3 Patent on Design

A design patent allows healthcare firms and experts to protect their AI-designed gadgets, goods, and equipment. Design patents provide the owner authority over a product's visual, aesthetic, and kinematic features. This will preserve the creator's rights in everything from the whole color scheme of an AI-based user interface, information layout to the look, and operation of a wearable activity monitor's touch notifications.

8.4.4 *Trade Secrets*

Because few individuals have the technological know-how to reverse engineer these systems, firms usually rely on trade secrets to protect their concepts. The WIPO Copyright Treaty of 1996 makes it illegal to circumvent a technical protective measure to access copyrightable computer code.

8.4.5 *Solutions for RPA to Meet up Challenges*

While the "robot" part of robotic process automation receives much attention, successful RPA deployment focuses on affected people and processes. The expenses of developing and deploying a sophisticated network of bots can be in the millions of dollars, depending on the scale; therefore, failure and abandonment of the operation can be costly. Let's look at a four-phase method to incorporate RPA because a lack of preparation most commonly causes loss.

8.4.5.1 Planning

The procedures that will automate and the logistical difficulties accompanying their implementation are identified, with an eye toward compatibility with current processes and systems.

8.4.5.2 Preliminary Development

The planning phase establishes automation processes to select automation candidates and identify potential hazards correctly.

8.4.5.3 Deployment and Testing

The product has been monitored to detect outages and defects during this phase. Bots can then be scaled and deployed when this step has been accomplished.

8.4.5.4 Support and Maintenance

The fully deployed product is regularly updated across the database to maintain productivity.

8.5 Conclusion and Future Scope

The authors finally concluded the chapter where authors try to portray the necessities of RPA Utilization in healthcare during Covid-19 and also how Intellectual Property Law, which has a scientific-legal sense, is enabled to safeguard the assets of RPA with various examples. As it is a new field, therefore, it needs a lot of improvement in future about the advancement This new technology has the appetite to take over the entire healthcare sector with the help of automation and artificial intelligence. But if there is one thing that is proving as a hindrance to this technological advancement is their cost–benefit analysis. It is something that every organization may look upon if they want to embellish a revolution.

References

1. Zibulewsky, J.: The emergency medical treatment and active labor act (EMTALA): what it is and what it means for physicians. Proc. (Bayl. Univ. Med. Cent) **14**(4), 339–346 (2001)
2. Chatila, R., Havens, J.C.: The IEEE global initiative on ethics of autonomous and intelligent systems. In: Aldinhas Ferreira, M.I., Silva Sequeira, J., Singh Virk, G., Tokhi, M.O., Kadar, E.E. (Eds.), Robotics and Well-Being, vol. 95, pp. 11–16. Springer International Publishing, Cham (2019). https://doi.org/10.1007/978-3-030-12524-0_2
3. Bhatnagar, R., Jain, R.: Robotic process automation in healthcare-a review. Int. Robot. Autom. J. **5**, 12–14 (2019). https://doi.org/10.15406/iratj.2019.05.00164
4. Sriram, R.D., Subrahmanian, E.: Transforming health care through digital revolutions. J. Indian Inst. Sci. **100**(4), 753–772 (2020). https://doi.org/10.1007/s41745-020-00195-0
5. Barbieri, D., Giuliani, E., Del Prete, A., Losi, A., Villani, M., Barbieri, A.: How artificial intelligence and new technologies can help the management of the COVID-19 pandemic. IJERPH **18**(14), 7648 (2021). https://doi.org/10.3390/ijerph18147648
6. Basant, R., Srinivasan, S.: Intellectual property protection in India and implications for health innovation: emerging perspectives. IEH **3**, 57–68 (2016). https://doi.org/10.2147/IEH.S56236
7. AI interns: Software already taking jobs from humans | New Scientist. https://www.newscientist.com/article/mg22630151-700-ai-interns-software-already-taking-jobs-from-humans/. Accessed 5 Sep 2022
8. Robotic Automation Emerges as a Threat to Traditional Low-Cost Outsourcing—HFS Research. https://www.hfsresearch.com/research/robotic-automation-emerges-threat-traditional-low-cost-outsourcing/. Accessed 5 Sep 2022
9. Nine likely scenarios arising from the growing use of robots, p. 2
10. Robotic Process Automation Market Size Report (2030). https://www.grandviewresearch.com/industry-analysis/robotic-process-automation-rpa-market. Accessed 5 Sep 2022
11. Klubnikin, A.: RPA in Healthcare: A Path to Intelligent Automation. ITRex (2021). https://itrexgroup.com/blog/rpa-in-healthcare/. Accessed 30 Dec 2021
12. Robotic Process Automation Market Size Report, 2021–2028. https://www.grandviewresearch.com/industry-analysis/robotic-process-automation-rpa-market. Accessed 4 Jan 2022
13. so_robotic-process-automation-for-healthcare_en.pdf. Accessed: Jan. 04, 2022. [Online]. Available: https://www.kofax.com/-/media/Files/Solution-Overview/EN/so_robotic-process-automation-for-healthcare_en.ashx
14. F.-Y. Wang, 'Parallel Healthcare: Robotic Medical and Health Process Automation for Secured and Smart Social Healthcares', IEEE Trans. Comput. Soc. Syst., vol. 7, no. 3, pp. 581–586, Jun. 2020, https://doi.org/10.1109/TCSS.2020.2995282

15. gx-lshc-medtech-iomt-brochure.pdf. Accessed: Jan. 04, 2022. [Online]. Available: https://www2.deloitte.com/content/dam/Deloitte/global/Documents/Life-Sciences-Health-Care/gx-lshc-medtech-iomt-brochure.pdf
16. Siderska, J.: The adoption of robotic process automation technology to ensure business processes during the COVID-19 Pandemic. Sustainability **13**(14), Art. no. 14, Jan. 2021. https://doi.org/10.3390/su13148020
17. RPA and COVID-19: can automation help businesses to return to the "new normal"?. Nividous Intelligent Automation Company, Apr. 28, 2020. https://nividous.com/blogs/rpa-helps-businesses-return-to-new-normal-covid-19. Accessed 29 Dec 2021
18. Torgal, A.: Robotic Process Automation Revolutionizing Healthcare Industry during COVID-19. https://www.ceoinsightsindia.com/industry-insider/robotic-process-automation-revolutionizing-healthcare-industry-during-covid19-nwid-3305.html. Accessed 29 Dec 2021
19. Peterson, T.: 12 Reasons Why Automated Care Is Helpful in the Healthcare Industry. https://www.adsc.com/blog/reasons-why-automated-care-helpful-in-healthcare-industry. Accessed 4 Jan 2022
20. Lutz, T.: Intellectual Property and Healthcare in 2020. ReferralMD, Dec. 21, 2019. https://getreferralmd.com/2019/12/intellectual-property-and-healthcare-in-2020/. Accessed 4 Jan 2022
21. World Intellectual Property Organization—2008—WIPO Intellectual Property Handbook.pdf. Accessed: Mar. 08, 2022. [Online]. Available: https://www.wipo.int/edocs/pubdocs/en/intproperty/489/wipo_pub_489.pdf
22. Emsley, L., Rachel, T.: Be Competitive: Patent Planning for Robotics Companies | Articles | Finnegan | Leading Intellectual Property Law Firm. Be Competitive: Patent Planning for Robotics Companies. https://www.finnegan.com/en/insights/articles/be-competitive-patent-planning-for-robotics-companies.html. Accessed 30 Dec 2021
23. Making Your Robotics Company a More Attractive Investment—Robotics Business Review, Oct. 21, 2012. https://www.roboticsbusinessreview.com/unmanned/making_your_robotics_company_a_more_attractive_investment/. Accessed 30 Dec 2021
24. Sunarti, S., Fadzlul Rahman, F., Naufal, M., Risky, M., Febriyanto, K., Masnina, R.: Artificial intelligence in healthcare: opportunities and risk for future. Gaceta Sanitaria **35**, S67–S70 (2021).https://doi.org/10.1016/j.gaceta.2020.12.019
25. Fullman, N., et al.: Measuring performance on the healthcare access and quality index for 195 countries and territories and selected subnational locations: a systematic analysis from the global burden of disease study 2016. The Lancet **391**(10136), 2236–2271 (2018). https://doi.org/10.1016/S0140-6736(18)30994-2
26. Reddy, S., Allan, S., Coghlan, S., Cooper, P.: A governance model for the application of AI in health care. J. Am. Med. Inform. Assoc. **27**(3), 491–497 (2020). https://doi.org/10.1093/jamia/ocz192
27. Gopal, G., Suter-Crazzolara, C., Toldo, L., Eberhardt, W.: Digital transformation in healthcare—architectures of present and future information technologies. Clin. Chem. Lab. Med. **57**(3), 328–335 (2019). https://doi.org/10.1515/cclm-2018-0658
28. Interns, I.: Artificial Intelligence in Healthcare and Role of IP | Intepat. Intepat IP, Mar. 24, 2020. https://www.intepat.com/blog/intellectual-property/artificial-intelligence-in-healthcare-and-role-of-ip/. Accessed 4 Jan 2022

Chapter 9
RPA Revolution in the Healthcare Industry During COVID-19

Nilesh Harshit Barla, Shaeril Michael Almeida, and Michael Sebastian Almeida

Abstract Over the last year, the evolution in Robotic Process Automation (RPA) has been staggering. The automation it brings to applications has yielded efficiency, reduced operating costs, and decreased the time of research, development, and production. Industries have already integrated RPA into their workflow and are profoundly transforming into an intelligent automated industry with minimum human intervention, calling this the fourth industrial revolution. In this race of transformation, the healthcare industry is quite ahead of many other industries. It stood the test of time when COVID-19 was spreading rapidly and was also resilient against all odds. The system did experience an unprecedented crisis that depicted its weakness, fragility, and unpreparedness. The healthcare system was forced to adapt to a new paradigm. And though there was the loss of life and economy, we learned to evolve as a community to tackle this crisis. This chapter sheds light on the role of RPA and covers how these technologies can assist healthcare workers in their day-to-today activities, reviewing what the fourth industrial revolution would look like in the healthcare sector. The intelligent, automated system would provide a seamless experience of gathering information by various means, processing, and assisting healthcare workers to deliver quality treatment.

N. H. Barla (✉)
PerceptronAI, HSR Layout, Bangalore 560102, India
e-mail: nielspace26@gmail.com

S. M. Almeida
CHRIST (Deemed to be University), Hosur Road, Bangalore 560029, India
e-mail: shaeril.michael.almeida@christuniversity.in

M. S. Almeida
Oracle India Pvt. Ltd. Bannnergahatta Road, Bangalore 560029, India
e-mail: mikevolcano@gmail.com

S. Bhattacharyya et al. (eds.), *Confluence of Artificial Intelligence and Robotic Process Automation*, Smart Innovation, Systems and Technologies 335,
https://doi.org/10.1007/978-981-19-8296-5_9

9.1 Introduction

The emergence and growth of new technology led to the advent of Industrial Revolution 4.0 and carries the potential to revolutionize the existing technological mode. The global economic crisis led to the bankruptcy of many companies, which in turn led to the development of new theories of innovation. The need to use innovative technologies was felt to maintain a unique competitive advantage. The industrial revolution of the twenty-first century led to research that started with the hypothesis related to the innovational development of economic processes. The aim of this type of research was to concentrate on the activation of the development of high-tech spheres of the national economy.

Hence, the industrial revolution of the twenty-first century is Industry 4.0, which is all about sustainability and balance.

The healthcare sector is also not untouched by industrial revolution 4.0. There is a shift from hospital-centric to virtual care. The high-tech technologies have enabled the healthcare sector to explore areas like 3D printing of tissues and implants, home care, modern payment methods, etcetera. Increased use of automation, robotics, and artificial intelligence approaches like deep learning, and machine learning will change the face of healthcare. Industrial revolution 4.0 has made the healthcare sector more data-driven and robotic-centered. By the year 2030, there will be a revolution in the healthcare sector due to AI [1].

In this chapter, we have discussed how data-driven technologies like intelligent software and automation systems can have an impact on the healthcare industry. We review the current trends of robotic process automation and artificial intelligence and explore how they are being used in current healthcare settings. We have also discussed how these systems can be combined and used in clinical settings.

The chapter is divided into 5 sections, each providing information about the latest trends and applications of robotic process automation and artificial intelligence. In the first section, we have discussed the background of big data and its impact and data-driven technologies like RPA and AI. In the second section, we have explored the subject of robotic process automation, its type, advantages, the role of AI with RPA, and its applications in healthcare. The third section explores the value RPA creates and the ethical challenges it might face. In the fourth section, we have discussed the real-world use cases of RPA and AI. Finally, we conclude the whole chapter by reasoning out how the current healthcare setting will fail if these technologies are not integrated instantly.

9.2 Background

Today the world is loaded with data from every stream. This voluminous data is available from social media websites, clickstream data from the websites, video data from streams, call center voice recordings, genomic and proteomic data from medicine and life sciences, and data from sensors and radio frequencies. Organizations can make use of this data and analyze it to find solutions to problems, develop new products and services, or study any phenomenon in detail. Data scientists work very closely with big data and derive useful conclusions. Data-driven decision-making can be the solution in the future, but it comes with a lot of controversial questions and challenges. Acquisition of big data is not easy as it can cause a flow of a lot of staggering information and filtering of the same can result in loss of what is important. The next challenge is the automatic generation of metadata, which is not an easy task. Downstream analysis needs metadata. The information extraction process can convert the data into a structured form ready for analysis, but this also can present wrong facts, if not done very scientifically. Analysis and mining of big data is a tedious and tough process. Despite all these challenges, big data is a huge ray of sunshine to the world if managed properly. In the health sector, big data expands the levels of knowledge generation and knowledge dissemination which can help medical professionals to design better treatments. Genomics research is made possible by using natural language processing to phenotype patients, by using the Electronic Medical Records and Genomics Network. Big data can easily improve the quality of care by analysis of everyday information generated at medical facilities [2]. Big data can help in improving the quality of healthcare services, managing the large population flow, early detection of diseases, and improving decision-making, all at a lower cost. However, the challenges of managing data structure, security, and governance remain in the health sector [3].

With the beginning of a new Industrial revolution 4.0, cognitive technologies are operating at various levels. As depicted in some of the popular movies, smart machines can be a threat to humans if not managed properly. The key question is how a machine can be classified as a smart machine. The answer here is that smart machines can be defined in terms of cognitive technology on a lower and a higher level. On a lower level, they respond to human questions and form their own objectives and on a higher level, they perform digital and physical tasks with the help of numerical analysis. Hence, Smart machines, which are a combination of multidimensional technologies, come into the picture. Smart machines can perform the 4 cognitive tasks of analyzing numbers, analyzing words, images, and videos, and performing digital and physical tasks. On various levels, smart machines act as support for humans or engage in repetitive task automation. Smart machines have contextual awareness and learning but no self-awareness. This quality can have a positive effect as it can reduce or eliminate variations in results that rise in humans because of emotional, mental, and physical stress.

Robotic Process Automation or RPA is part of the spectrum of emerging artificial intelligence tools, including virtual agents, machine learning, computer vision, natural language processing, and reinforcement learning. The move to artificial intelligence technologies can have many applications, for instance, insurance, image classification that can be leveraged for claims, and text analytics for servicing customer queries. These new technologies will further drive automation and augmentation of insurance processes.

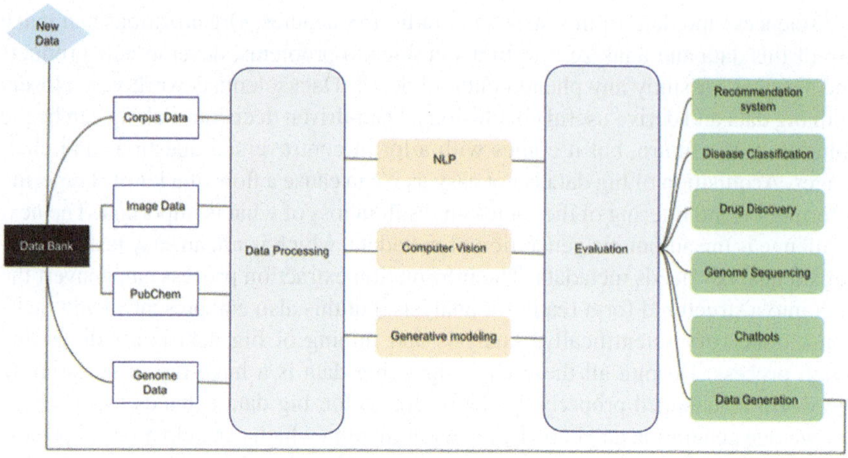

Fig. 9.1 An overview of an intelligent automated system

The healthcare sector is constantly changing and evolving with better services. The current pandemic has taught the world the unpreparedness that the health sector had to face and the need to bring about a complete revolution in the healthcare sector. The winds of change are seen, and especially private healthcare is becoming more and more advanced by adopting a combination of artificial intelligence, big data, and RPA. The usage and benefits of RPA in the healthcare sector are numerous. Digitization and automation in the healthcare sector have made RPA very effective in this sector. RPA results in value creation in the health sector. Research has found [4] that RPA can be successfully used in the main departments of healthcare as well as in the ancillary departments such as finance and billing, human resource department, and IT department as well. RPA can be best used in healthcare in the areas of processes that are repetitive and rule-based, processes that are prone to human error, activities that need out-of-office support, and that lack integrated applications (Fig. 9.1).

9.3 Robotic Process Automation

The term Robotic Process Automation (RPA) may seem to reflect another attribute related to robotics. Rather, it is related to repetitive tasks that are rule-based, and the processes contain some inputs and outputs. The tasks could be logical workflows that would typically have human intervention [5]. With the advent of RPA, these tasks would be executed via different applications and systems, but without human mediation. The term "robotic" in RPA needs to be visualized with a softer version, labeling it as "software bots" performing rational tasks. But in any case, it could also leverage a physical non-human entity to perform certain repetitive tasks replacing humans for instance a robotic arm.

9.3.1 Types of RPA

RPA can be broadly classified as traditional and cognitive. While the traditional type deals with basic tasks that do not need any decision-making or reasoning [5]. It is a type of automation process that relies heavily on a certain set of rules. Rules can be defined as a set of instructions that are required to achieve a certain task. These rules are generally defined in a framework, workflow or sometimes it is hard coded using a programming language like Python, JAVA, C++ et cetera. For example, if you want to scrap a website to build a medicines database, then you can simply enter the names of all the websites in a particular framework and the framework will extract all the medicine names and details and save them in different categories. There are limitations to this type of RPA as it is not equipped to handle changes in UI automatically or recognize human speech.

The Cognitive RPA on the other hand is equipped to recognize images, handwritten text, and human speech. As these are unstructured data, cognitive technologies such as speech recognition, natural language processing, and computer vision enable Cognitive RPA to extract and process the information further and derive meaningful conclusions and interpretations based on the task it is assigned to. Since the same data can have different interpretations for various tasks, objectives must be defined before the Cognitive RPA is put into action. In Sect. 9.3.3 and onwards, we will discuss more elaborately these topics as to how cognitive RPA can be used in healthcare settings.

9.3.2 Advantages of RPA

Now, let's discuss the general advantages of RPA in healthcare. These advantages are discussed in contrast with manual work.

RPA permits healthcare to work smarter. RPA is a type of business process automation that employs software robots, commonly referred to as automation or virtual/digital labor. These digital workers engage with a system's "backend" and "frontend"; this permits certain repetitive tasks automatically just as a human would have done. A front-end digital worker does the tasks of opening the documents, typing and clicking on the document, allowing the transfer of the data between systems. All this is done with the help of certain predefined systems and procedures. This allows the task to be done on a 24/7 basis automatically without any stoppage in between. Due to this, the time spent by humans on these monotonous automatic tasks can be used effectively somewhere else [6]. Thus, RPA helps in managing patients' appointments easily and also in reducing operational costs.

Completion of specific tasks. Robots are being used in healthcare sectors to complete specific tasks as they recognize medications using barcodes which enables the automation of prescription-filling processes. A set of parameters which include robots, automated pharmacy systems, barcodes, computerized medication administration records, and computerized discharge prescriptions and instructions is put into place by using AI and data management, and the task is automated [7].

RPA in pharmaceutics. The workers in the Pharmaceutical industries had to spend a lot of time in the lab, indulging in the mundane task of evaluating and generating new medicine. The task of generating a new drug and testing its efficiency is not only time-consuming, but also involves a set of repetitive processes which causes moral hazards to the human mind due to multiple failures. This drawback initiated the usage of RPA in pharmaceuticals and changed the entire scene. Robots are now used to perform these repetitive tasks at a more efficient level. For example, an automatic Perkin Elmer system not only captures 200,000 images a day, but also performs significant scientific experiments such as cultivating cells, dispensing substances, and rearranging them all at the same time, which would have required two full-time personnel to complete [8].

Quick diagnosis of problem. The world is facing a shortage of medical skilled labor force, for example, the availability of doctors, nurses, and radiologists is extremely less; which leads to longer waiting times and slow diagnosis of the problem. RPA in combination with machine learning and content flow can quickly diagnose any medical case. Content flow is a 3D image-based search engine that makes use of deep learning and further uses the recorded knowledge of medical images [9].

Better utilization of the skill. RPA promotes automation and quick services performed by robots. This allows the healthcare industry to use the skills which humans have in a better way. The nurses and doctors have more time now to spend with patients and serve them better. The experience of the patients improves as their doctor and nurse can give them more quality time [4].

Improvement in the quality of work. Humans tend to make more mistakes due to unavoidable human errors. These mistakes in the healthcare sector can be life-threatening. RPA offers good scalability and commits fewer mistakes. The quality of the work thereby increases. RPA with a set of designed instructions makes no mistakes or is very negligible.

Enhanced patient experience. RPA with a combination of AI makes processes for the patients in healthcare easier. The availability of doctors and nurses increases for the patients. Appointment booking becomes easy, and the FAQs are quickly clarified. This makes the patients have a better experience during any medical treatment.

Accuracy in the management of processes. As the concept of RPA is associated with the recurrence of tasks, it is imperative to note that accuracy automatically becomes a by-product. With predefined processes in place, it is easy to achieve the accuracy needed to ensure the smooth functioning and handling of various processes.

Elimination of operational risk. As the processes are well defined, there is no room for the tasks to be done in the wrong manner or for not having the required knowledge to perform a particular task. This is an added advantage, as the management can be free from regular operations and use the bandwidth to focus on more complex operations.

Data security. Every organization has the responsibility of protecting its data. There are many policies and laws to protect data. However, in the case of RPA, each bot performs a single task which in turn ensures that there is no leakage of information from one part to another; this is a very important benefit.

Enriched analytics. Data without analytics is not very helpful. While data analysts enable the use of data for better results, RPA also has a mechanism for assessing the performance of the workflows, streamlining the various functions, and resolving issues.

Seamless integration. When the functionality of a system requires change, or certain additions are to be made, there is a lot of apprehension from the management perspective as the cost of the impending change, and the related effect of the change on the existing system is getting affected. However, in the case of RPA, there is no disruption to the existing systems, as the integration is smooth. The RPA bots interact with the system at the User Interface level, as the way humans would, and hence there is no complication.

Improved use of human resources. There are disagreements among the various organizations that humans will lose out on their jobs if the bots take over tasks that were predominantly performed by humans. However, the fact is that human intervention is very important to handle very complex processes. Also, organizations use human resources to focus on work that requires higher intelligence and discerning powers which only humans can perform. While the bots can take care of mundane tasks, the more intricate and sensitive business aspects can be taken care of by humans.

RPA and COVID-19. Many countries faced a variety of challenges during this COVID-19 Pandemic. Challenges like shortage of medical staff, delay in healthcare services, the immediate requirement for patients' treatment, the need for making quick decisions, and the implementation of the same, were faced across the globe. Many countries successfully implemented RPA during these tough times of pandemic and reaped the benefits of immediate response at a scalable result [10]. The following processes were automated during this pandemic.

Test results. Robots were used to study and process the covid test results. The results were sent to the concerned people through SMS and in case the result was positive, the robot further advised the patients to stay home and wait till the doctors contact them. The robot also updated the results on the website to be checked by the medical professional [11].

Self-assessment of symptoms. RPA-enabled websites allowed people to perform the self-assessment of symptoms. RPA would guide the person to take a lab test if the symptoms showed the need. An appointment with the lab will also be booked for the person who needed one [12] with the help of RPA.

Vaccination drive. RPA enabled the vaccination drive for the COVID-19 vaccine in many countries. The robot will prepare the list of vaccination requirements on the basis of age, place, and health conditions and will update the health center and the public. The appointments for vaccinations were also booked through RPA.

Traveler's test. Research [13] showed that in the case of an international traveler arriving in a country without a valid RT PCR, a robot will help in the generation of a temporary patient identity to get a quick lab test taken. This would help in the quick tracking of the patient.

RPA has brought significant benefits to the processes. Some of the processes are also handled by robots, which were earlier handled by humans [14].

9.3.3 Cognitive RPA

While RPA has many advantages, it is not without certain limitations. To overcome the constraints of RPA, organizations would need to draw the power of Cognitive RPA which relies on technologies such as Optical Character Recognition (OCR), text analytics, pattern recognition, representation learning, and machine learning—all part of Artificial Intelligence (AI) [14]. The outcome is aimed at enhancing the customer experience as well as the workforce in an organization (Fig. 9.2).

Cognitive RPA enables organizations in many ways which include automating processes involving unstructured data sources in the form of voice recordings, emails, scanned documents, etcetera. The added advantage is that it helps automate very

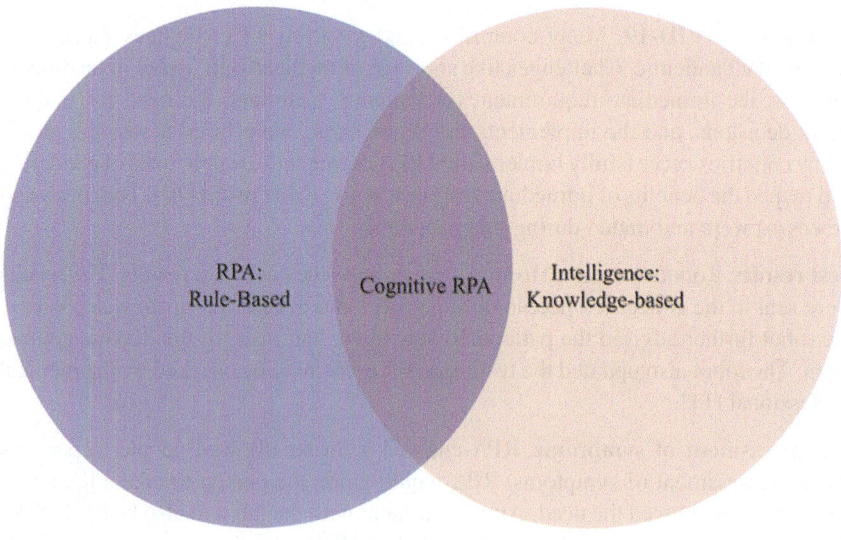

Fig. 9.2 Venn diagram of RPA and cognitive RPA

complex and less rule-based tasks. It enables the regular RPA system by providing structured data from unstructured data using natural language processing (NLP) and text analytics.

Organizations need to focus on the capabilities of RPA as it is possible to integrate it with various communication channels, such as messaging, chat, interactive IVR, etc. This integration equips the customers to do more without the need to have any human intervention. For instance, consider a chatbot in a medical center or hospital that automates the entire process of registration for a patient. The verification process might need human intervention in case of some discrepancies. The next time when the verification process is in progress, the system would validate the patient's identity based on the process used by the human. This is possible with the techniques in machine learning. Consider the scenario of a patient's experience in getting all of the medical processes, beginning with the registration, billing, and related work, complete with the discharge formalities–all of this done without human intervention. This might seem futuristic, but the possibilities are very encouraging.

In another scenario, when a patient is admitted to a hospital in a critical condition, it is a very crucial time for the doctors and nurses attending to such patients. While the doctors and nurses are busy treating the patient, the family members of the patient would be required to fill in the required admission forms and make the necessary payments related to the admission [15]. The nurses and other medical assistants would ideally pitch in to help the person attending to the patient, in completing the processes.

These healthcare scenarios can be managed better with the help of RPA. Here are a few areas concerning healthcare that are highlighted:

Digitizing data. The conversion of data from paper to electronic health record system (EHRS) is a time-consuming process as it must be done manually. Technology, in the form of optical character recognition (OCR) and computer vision, provides a solution with intelligent scanning from handwritten forms, and loading it automatically into the software systems.

Online appointments. Bots are programmed to send notifications to patients about their appointments with the doctor [15]. The COVID pandemic made hospitals turn to online appointments for most patients and even made provisions for telemedicine.

Discharge guidelines. It is very important to have follow-ups for patients, especially those who are critical. RPA helps in the routine checkup details that need to be sent as notifications to patients. Also, necessary updates about medicines to be taken on a regular basis post-hospitalization also can be handled.

Remote patient monitoring. Technology has made life better for patients in one way or the other. While traveling to the hospital can be strenuous for the patient, the COVID pandemic has also made it necessary to focus on remote patient monitoring [15, 16]. A lot of health data, from heart rate to blood pressure can be captured with modern remote patient monitoring devices, without the supervision of a health-care provider. The Food and Drug Administration (FDA) has authorized hospitals, remotely, to use wireless, noninvasive tools measuring common physiological parameters, until the COVID pandemic ends. Patients can use the mobile application with their data from wearables and smart sensors to send it to the healthcare staff of the hospital [15, 16]. The raw patient-generated data from the mobile app is stored in a cloud repository. In some cases, however, the systems provide direct-to-cloud connectivity for devices. This eliminates the need for patients to download a dedicated mobile application, as the data is captured in the cloud. The Hospital-side web software must comply with Health Insurance Portability and Accountability Act (HIPAA) standards and adhere to the interoperability of different systems and applications to exchange data and prevent the isolation of information.

Automated infection control system. Patients can use the mobile application with their data from wearables and smart sensors to send the information to healthcare staff. The mobile app provides the patients with a lot of features that keep them updated on details such as medication schedules, interactions with doctors, access to medical information, and related data submitted by them. The role of the Automated infection control system is huge when the emphasis is given to the overall impact on the global society [16]. To sense oncoming infections, the hospital data is used by automated infection control surveillance technologies, analyzed thoroughly and relevant reports are generated in real-time. The impact of the Automated infection control system can be seen in the following categories:

Medical

- Mortality rates have reduced
- Accuracy of the measurement of infections has improved
- Comprehensive identification of infections is possible
- Measuring Urinary tract infections (UTS) and less acute cases have increased
- Hospital infection rates have reduced.

Functional

- The time taken by ICPs to review charts has reduced
- The depth and scope of reporting have expanded
- Collaboration with the staff of all departments using specific goals and measurable reports have improved.

Economical

- The added cost of care has reduced
- The overall hospital antibiotic expenditures have decreased
- By decreasing UTIs by 55% to 60%, more than $1 million saved.

The Table 9.1 gives an insight into the benefits of the automated system, compared to the traditional method.

While the focus is on eliminating human intervention in repetitive tasks, some express concern about losing jobs. However, this should not be considered as a threat, but rather the human resources can be utilized for more useful work that involves better decision-making. The potential of humans cannot be replaced by

Table 9.1 Automated system compared to the traditional method

Traditional surveillance	Automated surveillance
Labor intensive, limited to 10% of patients at high-risk infection	Replaces the manual chart review process, enabling the Infection Control Practitioners (ICPs) to focus their efforts on prevention
Infection Control Practitioners (ICPs) focus clinical guidance only on a few high-risk patients	Alerts healthcare providers about potential infections and guides the selection of treatment
Underreports infection rates for some types of infection	Combines reports along with guidance on antibiotics and related applications to reduce errors

bots, especially in making decisions that affect the organization's overall growth. We will cover the topic in Sect. 9.3.7.

9.3.4 *RPA and Artificial Intelligence*

As mentioned previously to overcome the limitations of RPA, we need to leverage the cognitive abilities of systems such as Artificial Intelligence. In this section, we will explore how AI systems can modify RPA and enhance the ability of the current healthcare system.

The AI system is designed to mimic similar functionalities of the brain such as reasoning, decision-making, cognition, prediction, classification, understanding, interpretation, etcetera. The system can be purposefully used to acquire knowledge, extract patterns, and meaningful information, and derive interpretation from a given distribution without being explicitly hard coded. As data is readily available from all spaces, information can be extracted and made use of in a very promising manner. Data contains a lot of trends and hidden information which can reveal a lot of patterns in terms of human behavior, and the evolution of human health not only that it can also reveal the evolution of various diseases and approximate the new ones. These patterns are not easy to find. AI systems, especially deep learning algorithms have recently established themselves as one of the most efficient systems in the exploratory analysis [17]. With these capabilities, the field of AI is evolving rapidly, and it is widely adopted in various industries; it is one of the most influential fields in the history of mankind.

In the field of healthcare, the adoption of AI is quite rapid and rigorous as well. With its success in other industries and the rapid development of accelerated hardware like central processing units (CPUs), graphic processing units (GPUs), and tensor processing units (TPUs), researchers in the field of biomedicine are actively seeking opportunities to integrate AI techniques into the healthcare systems to intelligently automate various tasks. The AI systems are foreseen to be beneficial in the overall development of the healthcare system in terms of medical analysis, diagnosis, treatment, prognosis, research, and development [18].

Some of the areas where AI heavily has been used are clinical diagnosis and treatment, predicting protein structures and amino acid sequences, clinical decision-making, and much more [19]. The AI systems are also helping researchers and doctors to discover new information buried beneath the existing data, such as handwritten clinical notes, genomic data, MRI, and CT scans which can be leveraged for discovering unknown diseases and viruses, and enhance research. This information can also help them to develop precise drugs in less time. Not only that, but these systems can also be used to understand the working of the brain, which can be beneficial to cure diseases like dementia and other neurological disorders [20], and the possibilities are endless.

One of the key properties of AI as discussed earlier has been discovering trends, and patterns, and understanding new and unknown diseases, viruses, bacteria, etcetera, which companies like DeepMind are pioneering at. By integrating AI systems with RPA in the healthcare industry, we can automate the process of research and development through data mining on various and large datasets. The automated data mining process will enable researchers to study the extracted information and design a new automated pipeline that can develop new drugs for new diseases (based on the study) and conduct a clinical trial on a virtual cell [21]. Essentially the drug composition that performs well on a virtual cell can be further moved to the final clinical phase before the phase involving regulatory approval and mass production. Finally, once the individual process is approved, then they can be automated together by creating a pipeline to reduce the workload of both doctors and medical and non-medical staff.

With such an intelligent system, we can easily anticipate the revolution it can bring to the healthcare industry. But these systems must be explored much more carefully in order to understand their full potential. For instance, the following question lingers around the minds of many when it comes to integrating such a powerful tool into the healthcare system.

As AI systems demand large quantities of data, who will take responsibility for the unethical use of patient information? Such as harassment, and unwanted financial gains [22].

Since AI systems automate most of the routine and tedious work, who will take responsibility for the rise in the unemployment level [22]?

Another important concern is the representation bias which occurs during training when the data is skewed toward one end of the spectrum. This skewness in the data can lead to false positives which can be detrimental [23].

These points are a matter of concern and must be tackled. In the following sections, we will discuss how RPA and AI can create an efficient working pipeline to tackle different issues, mainly the issue of privacy.

9.3.5 Applications of AI-Enabled RPA

The greatest opportunity offered by AI is not reducing errors or workloads, or even curing cancer: it is the opportunity to restore the precious and time-honored connection and trust.
Deep medicine: Eric Topol

In recent years, we have witnessed RPA and AI growing at an exponential pace and enabling us to intelligently automate a lot of processes. It is how valuable they are for the present times and for the times preceding. The combination of both technologies can help healthcare experts to save time, reduce errors, and attend to many patients at the same time. These technologies will not only benefit the patients, but the doctors as well who suffer from mental fatigue by attending too many patients continuously causing them to make errors in their decisions. Let's discuss a few areas in healthcare where cognitive AI can be successfully used and can yield good and promising results.

Pharmaceuticals. RPA is proving to be a great solution for different types of industries as it helps organizations to focus on the quality of the products or services that it is engaged in [24]. The Pharmaceutical industry is benefiting from this technology as it caters to a variety of its needs.

The medical data generated by various hospitals, healthcare providers, and the larger medical fraternity has posed the challenge of managing huge volumes of data. When technology provides much-needed help in undertaking routine clinical procedures with accuracy, hospitals, and medical centers get the time to focus on areas that need more human interventions [24], such as areas like emotional support and counseling. And this is especially true in complex tasks such as in the case of radiology and monitoring patients in critical conditions because at times measurement of the heart rate or oxygen level won't lift up the morale of the patients, but rather human-to-human emotional support will. Some of the tasks where RPA or rather a cognitive RPA can be beneficial are:

Randomized Clinical Trials (RCT). The designing and completing of an RCT is error-prone and time-consuming as it involves the collection of huge volumes of data and utilizing them effectively. AI techniques such as machine learning and deep learning can help to extract information and can structure the data in the proper readable format making it easy for doctors and nurses to follow through.

Managing trial master file. As data needs to be entered and managed manually in a traditional setup, using RPA to manage the same, and other activities related to clinical trials can be done automatically with appropriate data structuring. AI-enabled bots can be trained or programmed to do data verification and check for missing data. In any case, these systems can also fill the missing data with proper statistical strategies.

Adhering to compliance and regulatory conditions. The employees of pharmaceutical companies coming under Regulatory Affairs use spreadsheets to manage data [24]. However, this time-consuming work hinders effective adherence to regulatory submissions and other compliance standards. RPA tools facilitate faster access to data and accurate methods to compile data, thus making it useful, whenever needed.

Estimating the precise usage of medicines. The automation of tasks related to collecting, managing, presenting, analyzing, and applying data will help in the precise forecasting of future usage of drugs. This is possible with the use of RPA software and improved integration with Artificial Intelligence (AI). For instance, in a certain hospital, a system can be installed or developed that can monitor the use of medicines or drugs. The system can predict the number of medicines required based on the usage and number of patients getting admitted for that common health issue. The system can inform the concerned authorities involved in that department so that they can place an order before the medicines run out of stock. Because this system is AI-enabled, it can keep track of the trend in that hospital or at best the whole town, district, or even city. These AI-enabled hospitals could communicate with each other and could estimate the number of drugs and forecast the rise in the price of the same therefore placing orders automatically.

Forecasting medical emergencies. Elaborating on the previous point, hospitals powered by intelligent systems can also forecast medical emergencies. These systems can be trained on news and Twitter trends which can enable them to do the following:

- Alert doctors and concerned authorities to be mentally prepared.
- Get an ample supply of medical facilities like oxygen supply, medicines, blood, etcetera.
- Preparing the infrastructure like extra health monitoring systems, beds, et cetera.
- Hiring vehicles and drivers for ambulance services.
- Proper work-shift planning so that medical personnel may get enough rest so that they function properly.

Managing supply chain. Pharmaceutical companies rely on various vendors for the supply and distribution of drugs. RPA provides flexibility to companies for automating numerous regular processes related to the supply chain, enabling the reduction of operational costs and improving overall efficiency.

Telemedicine. One of the exciting and intriguing disciplines in healthcare is telemedicine. The term basically means distance healing. In telemedicine, the patient and doctor use a communication system such as a video call to diagnose health-related issues. Telemedicine is growing quite profoundly in recent times, and it is evolving quite rapidly. This is mostly because of systems like RPA and AI. As such, apps are being developed which can remotely diagnose patients in the comfort of their homes. For instance, the flashlight at the back of the smartphone can be used to measure the heart rate. There are apps that can access the flashlight of the smartphone and transform it into an electrocardiogram or ECG device. This can allow doctors to diagnose a lot of heart-related issues [25]. With such apps, patients' data can be collected easily, and a database can be generated using RPA. It can also be leveraged for intermediary and analysis of data, based AI systems and medicines can be recommended to doctors based on the patient's condition [26]. Doctors can select, add if needed and then approve the medicines for the patients. Once approved and payment is done, the system can locate nearby drug stores possessing the recommended medicines by the doctors by sifting through the database, assigning a delivery partner, and delivering the medicine.

RPA plays a major role in collecting and sifting the data in existing, and AI systems play a major role in finding patterns and correlating them with the patient's condition, thereby assisting and recommending probable medicines, diets, and treatments to the doctor.

In many cases, telemedicine can recommend suitable therapy as well. For instance, there are apps that are designed for mindfulness, and psychiatry help. Some of the apps offer AI-enabled solutions whereby the patient needs to fill out a questionnaire and the app will recommend a proper plan. These recommended plans are even monitored by the assigned psychiatrist and physician which is again possible through RPA and AI. This type of intelligent system allows both the patients and doctors to achieve the goal of living a healthy life and helping as many patients as possible respectively.

Surgery. Surgical robots are quite popular in the healthcare industry and have been used since 1921 [27]. The aim of these robots is to automate the surgical process and get the job done with acute preciseness and accuracy. These may be faceless white cylindrical arms attached to a table with various buttons and a screen, but it is quite reliable and doesn't carry any emotion. It just executes the instructions given to it. As of 2019, there are more than 200,000 deaths recorded during surgery [28]. All because of an error in judgment and a slight mishap. The Robotic surgeon can reduce the error to 0 and can provide a safe and error-free surgery.

Usually, the robot is loaded with a set of instructions. These instructions enable the robot to automatically perform all the actions without any human interference. The robot is loaded with sensors and cameras that allow it to efficiently navigate through the desired area. Most robotic surgery is semi-automatic where a human surgeon controls the robot. These days the robots are powered with AI algorithms that enable the robot to quickly identify the affected region and perform the surgery.

These robots are initially made to perform surgery in a simulated environment. The one commonly used is the Da Vinci Robotic system [27–29]. It is a training simulator used to learn to operate and practice surgical skills for robotic surgery. The progress of AI-enabled robotic surgery is very slow mostly because of the lack of data, and a realistic working environment. In such cases, the techniques like reinforcement learning or RL [30] come into the picture. In this experiment [29], the authors presented an RL technique called model-free reinforcement learning [31] to train a surgical robot in a simulated environment. The model-free RL was integrated into the simulation which allowed the robot to perform surgical tasks using images or state-action data. Authors in [29] have demonstrated how techniques like Q-learning enhanced with long-term replay, which they call hybrid-batch learning, can train robots for surgery in less time and are much more efficient.

This experiment [29] has shown promising results of AI playing an important role in the development of robotic surgery. But whenever we replace humans with a non-human entity, questions regarding safety and reliability come into the picture, few of them are:

- In case of a software malfunction if the robot makes a wrong move, then who will be responsible [28]?
- Cost of surgery.
- Time taken for preparation of the surgery is more since the DL model needs to be fine-tuned followed by rehearsing it on the simulation before the surgery.
- Requirement of technical staff is required who can program complex surgeries.

Prosthetics. In healthcare, prosthetics plays a major role in the rehabilitation of patients who lost their body parts in accidents or even due to amputation by providing them with an artificial body part. This is very helpful and some of these parts can restore the full functionality as well.

Developing prosthetics in recent years has seen a profound evolution due to the invention of 3D printing. 3D printing or additive manufacturing is digital manufacturing technology [32] invented in 1984 [33]. It can essentially print or manufacture any design from simple to extremely complex. 3D printers can manufacture an object from a computer-aided design, a software used for designing 3D shapes. These printers can also work with different materials like polymers, metals, Ceramics, Composites, Smart materials, and Specials materials [32]. In the healthcare sector, 3D printing is used for various purposes such as 3D skin, tissues, bone and cartilage, prosthetics, bone and cartilage, etcetera.

One of the biggest advantages of 3D printers is that it has automated the full process of manufacture of prosthetics and reduced manual labor. All it requires is the design that these printers can use as an instruction. A few other advantages are:

- Reduce the cost of manufacturing [34].
- Patient-orientated [34]
- Rapid prototyping [34]
- Reducing the risk of error [35].

Combining AI with 3D printing can enable much more intrinsic designs with much preciseness and accuracy. Since AI systems are trained on a huge amount of data to learn feature representations, distributions, and patterns, on-demand features of a particular patient can be fed into the system to get an accurate 3D design of the body part. Once the design is generated, AI can self-evaluate the design by correlating its dimensions with other designs to check if the recommendation is error-free. Finally, once the design is approved by the doctors, it can be sent to a 3D printer to get the results. In addition to that, the final design can also be stored in the database with various other variations in the original design to increase the database.

Cell engineering. Cell and cell-based tissue engineering have been at the forefront of medical research and development, and clinical trials and it displays a great feat of technical and innovational advancements that both engineering and medicine have achieved together. But cell growth is not an easy task to achieve. Unlike other departments of healthcare which can have a general output or production, cell growth via engineering requires both unique and intrinsic solutions as well as a general and reproducible solution for the masses. One of the major hindrances in cell growth is that cell culture requires a sufficiently long time for growth in a favorable (artificial) environment, and if the environment differs slightly in any setting, then there can be a phenotype drift that can produce variability. In any case, cell engineering is hard in terms of reproducibility, scalability, economic stability, and regulatory concerns [36]. But it is also true that cell engineering has theoretically and also practically offered promising results such as offering therapeutic solutions to non-curable tissues.

Cell growth falls under two categories: autologous therapy which is personalized i.e., the patient is treated with his/her own cell and does not require mass production and other is allogeneic therapy which is aimed to scale up the production for masses. But cell engineering can be possible only on a small scale because of the phenotypic drift. To address phenotype drift, engineers are compelled to use intelligent automated systems such as AI-enabled RPA [37–39].

In cell engineering, an AI-enabled RPA system can be considered a full lifecycle. It can be broadly classified into four phases: bioprocess, monitoring, screening, and production. In bioprocess, an automated pipetting system is generally used to automate the whole process of culture automation which increases productivity and accuracy and produces high throughput in biomedical labs. With an automated pipetting system, operations like monitoring flow rate, i.e., the process of maintaining the shear forces during the media change, becomes feasible.

In the screening process, the morphology of the cell is evaluated. The morphology of the cell is the reflection of the previous two steps, i.e., bioprocess and monitoring. If these two processes are carried out precisely, then the desired cell morphology is achieved. Cell morphology is also the by-product of phenotype. In the screening process, machine learning algorithms may include computer vision and reinforcement techniques to evaluate cell morphology. The machine learning systems can:

- Detect faulty structures within the cell
- Improve intrinsic process
- Optimize the process's whole lifecycle by providing other meta information for better growth culture like temperature, light exposure, humidity, etcetera.
- Predict function-specific gene-phenotype [40].

Radiology. This is identified as the discipline in medicine that deals with medical imaging to diagnose diseases. This discipline is one of the backbones of modern-day healthcare and medicine. But given the number of trained readers available, imaging data is available disproportionally [41]. This increases the workload of the radiologists. For instance, an average radiologist on an 8-h shift must interpret an image for 4 s [42]. It is enormous pressure, especially if there are several patients waiting in the queue. The increase in workload usually leads to impairment in decision-making which can be detrimental to the patients [43]. In such cases, RPA and AI can be of tremendous help.

Radiologists primarily rely on their visual perception and their decision-making to diagnose and prescribe treatments. The job of the radiologist is essentially to find and identify the region of interest in each image; these tasks leverage tools such as annotations, bounding boxes, segmentation masks, etcetera, for the same. A seamlessly integrated AI system can be designed to achieve the same task of identifying the regions of interest with higher-level efficiency and reduced error. This system can be proposed as a supervised learning task where it is provided with both input and output for training. Once it is trained, it would be able to classify all the anomalies in each picture, thus reducing the workload of the radiologists. Similarly, RPA can be used to facilitate some of the mundane tasks that are both repetitive and tedious. Some of the areas where RPA and AI can be put into effective use are:

Data curation. Although the availability of data is not a problem these days, the quality of data is. When it comes to medical images like CT and MRI scans, the images can be difficult to interpret at times, and training an AI system on such data can lead to errors and inaccuracies in classification tasks. Properly evaluated and curated data against a certain benchmark score will certainly increase the model's accuracy and robustness against the anomalies and will yield good results. RPA can be leveraged to collect data and for the curation process. The process itself consists of: ethical approval of the data, accessing data from the database, querying data, data de-identification, data transfer, quality control, and data preprocessing such as resizing and labeling data [44].

Selecting tasks-specific algorithms. AI systems can have different algorithms, and even though they are powerful, they can be misapplied. It is key to understand that AI systems model the distribution of data. A simple algorithm cannot model a complex data distribution. AI algorithms must also be properly curated based on the task and data in hand. In general, medical imaging covers a much broader spectrum of specialties, which includes ophthalmology, pathology, cardiology, and radiation oncology [45], and each of these specialties contains a different type of distribution differing from the other. One algorithm can work on one category but may not work on another. Also, the type of task we want to achieve can differ. For instance, do we want segmentation or instance segmentation, or both? In such cases, RPA can be used to determine which AI tools and algorithms must be applied based on the specialization, data, and task at hand. Once necessary details are determined, the data can be fed into the training pipeline.

Annotations. Annotation or Image annotation is the task of labeling certain specific parts of the image for explanatory and descriptive information [46]. Radiologists rely on labeling images to interpret them better. Annotations can be of certain types including bounding box, instance segmentation, and semantic segmentation and each of these annotations serve well in their respective task. The issue with annotations is that it is extremely tedious, and it takes hours to annotate image data. An AI-based RPA can quickly collect data and annotate new and similar data for training new models, and at the same time, store the data in the database.

Clustering. In AI and machine learning, clustering, or image-clustering is an unsupervised learning task of grouping similar images together. Similarity can be defined as the quality of the images as well as domain specific. In the context of radiology, this is quite helpful because AI and RPA combined can now automatically separate low-quality images and high-quality images and also group domain-specific images. For instance, all the cardiology images could be clustered together, likewise, all the neurology images can be clustered together, and so forth. Integrating these technologies into the production pipeline can make the data-management process efficient and reliable.

Treatment recommendations and report generation. Generating radiology reports is not an easy task, it requires expertise, it is time-consuming, and it increases the workload as well. A successful report generation is required to have a correlation between the visual contents or the annotations of the image with the natural language description [47]. In such cases, RPA is usually combined with AI or deep learning (DL) algorithms capable of natural language processing. Now, RPA combined with a DL algorithm can take an input image, process i.e., generate necessary annotations and based upon those annotations generate a full medical report leveraging another DL algorithm that deals with natural language. Similarly, the same process can also be used to recommend medicines, therapy, and treatments, and the concerned doctor can analyze the best option and move ahead with the treatment process.

Follow-up reminder. Most of the time, patients are required to have follow-ups with the doctor to make sure that everything is fine. On a normal day, the patient must go to the reception and ask for another appointment on a certain recommended day. But with smart and intelligent automation processes, patients can book follow-up appointments along with reminders through smart effective software automatically. The software can use the generated report of the patient to generate recommended appointments, if both the doctor and patient are available, then they can book a slot and a reminder can be sent to both.

Drug discovery. This is a process of identifying new medicine leveraging disciplines such as chemistry, biology, and pharmacology. But developing a new drug, understanding its effects and side effects, testing its reliability, and putting it into production is both time-consuming and expensive [48]. Developing a single drug takes around 10–15 years [49]; it is also very complicated as it requires finding a combination of molecules that can precisely target a particular disease. Generally, there can be around 5000–10,000 potential candidates and each of them is tested against the target disease; it is a very rigorous and tedious process that can include various screening methods. This motivates the use of RPA and AI because they can both sift and process large information of data yielding reliable results as well as automating the whole screening process before approval and production.

Firstly, RPA can process huge amounts of unstructured data by collecting valuable information, labeling them, and transforming it into a structured format for the purposes of AI training. Secondly, AI can make use of that data to find patterns and correlations that can enable them to predict relevant molecular structures, properties, and molecular behavior for a particular type of disease. Once the relevant information is extracted from the algorithm, further steps can be taken to produce all the drugs of different combinations. For instance, RPA can be further used for sample analysis, testing, handling liquid samples, and even preparing samples for NMR (Nuclear Magnetic Resonance) spectroscopy and HPLC–MS (High-Performance Liquid Chromatography-Mass Spectroscopy) analysis [8, 50, 51].

Genome Sequencing. This is another area where both RPA and AI can effectively impact the field. We have discussed earlier how RPA can be used to collect information from unstructured data and transform it into a format that is clean and well organized. Similarly, in genome sequencing, RPA can be used to perform the same task but in a different way. It is important to understand that all data related to chemistry and biology are text corpus and transforming such data requires special algorithms; most of these algorithms belong to the area of natural language processing. These algorithms can transform the sequences into word embeddings [52]; word embedding is a term given to real-valued vectors which are the distributed representations of words. Creating word embedding can tend to become complex because of three reasons:

- Unstructured nature of the data
- Different sources
- Heterogeneity.

Data construction always remains an integral part of an AI system, and in this context, constructing a benchmark dataset is the key. Since the data is collected from various sources, a smooth and homogenous construction of the dataset is appreciated which includes removing unwanted information or noise as well as cropping the sequence length. Thus, a combination of algorithms must be used such that all the tasks related to data preprocessing are efficiently achieved, these methods can be described in these four steps: data construction, sequence segmentation, creating feature vectors, and AI framework [52], this is where the RPA comes into the picture, it can just automate the whole lengthy process and create an end-to-end pipeline. Once the dataset is constructed, it can be further moved ahead to create word embedding using algorithms like word2vec or C-BOW followed by creating feature vectors and then feeding it into the respective AI algorithms like CNN [53], RNN [54], or Transformers [55] to classify genome sequence. A reverse approach is also applicable to finding a genome sequence using a DNA structure.

Similar approaches are also true for finding protein sequence and protein structure.

Data Generation. So far, we've seen how AI and RPA can revolutionize healthcare by intelligently automating everything. But the fuel on which it runs is data. All AI systems become intelligent by finding patterns and representations within the data, in other words, the data itself is the source of shaping an intelligent system. But to train and produce new AI algorithms, we need data. When it comes to healthcare, data is highly sensitive, and there are many concerns regarding privacy. One way to tackle the issue of privacy is to create or generate realistic synthetic data. One approach is that the AI models must capture precise details of the actual real-world data which includes patterns, anomalies, structures, distribution, and linear and non-linear relationships [56].

Recent advances in AI have led to the development of state-of-art models that yield extraordinary results in the task of generating synthetic data. Two such examples are generative adversarial networks (GANs) [57] and autoencoders [58] (AEs). Both are classified as deep generative models and are capable of learning and preserving temporal information and feature representations [59]; this allows the models to generate good-quality synthetic data and are also capable of preserving patient-sensitive information. They achieve this by leveraging a mechanism called Differential Privacy which uses mathematical formulas to create representations for ensuring and quantifying the privacy of the system [59].

Generally, generating synthetic data can be divided into four phases:

- Preserving privacy
- Handling discrete data
- Quality test of the generated samples
- Temporal information.

The tasks mentioned above are complex and each of them needs a different algorithm. Essentially, all the algorithms used must produce a homogeneous result while preserving privacy and capturing the true statistical properties of the real data such that it is difficult to differentiate real and generated data. The use of AI can greatly benefit the task of data generation and can bridge the gap between diagnosing and treatment.

The complexity that arises in the four phases mentioned above can be combined using RPA. In essence, RPA can take care of the automation process and provide a seamless delivery of new data while preserving privacy. First of all, we must understand that GANs are good candidates for generating data that is secured because it is not reversible. GANs belong to the category of implicit models which also means that they are not deterministic but rather stochastic. But it is also true that GANs are poor at generating continuous data. This is where we can use autoencoders to produce continuous data [60].

Autoencoders are generally used to find latent variables and for the reconstruction of the original data. We can view latent variables as variables that cannot be observed but their properties can be inferred from the other variables. Thus, GANs can leverage the continuous feature space generated by the AE. At the same time, GANs can generate high-fidelity samples while preserving privacy. Thus, we can automate four processes:

- Generating continuous feature space of the original dataset using AE
- Using continuous feature space to train the GANs
- Using differential privacy to measure loss
- Updating the database once the model reaches the desired accuracy.

With this framework, we can tackle the issue of privacy and preserve patients' information while developing newer DL models.

Natural language processing for text analysis. Text analysis is one of the important areas within the healthcare industry which enables healthcare employees to search, analyze and interpret huge amounts of data within seconds. These data, however, are not fully structured, while some are in a proper format other aren't as they are present in the form of clinical notes. Other ways through which medical data is usually generated are electronic health records (EHR), which include clinical notes, research data, synthetic data, and chat records [61]. This data contains vital information that can be useful in developing precise drugs, treatments, monitoring, prediction of patient status over time, etcetera, and extracting this information is an important task. Fortunately, computational techniques such as natural language processing allow doctors and medical staff to do the same.

Recent works on NLP have led to the development of state-of-the-art language models, models like GPT-3 [62], that is, training on a huge corpus of unstructured internet data can find meaningful information within the data and then use that information for machine translation, sentiment analysis, natural language inference, and much more. Motivated by this, we can certainly assume NLP powered with deep neural networks can impact healthcare in a positive manner. Some of the applications are:

- Text Mining Applications in Healthcare
- Text Analytics for Clinical Decision Support
- Text Classification
- Hypothesis Generation and Knowledge Discovery.

9.4 RPA and Value Creation in Healthcare

RPA has made inroads into a whole range of areas in the healthcare sector. It has generated significant value in various terms [4]:

- Round-the-clock workflows: Although human intervention is necessary for crucial aspects of healthcare, normal workflows are certainly important. RPA has provided that as an added value, easing the burden on healthcare providers to a large extent.
- Channelizing data: The source of data is multi-layered, as it comes from different forms and multiple sources. This data generated through RPA and AI algorithms are stored systematically via clustering and can be directed in a logical format and used for multiple purposes in the cloud or in an in-build and local data center. The sources may be a physical medical bill, medical report of the patient, doctor's prescription, hospital discharge summary of the patient, a detailed medical history of the patient, images of scan, blood sample details, or composition test results, etcetera. Similar reports and other data relevant to the medical aspects might be generated via online sources as well, such as scanned reports. RPA enables the scrutiny of the data in a viable manner to ensure that only the relevant and important data is utilized to give a feasible, appropriate, and authentic output.
- Data management: Most of the data is unstructured; RPA can leverage AI systems to cluster similar data and then structure the data, making it efficient for searching, extracting, handling, and preprocessing for further use in the AI pipeline.

- Data security: In this chapter, we already saw how data can be privately stored in clouds and local databases. This assures the security of the patient's information from any kind of threats and unwanted extraction of money. We also saw how AI-enable RPA can be used to create synthetic data from real ones to further develop and train any deep learning algorithms.
- Research Enhancement: With the integration of RPA research in medicine, drugs, and treatment will be accelerated as most menial and repetitive tasks will be automated such as collecting the data, storing the data, analyzing, and extracting the data, etcetera. Now, researchers can focus more on studying the extracted information to build a new medicine, methodology, treatment, etcetera.
- Discovering uncharted territory: With data already available and with accelerated research, automated extraction of data through DL algorithms can possibly open doors to predict new bacteria, viruses, protein structures, and treatments, and can enhance the longevity of humans altogether.
- Expansion in the scope of healthcare: As RPA progresses, and with the introduction of newer technologies within the RPA framework, the scope of the usability of RPA in the healthcare sector increases exponentially. In the areas that were limited to human intervention earlier, the introduction of RPA has given rise to the induction of tech bots that can do more than humans with very less or almost error-free output [51]. This gives the much-needed push to the medical fraternity to enhance the quality of medical services, incorporating trendsetting medical advancements, including RPA as well. When the impetus is absolute, then there is more room for other areas of medical care, such as complicated surgeries, which could be noninvasive. The value addition is humungous, provided it receives the necessary acceptance from the relevant authorities, government bodies, and experts in ethical principles.
- Morale booster to the medical fraternity: As the healthcare providers and other medical practitioners, doctors, and nurses are focused on providing excellent healthcare to the patients, there will be a surge in their morale as they carry out their primary responsibilities, without worrying about the mundane work that would usually block their thought process. RPA would take care of the administrative work, along with some complicated jobs, and in turn, provide the much-needed respite to the medical fraternity.
- Job creation: Although many see automation as a threat to humanity in terms of creating unemployment, others can see it differently. As automated technology is growing, the need for personnel who can write code and structure and design these systems has rapidly gone up. Not only that, the reliance on RPA and AI in other sectors like chip manufacturing companies is also creating a lot of employment opportunities in these sectors as well. And since the repetitive and mundane tasks jobs are getting automated, opportunities in research and development is huge.

9.4.1 Ethical Challenges

Every area of healthcare is susceptible to ethical behavior from the doctors to the nurses, and other healthcare professionals [63]. While the challenges posed by the different diseases and sicknesses to the medical fraternity are a known fact, it is also important to focus on some of the ethical challenges posed by using RPA technology in the medical field.

- Testing on chronic patients: With the advancement in medical technology, it is important to conduct tests to declare the success of the technological process. However, it is unethical to test on elderly and chronic patients with very less or no hope of recovering from the ailment [63]. Hence, the tests can be termed as mere procedures that use humans as guinea pigs. In such cases, technological advancements are very much required. As mentioned previously, clinical trials of drugs on virtual cells can be very effective as they can be designed according to the requirements and specs and can be tested with various drug compositions.
- False hopes of improvement in health: The huge costs involved in some medical procedures and treatments, especially with major hospitals pose this question about affordability. This unethical method of increasing the costs makes the treatment of ailments unaffordable for the marginalized society. With digital technology or IoT, the government can monitor the hospitals by creating an audit committee to inspect the proper usage of the healthcare system. Such systems can control extortion and dishonest gains.
- Robots helping Humans is Inhuman: Although the intention of using robotic technology in the medical field is to enable better services to patients, it is a well-known fact that humans with medical conditions would prefer to have another real human to alleviate their problems, rather than having a robot to take care of their needs [63].
- Highly-priced medical care: Basic healthcare should not be deprived of the public. The huge medical charges made by hospitals and other medical care facilities are dubious. The government and medical authorities in consultation with manufacturers of drugs and medicines should work out a solution to this problem, as millions in the world go without proper medical care owing to huge medical expenses.

9.5 Case Studies

In this section, we will explore and discuss these technologies as they are used in the real world. This will give us an idea of how these technologies are evolving and are adding value to our lives. This section involves two case studies from various backgrounds.

9.5.1 Glaucoma Screening

According to the WHO, glaucoma is one of the major causes leading to vision blindness, which also makes it a leading concern for doctors around the world [64]. To prevent blindness and its relative symptoms like frail vision, early detection and treatment of glaucoma can be very crucial. With advancements in RPA and AI, this is very much possible, especially in the most remote and rural areas where advanced healthcare infrastructure has not been set up. In such cases, intelligent software can come to the aid. As such, this case study focuses on the use of telemedicine to diagnose glaucoma screening using RPA and AI.

So far, we've seen how efficient RPA can be in automating repetitive tasks which can be tedious, time-consuming and prone to error when handled by humans. In this case study [65], the authors leveraged the automation capabilities of RPA to the ocular screening process which remains the anchor point of the system. Apart from the screening process, additional tasks like registration, storing and gathering the data, medical appointment scheduling, and payment processing were also added to the pipeline. The screening process included facilitating early clinical examinations, in addition to comparing analysis results from previous and current examinations of the patient.

This experiment was conducted in a hospital based in Thailand. The workflow of the whole system is as follows:

- The first visit of the patient usually requires creating a patient record in the electronic resource planning software or ERP. This could be done with a simple mobile app and could be verified easily if the citizens have a unique identification number which could be found in their driving license, bank account number, etcetera. But in Thailand, identification verification requires face-to-face verification.
- Once the information has been entered into the database, RPA can automatically arrange the information with other similar information, and pull out a statistical comparison with the patients having the same concerns, i.e., do a partial analysis, etcetera, while doing a quick forecasting of your health.
- Apart from creating the account in the ERP, patients can also upload an image of the eye using a mobile camera which is then sent to the ML or DL pipeline for further preprocessing and investigation.
- Depending upon the results that the preliminary diagnosis shows, the patients are given medication, or in the worst-case scenario, they are appointed to a specialist.

Furthermore, after the diagnosis is confirmed by the specialist and relevant information on the ERP were collected, deep learning algorithms such as NLP can be used to summarize the whole screening, diagnosis, and treatment processes into a final report [66]. In this experiment, the patient was also notified about the final report, results, and an appointment for the next examination as well.

Such a system is extremely beneficial for assessing a patient's condition based on the information provided and yields relatable results in a short period of time. The result of this experiment shows that the screening system with integration of RPA reduces the average handling time per user by 75% making the overall satisfaction score of the application to be 8.10 out of 10 [65]. On top of it, the deep learning systems integrated with RPA take care of the preliminary stage of diagnosis, reducing the workload of the doctors and improving patient's experience. Further studies have also shown that teleophthalmology can also provide better eye care solutions and services to areas under COVID-19 pandemic restrictions [67].

9.5.2 AI-Enabled RPA During COVID-19

During the outbreak of COVID-19, WHO declared it to be a global healthcare crisis and it also led to a downfall of the whole medical system [68]. As it turned out, there was a shortage of medical personnel and medical and healthcare supplies. But fortunately, computer-automated services came to the rescue. Intelligent software like Deep learning and RPA achieved state-of-the-art performances in the areas such as diagnosis, risk assessment, monitoring, telehealth care, and most importantly research and development of vaccines. In this case study, we will explore how AI-enabled RPA can and has helped frontline workers during the COVID-19 health crisis.

Accelerating diagnosis. As we witnessed, the testing sites experienced a huge and overwhelming demand for testing results [10]. This led to people standing in long queues which ultimately prolonged their possible exposure to the virus and delayed the potential treatments as well. In such cases, RPA can be very handy as it can be used to:

- Automatically load the data and vital information of the patients to EMR records
- Cluster records with similar information, for example, separate patients with respect to a certain age group, recent travel history et cetera.
- Analyze the patient's information using deep neural networks and yield results based on the information provided.
- The results can be further delivered through a mobile application which can reduce the waiting time and manual errors.

In such cases, the data collection, storage, and retrieval are fully taken care of by RPA, while privacy, preprocessing, and prediction being taken care of by the AI system. Furthermore, these systems can work 24/7 without any hindrances. Because these systems follow a sequence of commands, it is quite efficient and easy for all the doctors, and other staff to understand what is going on and get accustomed. Thus, heterogeneity and miscommunication are eliminated in communication among the medical personnel in varying shifts and the process runs seamlessly.

Monitoring patients. With the shortage of medical personnel, especially nurses, monitoring patients with critical conditions was extremely difficult. Screening bots can be used to monitor the health condition of the patients and continuously monitor their health condition. This can be done by keeping track of the temperature, oxygen level, blood pressure, etcetera. If any of the concerned parameters crosses a certain threshold, then the nurses and doctors can prioritize that patient and attend them.

9.5.3 Baylor Scott and White Health|Revenue and Financial Stabilization

Throughout the chapter, we have seen methods and ways in which AI-enabled RPA can help doctors and healthcare workers to efficiently diagnose and treat patients, and we also explored the research and development part of drugs and medicine. This case study will focus on the economic side of the entire healthcare system. As it turns out, the administrative cost takes up to one-third to one-fourth of the total expenditure in the healthcare system [69]. If the administrative cost can be lowered, then the wealth can be used in other areas to improve the quality of service.

One such hospital is Baylor Scott and White Health (BSWH) [69]. They use AI-enabled RPA services to automate most of the administrative tasks. They focus on three major points in revenue generation:

Improving the patient's financial experience through automated services. One of the ways in which BSWH is improving the patient's financial experience is through transparency in pricing. They use machine learning-based tools that can estimate the price of the patients' treatment before they even receive the hospital's care and facility. Creating price estimates usually contains combining various information from numerous systems. Such systems can fetch information automatically like real-time eligibility and benefit data from patients' insurance companies and process the whole information to give patients' a better experience. With RPA and machine learning technology, BSWH can generate price estimates within seconds which earlier took 5 to 7 min when done manually.

Reducing the cost to collect insurance claims. Collecting insurance is another benefit that BSWH provides. With AI-enabled RPA, BSWH can log into multiple payer websites and perform tasks like screen-scrapping to collect claim status from the payer. If the claim status is accepted, it is then scheduled to be paid automatically to the patient. If the claims are denied, only then human intervention is required, otherwise, the whole pipeline is taken care of automatically.

Net revenue optimization. The goal of any business is to be profitable which includes healthcare businesses as well. In all businesses, optimizing revenue takes a lot of effort because it requires pulling out documents for analyzing the business expenditure. In large-scale companies, such as hospitals and healthcare centers, analyzing documents can be hard and tedious even after having advanced data structures and business intelligence software. These software rely on data and data collection remains an integral part of the whole pipeline. Humans can make an error during data retrieval which can lead to optimization error. In such cases, an RPA-enabled data retrieval pipeline can be created that fetches information from various departments. With AI technologies like OCR and NLP, information can be extracted and processed which can help to find anomalies in the current pipeline. Experts can then use this information to bridge the gap between the supply and demand services, cut out unnecessary expenditures, and introduce new methods that can increase the productivity of the doctors, improve patients' experience, and attract more third-party companies and investors to become part of the business, thereby increasing the net revenue.

9.6 Conclusion

The world is facing numerous problems along with the pandemic, the UK sits in a painful deadlock over Brexit, the Russia–Ukraine war is posing new challenges to the world, and there is a fear of new variants of coronavirus constantly hovering over humankind. With all these challenges, the healthcare industry will severely suffer a blow in the coming days. The chances of healthcare facing a crisis are high. The current situation in the research of healthcare is that the algorithms that feature prominently in this sector are not executable at the frontlines of clinical practice. The addition of AI to a fragmented system cannot ensure sustainability in this sector. Healthcare delivery is a combination of factors that are very complex and are influenced by political, social, and economic situations. Healthcare is highly influenced by medical norms and applying RPA sometimes in a restricted industry becomes tough. The commercialization of healthcare, however, has enhanced RPA application. Today, most hospitals lack the infrastructure to record and organize data to build better algorithms. The entire data cannot be used globally as patients across the globe are unique in terms of their response to medication and other healthcare-related practices. Fundamental issues related to data holding, data handling, and policymaking should be tackled in the first place. The government will play a major role in policymaking related to the same, as the regulations related to data can bring about a revolutionary change in this sector. However, Cloud computing, predictive analytics together with deep learning and machine learning hold a promising future for the usage of RPA in healthcare.

References

1. Wehde, M.: Healthcare 4.0. IEEE Eng. Manag. Rev. **47**, 24–28 (2019)
2. Murdoch, T.B., Detsky, A.S.: The inevitable application of big data to health care. JAMA **309**(1351) (2013)
3. Kruse, C.S., Goswamy, R., Raval, Y., Marawi, S.: Challenges and opportunities of Big Data in health care: a systematic review. JMIR Med. Inf. **4** (2016)
4. Ratia, M., Myllärniemi, J., Helander, N.: Robotic process automation—creating value by digitalizing work in the private healthcare?. In: Proceedings of the 22nd International Academic Mindtrek Conference (2018)
5. Lacity, M., Willcocks, L.P.: Innovating in service: the role and management of automation. Dyn. Innov. Outsourcing, 269–325 (2018)

6. Seuwou, P. (2021)
7. Wiljer, D., Hakim, Z.: Developing an artificial intelligence–enabled health care practice: Rewiring health care professions for better care. J. Med. Imaging Radiation Sci. **50** (2019)
8. Bhatnagar, N.: Role of robotic process automation in pharmaceutical industries. Adv. Intell. Syst. Comput. 497–504 (2019)
9. Dubey, A.: Bioinformatics & Proteomics open access journal committed to create value for researchers showcasing the. Impact (2020)
10. Sarker, S., Jamal, L., Ahmed, S.F., Irtisam, N.: Robotics and artificial intelligence in healthcare during COVID-19 pandemic: a systematic review. Robot. Auton. Syst. **146**, 103902 (2021)
11. Jerry, J.: Use Of Robotic Process Automation (RPA) For Rapid Analysis And Interpretation Of Multidrug Resistant Organisms And COVID-19 Results | Antimicrobial Resistance And Infection Control; 10(SUPPL 1), 2021. | EMBASE. [Online]. Available: https://pesquisa.bvsalud.org/global-literature-on-novel-coronavirus-2019-ncov/resource/en/covidwho-1448352. Accessed 25 Jun 2022
12. Doğuç, Ö.: Robotic Process Automation (RPA) applications in covid-19. Contrib. Manag. Sci., 233–247 (2021)
13. Kedziora, D., Smolander, K.: Responding to healthcare emergency outbreak of COVID-19 pandemic with Robotic Process Automation (RPA). In: Proceedings of the Annual Hawaii International Conference on System Sciences (2022)
14. Kaur, J.: Getting started with cognitive robotic process automation, https://www.xenonstack.com/blog/cognitive-rpa
15. Editor: Remote patient monitoring systems: Components, types, vendors, and implementation steps, https://www.altexsoft.com/blog/remote-patient-monitoring-systems/
16. Wright, M.O., Perencevich, E.N., Novak, C., Hebden, J.N., Standiford, H.C., Harris, A.D.: Preliminary assessment of an automated surveillance system for infection control. Infect Control Hosp Epidemiol. **25**(4), 325–332 (2004). https://doi.org/10.1086/502400. PMID: 15108731
17. Sarker, I.H.: Machine learning: algorithms, real-world applications and research directions. SN Comput. Sci. **2**, 160 (2021). https://doi.org/10.1007/s42979-021-00592-x
18. Secinaro, S., Calandra, D., Secinaro, A., Muthurangu, V., Biancone, P.: The role of artificial intelligence in healthcare: a structured literature review. BMC Med. Inform. Decis. Mak. **10**(21), 125. https://doi.org/10.1186/s12911-021-01488-9
19. Burgos, N., Bottani, S., Faouzi, J., Thibeau-Sutre, E., Colliot, O.: Deep learning for brain disorders: from data processing to disease treatment. Brief. Bioinform. **22**(22), 1560–1576. https://doi.org/10.1093/bib/bbaa310
20. Hassabis, D., Kumaran, D., Summerfield, C., Botvinick, M.: Neuroscience-inspired artificial intelligence. Neuron **95**, 245–258 (2017)
21. Ulman, V., Svoboda, D., Nykter, M., Kozubek, M., Ruusuvuori, P.: Virtual Cell Imaging: a review on simulation methods employed in image cytometry. Cytometry A **89**, 1057–1072 (2016)
22. Floridi, L., Luetge, C., Pagallo, U.: Key ethical challenges in the European medical information framework. Mind. Mach. **29**, 355–371 (2019)
23. Mehrabi, N.: A Survey on Bias and Fairness in Machine Learning. [Online]. Available: https://arxiv.org/pdf/1908.09635.pdf. Accessed 25 Jun 2022
24. Andruszkiewicz, D.: RPA application in the Pharmaceutical Industry
25. Mahmoodzadeh, S., Moazenzadeh, M., Rashidinejad, H., Sheikhvatan, M.: Diagnostic performance of electrocardiography in the assessment of significant coronary artery disease and its anatomical size in comparison with coronary angiography. J. Res. Med. Sci. **16**(6), 750–5 (2011). PMID: 22091303; PMCID: PMC3214392
26. Ebrahimi, Z., Loni, M., Daneshtalab, M., Gharehbaghi, A.: A review on deep learning methods for ECG arrhythmia classification. Expert Syst. Appl. X **7**, 100033 (2020)
27. Kakar, P.N., et al.: Robotic invasion of operation theatre and associated anaesthetic issues: a review. Indian J. Anaesth. **55**(1), 18 (2011)

28. Barnoy, Y. et al.: Robotic Surgery with Lean Reinforcement Learning (2021). arXiv.org. Available at: https://arxiv.org/abs/2105.01006. Accessed 22 August 2022
29. Arulkumaran, K. et al.: A brief survey of Deep Reinforcement Learning (2017). arXiv.org. Available at: https://arxiv.org/abs/1708.05866. Accessed 22 August 2022
30. Swazinna, P. et al.: Comparing Model-Free and Model-Based Algorithms for Offline Reinforcement Learning (2022). arXiv.org. Available at: https://arxiv.org/abs/2201.05433v1. Accessed 22 August 2022
31. Shahrubudin, N., Lee, T.C., Ramlan, R.: An overview on 3D printing technology: technological, materials, and applications. Procedia Manuf. **35**, 1286–1296 (2019)
32. Savini, A., Savini, G.G.: A short history of 3D printing, a technological revolution just started. 2015 ICOHTEC/IEEE International History of High-Technologies and their Socio-Cultural Contexts Conference (HISTELCON) (2015)
33. Dodziuk, H.: Applications of 3D printing in Healthcare. Polish J. Cardio-Thoracic Surg. **3**, 283–293 (2016)
34. Banerjee, A. et al.: Artificial Intelligence in 3D printing: A revolution in health care. Lecture Notes in Bioengineering, pp. 57–79 (2021)
35. Doulgkeroglou, M.-N. et al.: Automation, monitoring, and standardization of cell product manufacturing. Fronti. Bioeng. Biotechnol., 8 (2020)
36. Peroglio, M., Gaspar, D., Zeugolis, D., Alini, M.: Relevance of bioreactors and whole tissue cultures for the translation of new therapies to humans. J. Orthop. Res. **36**, 10–21 (2018)
37. Costariol, E., Rotondi, M., Amini, A., Hewitt, C., Nienow, A., Heathman, T., et al.: Establishing the scalable manufacture of primary human T-cells in an automated stirred-tank bioreactor. Biotechnol. Bioeng. **116**, 2488–2502 (2019). https://doi.org/10.1002/bit.27088
38. de Sousa Pinto, D., Bandeiras, C., De Almeida Fuzeta, M., Rodrigues, C., Jung, S., Hashimura, Y. et al.: Scalable manufacturing of human mesenchymal stromal cells in the vertical-wheel bioreactor system: an experimental and economic approach. Biotechnol. J. (2019)
39. Hamad, S., Derichsweiler, D., Papadopoulos, S., Nguemo, F., Šarić, T., Sachinidis, A., et al.: Generation of human induced pluripotent stem cell-derived cardiomyocytes in 2D monolayer and scalable 3D suspension bioreactor cultures with reduced batch-to-batch variations. Theranostics **9**, 7222–7238 (2019). https://doi.org/10.7150/thno.32058
40. Sailem, H.Z., Rittscher, J., Pelkmans, L.: KCML: a machine-learning framework for inference of multi-scale gene functions from genetic perturbation screens. Mol. Syst. Biol. **16**, e9083 (2020)
41. Hosny, A., Parmar, C., Quackenbush, J., Schwartz, L.H., Aerts, H.: Journal artificial intelligence in radiology. Nat. Rev. Cancer **18**, 500–510 (2018)
42. McDonald, R.J., et al.: The effects of changes in utilization and technological advancements of cross-sectional imaging on radiologist workload. Acad. Radiol. **22**(9), 1191–1198 (2015). PMID: 26210525
43. Fitzgerald, R.: Error in radiology. Clinical Radiol. **56**, 938–946 (2001)
44. Willemink, M.J. et al.: Preparing medical imaging data for machine learning. Radiology. **295**(1), https://doi.org/10.1148/radiol.2020192224. Epub, pp 4–15
45. Krupinski, E.A.: Current perspectives in medical image perception. Atten. Percept Psychophys. **72**(5). https://doi.org/10.3758/APP.72.5.1205. PMID: 20601701; PMCID: PMC3881280, pp. 1205–17
46. Channin, D.S., Mongkolwat, P., Kleper, V., Rubin, D.L.: The annotation and Image Mark-Up project. Radiology **253**, 590–592 (2009)
47. Yuan, J., Liao, H., Luo, R., Luo, J.: Automatic radiology report generation based on multi-view image fusion and medical concept enrichment. In: Lecture Notes in Computer Science, pp. 721–729 (2019)
48. Paul, D., Sanap, G., Shenoy, S., Kalyane, D., Kalia, K., Tekade, R.K.: Artificial Intelligence in drug discovery and development. Drug Discov. Today **26**, 80–93 (2021)
49. Lavecchia, A.: Deep learning in drug discovery: Opportunities, challenges and future prospects. Drug Discov. Today **24**, 2017–2032 (2019)

50. Hurst, W.J., Mortimer, J.W., V.C.H. Publishers, Inc.: Beach deerfield, and USA FL, "US$24.95," J. Chemometrics **2**, 298–298 (1987)
51. Ward, K.B., Ann Perozzo, M., M.Zuk, W. : Automatic preparation of protein crystals using laboratory robotics and automated visual inspection. [Online]. Available: https://doi.org/10.1016/0022-0248(88)90328-4. Accessed 25 Jun 2022
52. Wu, F., Yang, R., Zhang, C., Zhang, L.: A deep learning framework combined with word embedding to identify DNA replication origins. Scientific Reports, vol. 11 (2021)
53. Lawrence, S., Giles, C.L., Tsoi, A.C., Back, A.: D.: Face recognition: a convolutional neural-network approach. IEEE Trans. Neural Netw. **8**, 98–113 (1997)
54. Sherstinsky, A.: Fundamentals of Recurrent Neural Network (RNN) and long short-term memory (LSTM) network
55. Park, S. et al. C.: [PDF] Vision Transformer for Covid-19 CXR diagnosis using chest X-ray feature corpus: Semantic scholar
56. Tucker, A., Wang, Z., Rotalinti, Y., Myles, P.: Generating high-fidelity synthetic patient data for assessing machine learning healthcare software. npj Digital Medicine (2020)
57. DuMont Schütte, A.P. et al.: Overcoming barriers to data sharing with medical image generation: a comprehensive evaluation. npj Digital Med. (2021)
58. Bank, D., Koenigstein, N., Giryes, R.: Autoencoders
59. Torfi, A., Fox, E.A., Reddy, C.K.: Differentially private synthetic medical data generation using convolutional Gans. Inf. Sci. **586**, 485–500 (2022)
60. Kingma, D.P., Welling, M.: Auto-encoding Variational Bayes, ArXiv preprint arXiv: 1312.6114 (2013)
61. Feldman, K., Hazekamp, N., Chawla, N.V.: Mining the clinical narrative: All text are not equal. In: 2016 IEEE International Conference on Healthcare Informatics (ICHI) (2016)
62. Brown, T.B. et al., Language models are few-shot learners
63. Lehmann, L.S.: Ethical challenges of integrating AI into healthcare. artificial intelligence in medicine. In: Group of Young: 8 Ethical Issues in the use of Robots in Healthcare, pp. 1–5 (2021)
64. Weinreb, R.N., Aung, T., Medeiros, F.A.: The pathophysiology and treatment of glaucoma: a review. JAMA **14**(311)(18), 1901-11(2014).https://doi.org/10.1001/jama.2014.3192. PMID:24825645;PMCID:PMC4523637
65. Thainimit, S., Chaipayom, P., Sa-arnwong, N., Gansawat, D., Petchyim, S., Pongrujikorn, S.: Robotic process automation support in telemedicine: glaucoma screening usage case. Inf. Med. Unlocked **31**, 101001 (2022)
66. Sorin, V., Barash, Y., Konen, E., Klang, E.: Deep Learning for Natural Language Processing in radiology—fundamentals and a systematic review. J. Amer. College Radiol. **17**, 639–648 (2020)
67. Lam, P.Y., Chow, S.C., Lai, J.S., Choy, B.N.: A review on the use of telemedicine in glaucoma and possible roles in covid-19 Outbreak. Surv. Ophthalmol. **66**, 999–1008 (2021)
68. Mahase, E.: Covid-19: Who declares pandemic because of "alarming levels" of spread, severity, and inaction. BMJ. m1036 (2020)
69. Mofatteh, M.: Neurosurgery and artificial intelligence. AIMS Neurosci **2021**(8), 477–495 (2021). https://doi.org/10.3934/Neuroscience.2021025

Chapter 10
Importance of Artificial Intelligence (AI) and Robotic Process Automation (RPA) in the Banking Industry: A Study from an Indian Perspective

Sunita Kumar, Shivi Khanna, Nabanita Ghosh, and Shiv Onkar Deepak Kumar

Abstract The paper's primary focus is intelligence creation by AI and fast implementation by RPA. It starts with the applicability of the Turing test propounded by Alan Turing, the father of Artificial Intelligence (AI). At the backdrop lies various events, namely the recent motivation addition of various bank account holders. These factors fuelled the demand for AI and RPA implementation in the banking industry. It pitched how AI and RPA work in real-time scenarios such as financial fraud and money laundering. It discusses how AI builds the knowledge graph and recommends products and services for each customer. This knowledge is implemented and delivered using RPA. The AI application gained prominence in every banking business segment, such as equity, personal, investment and loan. The application of RPA is present in all business segments, although the percentage is increasing yearly. The AI and RPA can help banks to convert the challenges to opportunities. There have been various challenges, and the application of AI and RPA combinations is the key to solving the inefficiencies. Advanced analytical techniques on open-source data have been used in this paper.

S. Kumar (✉) · S. Khanna
School of Business and Management, Christ (Deemed to be University), Bangalore, India
e-mail: sunita.kumar@christuniversity.in

S. Khanna
e-mail: shivi.khanna@christuniversity.in

N. Ghosh
School of Commerce, Finance and Accountancy, Christ (Deemed to be University), Bangalore, India
e-mail: nabanita.ghosh@christuniversity.in

S. O. D. Kumar
Head of Data Science, Straive, Chennai, India
e-mail: shiv.onkar@straive.com

S. Bhattacharyya et al. (eds.), *Confluence of Artificial Intelligence and Robotic Process Automation*, Smart Innovation, Systems and Technologies 335,
https://doi.org/10.1007/978-981-19-8296-5_10

10.1 Introduction

The importance of AI can be understood from the words of Alan Turing, the father of AI. He proposed the Turing test in his research paper "Computing Machinery and Intelligence" in 1950. The analogy in the banking industry is that the customers would get a seamless service experience without the intervention of the service agent (computer or human) [1, 2] (Fig. 10.1).

The RPA was visible on the ground since early, 2000 and development started a decade earlier. It can be traced back to 1959, when Arthur Samuel invented ML (Machine Learning) [3].

The banking industry has always been a front runner in accommodating the latest technology to enhance customer serviceability and the faster accomplishment of day-to-day administrative work. The future of the banking industry would have reduced footfalls through the judicious application of AI and RPA.

10.1.1 Background of the Study

The study has been taken based on different factors (one of such significant factors is motivation) influencing the implementation of AI in the banking industry. Various websites, conferences and journals have identified the use cases. The corresponding solution has been generated from open-source data from public sites.

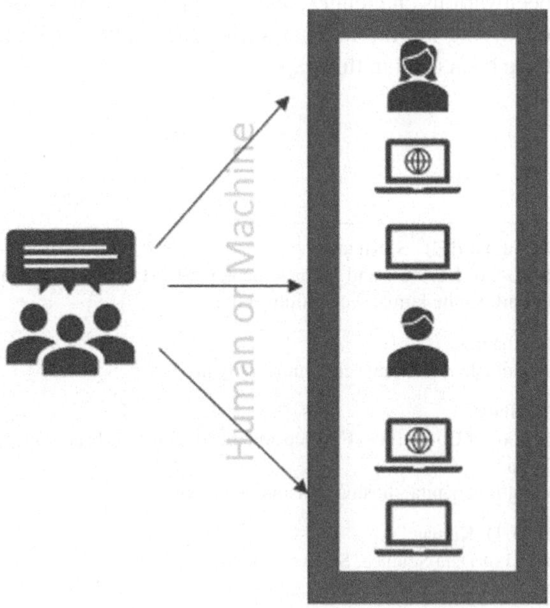

Fig. 10.1 Turing test to depict "Human or Machine"

10.1.2 Motivation

The motivation for implementing AI in the Banking industry: As per official data published by Reserve Banks of India (RBI) (Banks in India, 2021), below is the group of banks and the respective counts as on April 2021 [4, 5].

- Public-Sector banks (Majority stake owned by the government): 12
- Private banks: 21
- Regional Rural bank: 43
- Payments banks: 6
- State Co-Operative banks: 32
- Urban Co-Operative banks: 53.

The above banks provided individual accounts to India's population of more than 80% in 2018 (RBI 27-Jun-2019), and around 66% (opened between 2014–2021) were "Jan Dhan" (zero balance account opened for authentic citizens) accounts [4, 5].

The activities undertaken by the above banks are manifesting the following **challenges**:

- Volume: The count of the bank account holder has multiplied.
- Variety: The depositors are from all sorts of financial statuses.
- Diversification: Depositors are from all around the geographical corners.
- Non-internet access: Many "Jan Dhan" account holders still use a basic essential without internet access.
- Physical accessibility: Many account holders stay far from the branch and rarely visit branches.
- Operational costs: The costs are proportional to the number of customers.
- Risk: The risk is proportional to the number of customers.
- Customer experience: Each customer expects a privileged banking service even though they have a zero-balance account
- Regulatory compliance: These procedures are proportional to the number of customers.
- Fraud: The naiveness increases vulnerabilities, and customers (mostly with non-technical background) falls into the trap quickly.
- Loan and credit: The loan and credit facility are always disproportionate to the income generated.
- Investment: The depositors expect a high rate of investment.
- Ambiguity: Faced by the customers in interpreting the banking terms and connotations.
- Ethical constraint.

10.1.3 Various Ways to Handle the Challenges

The above challenges need intelligent systems to cater needs of the account holders. The system should have an intuitive understanding (without direct feed) of customers and their transactions. This allows banks to use existing or new financial products and services based on the transaction's patterns. The intelligent system predicts future trends during the customer journey, resulting in various recommendations like up-selling or cross-selling. The intelligent system enables banks to detect fraud and handle anti-money laundering [6].

10.1.4 Robotic Process Automation (RPA)

The concept of RPA consists of three meaningful words. The word "robotic" is an adjective for "robot", meaning an intelligent machine works like an innovative and robust human being. The processes convey a series of planned tasks. The automation does repetitive work many times with the same accuracy. The combination of these three words has been making much impact in all the enterprises targeting automation, with special mention of the banking industry. The effect of RPA gets accelerated with the presence of AI [3, 7–15].

10.1.5 Roadmap

The chapter's roadmap consists of 11 interconnected sections. The current Sect. 10.1 is about the introduction and briefly touches on various points. Section 10.2 highlights the methodology (high-level view of the solution's approach) used in this chapter. Section 10.3 peeps into the past and highlights various work done in these areas. There have been many types of research as well as practical implementations, and this section highlights a few of them. Section 10.4 describes the present situation and the current working scenario of AI and RPA in the banking industry. This section elaborated on two examples: equity and an individual's financial objective. Section 10.5 covered the various challenges in the introduction section and motivation section above. Till now, various sections have been forming platforms. Section 10.6 highlighted the real-time example of data of RBI and details how it helps in ATM's service management. Section 10.7 considered various use cases already happening in different banks. Section 10.8 showcased some use cases that correspond to those mentioned in Sect. 10.6. The use cases are Fraud Analytics, Anomaly detection, Stock market growth from new users' point of view, Customer 360: Understanding the Customer from all dimensions and Credit Card Customer Life Cycle. Section 10.9 analysed the future trends of AI and touched upon future's AI applications—Conversational AI, Personalised service with high accuracy, Data collection points, almost real-time Fraud detection, Ethical AI and big spending on AI. Section 10.10 questions whether AI and RPA are a panacea for all current problems. And finally, Sect. 10.11 concluded and suggested based on the study's findings.

10.2 Methodology

The paper has taken into consideration the following scientific steps. Firstly, understanding the real problem from multiple authentic sources. Then attempt to visualise the problem using the data from open source and publicly available sources. We have then focused on the multidimensional view of a given problem. Thereby Searching for the right sets of data to solve the problem. Thus, check the graphical visualisation of the problem by using the Exploratory data analysis. Then the research is done using the latest Machine learning techniques. We have then shared the outcome here for public benefit.

The secondary data-based analysis is being conducted simplified to enable the researchers to understand the application of AI and RPA in the banking industry.

10.3 Literature Review

Few questions are asked consistently in many seminars, and respective solutions are discussed. Is Fraud prevention important? How does the non-prevention of fraud impact organisations? What is the opportunity cost for non-prevention of Fraud?

The literature given below attempted to answer the questions mentioned above.

Pruitt, a certified fraud examiner from the Association of Certified Fraud Examiners (ACFE), requested the audience's opinion on how AI can help their organisation. Most of them (73%) mentioned fraud detection using the latest technology, say AI [16].

Fraud is being committed in almost all industries with different percentages and attributes. It is one of the misdeeds existing for ages. And prevention activities are also being done for generations. Both kinds of people are incredibly tech-savvy, highly motivated and multi-dimensional through process capabilities.

Standis [17] observes in his paper that "There are well-defined processes for fraud preventions. It requires best practices (consistently) and the latest technology (say AI and analytics) to build powerful tools enable the business model to extract significant results and significant ROI in anti-fraud activities". It was summarised in this way—"This war cannot be won because whatever intelligence is built, are for the short term because counterintelligence gets developed soon. The tools for sophistication are available to both sides. It gets advanced and outdated in due course of time. Hence, most of the defence is only for a while" [17].

The Social Network Analysis (also known as Network Link Analysis) uses AI to find "Fraud rings" or close groups committing Fraud. Various authors—Asaro and Poisson, Aberdeen, Gill and Lormel mention that Fraud is often committed in a group and hence detecting groups is also part of fraud analytics. These people form a team with different skills so that they commit forgery, manipulate claims and manipulate or duplicate identities. It was also studied how these people form a group or come together. It was observed that most Fraud-related activities require various sets of activities. These people search similar minded people who can complement each other and finally form a big group [18–21].

Can we predict the Fraud?—this question is also studied by various researchers—Asaro and Poisson, Aberdeen, Gill and Lormel. Considering the nature of Fraud, it is impossible to predict in any way or simply. There are a few ways suggested—like studying the trends, trying to group similar and non-similar activities, various essential factors contributing to Fraud and finally, providing an objective score for each step. It is evolving activity and will continue in the coming years [18–21].

The RPA helps simplify the challenges (mentioned in the paragraph above) posed by a combination of Volume and Variety. The AI can formulate techniques, and the RPA implements those techniques at the speed of light and brings down the operational costs per transaction. The transaction success results in customer success and hence a positive customer experience. Regulatory compliance needs the generation of various reports, and the RPA helps in generating those reports at the right time. The real-time usage can be seen in the online processing of loans and credit, where the customer gets an instant indication of loan amount. Similarly, for credit and overdraft facilities online, in real-time. Investment banking is another area where speed matters, and the customer gets an instant suggestion for investment as soon as the customer enters various required parameters [8–15].

The RPA helps banks in optimising costs and boost productivity. This combination helps optimise skilled resources. As per research by McKinsey, the RPA will perform 10% to 25% of overall capacity so that freed resources can focus on high-value tasks [8].

The Banking industry is growing at an unprecedented rate supported by technological advancements. Next, prominent questions come for virtual mode, speed, security and reliability of services. To address this, the banks must find a balanced way. As per Gartner, the RPA is expected to be more than a billion dollars by 2025. In India, Axis Bank uses RPA to a great extent. It is observed that RPA can reduce processing costs by 30% to 70%. The banks are using RPA in the following areas:

- Customer Service
- Regulatory Compliance
- Credit Card Processing
- Fraud detection
- Electronic KYC (eKYC)
- Report Automation.

The above points are a few areas, and many more will be done by 2025 [9].

The banks have many strategic business units (SBU). The implementation and impacts of RPA also depend on SBU functions. As per McKinsey, accounting SBU can have the biggest automation and can be automated up to 56%. Please see the link

[10] for various other SBUs' automation potential. As per Gartner, 80% of banks are already using RPA for various SBU [10].

A few examples areas are listed as follows:

- Allows to scale operations seamlessly.
- Saves time from 20 to 90%.
- Cuts down expenses by 30%, on average.
- Minimises IT interference.
- Ensures there are no extra infrastructure costs.
- Increases human efficiency by reducing human error.

One solution does not work for all SBU; hence planning and implementation have to be done differently for all SBU [10].

The RPA is becoming Intelligent RPA day by day. The new synonym of RPA is Intelligent Automation (IA). The IA learns by AI and gets intelligent day by day [12].

How IA is working in banks can be understood by the following points:

- IA enables banks to automate complex end-to-end processes.
- No New IT Infrastructure.
- Save Time and Money.
- Improve Processes.
- Augment Human Workers with a Digital Workforce.
- Better Regulatory Compliance.

Above are a few areas where IA and AI have good tango [12].

The article [13] mainly highlights the pace of growth for RPA. It mentions Gartner's prediction about the market size to reach $2.4 billion by the year 2022. It also points report from Fortune Business Insights, in which the prediction is to reach $6.81 billion by the end of 2026 [13].

It [13] suggests that RPA implementations are prominent in the following areas:

- Contact Centre Optimisation
- Trade Finance Operations
- Customer Onboarding
- Anti-money Laundering (AML)
- Bank Guarantee Closures
- Bank Reconciliation Process
- Loan Application Processing
- Automated Report Generation.

The article [13] details many areas for RPA growth.

The article [14] tries to answer various questions—Can the errors be stopped at the source instead of passing to the next downstream application? How to deploy digital solutions quickly? How respond to new demand quickly?

The article [14] presents a few statistics. It mentions that 75% of transactions, 40% of reporting, 60% of financial services can be automated [14].

The implementation of RPA differs from industry to industry. The banking and insurance implements 51% of RPA's solutions. The percentage implementation of other industries can be seen at the reference link [15].

10.4 How Do AI and RPA Work in the Banking Industry?

When we think about banks, then the concept of financial transactions comes into the picture. Either we deposit the money or withdraw the money or take financial products and services. The financial products may range from various types of loans to credit cards. When the users are associated with one or more products and services, then AI builds the knowledge graph (shown in Fig. 10.2, created after analysing the open-source fraud data set mentioned from [22–35] in the appendix) and recommends other products and services for each customer.

These aspects are also known as up-sell or cross-sell. There is a complex process to identify the right products (considering income suitability while minimising risks), and AI and RPA become handy. The AI builds knowledge, and the RPA does all complex processing and provides a final recommendation to the salesperson or customer relationship manager (CRM). The customer engagement with bank becomes more substantial, and the customer gets most of the banking needs under one umbrella. Examples are outlined below [36].

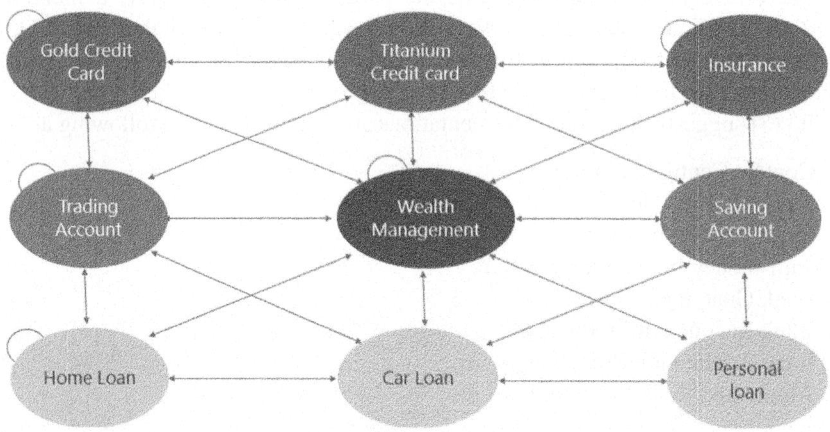

Fig. 10.2 The banking industry's knowledge graph

10.4.1 *Example: How AI and RPA Are Working in the Equity Segment of Banking Industry?*

The equity segment works on analysis of data from all possible dimensions. Besides the company's performance, a lot of macro-economic data also impacts the values of equity. Few mutual funds (IDFC in India, ETFMG in USA, Horizons ETF in Canada) have implemented the AI to manage the portfolio [37–39]. The following Fig. 10.3, created by author, tries to depict the work being done.

Example: How AI is helping in achieving the financial objective of individual

The above figure shows that there are multiple inputs before AI builds knowledge. The RPA is enabler and AI leverages those real-time data and runs on the ground.

An individual investor used to make the investment independently or take advices from a financial advisor. The main objective is to optimise the investment with minimum risks. He tries to achieve financial goals through portfolio diversification, depicted in the following Fig. 10.4, created after analysing open-source fraud data set mentioned from [22–35] in appendix.

The new advisor is also known as "Robo advisors" (equipped with complex algorithms). It is gaining immense popularity because it allows investors to build investment portfolios (primarily diversified) at a meagre cost. The Robo advisors shares the cost-free timely information in a legible and simplified manner as against the human advisor. The knowledge comes by AI and speed of processing is enabled by RPA.

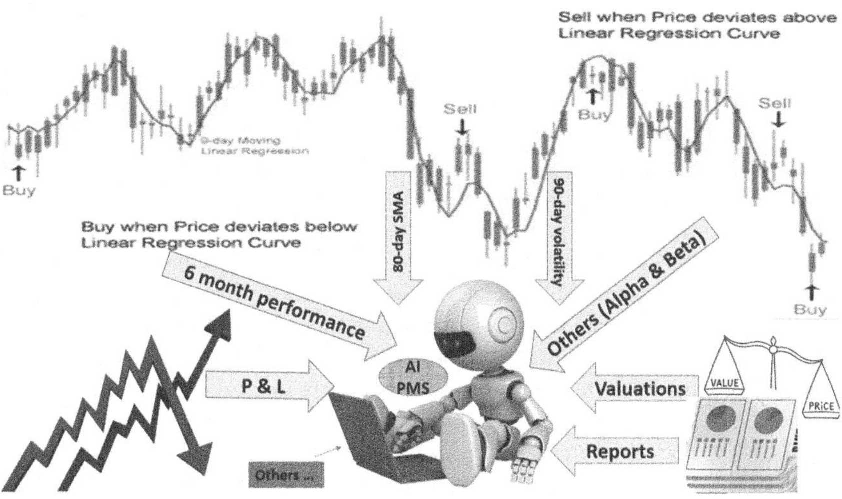

Fig. 10.3 The AI system can analyse a huge variety of information quickly

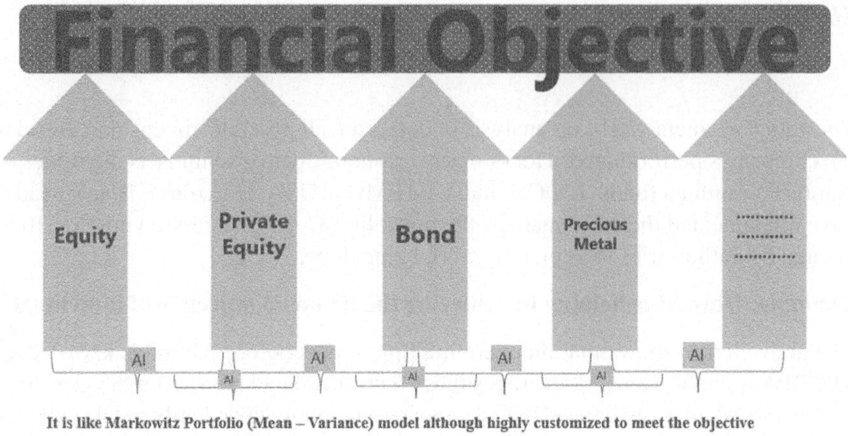

Fig. 10.4 The financial objective through diversification of portfolio

10.5 How AI and RPA Can Help in the Conversion of Challenges to Opportunity

The introduction sections outline various challenges existing in the banking sector. Let us explore how AI is making the inroad smooth for those challenges.

10.5.1 Volume

The customer count goes in million, and they have inherent groups based on certain specific heterogeneous aspects. To optimise product or service offers, the banks should have a meaningful and effective distinct group of customers based on criteria—demographic information, the transaction for the last few quarters, income declarations, various other products or services associated with the same bank or other banks, online or offline interactions to enhance cohesion in the segmentation process. The banks need AI to define better customer segments by analysing the volume and variety of data they have about their customers. When customers have been segmented into logical groups (making sense to business) based on transaction behaviour or their needs, in terms of banking products and services then it is feasible to conduct sales, promotion, and marketing campaigns accordingly. In the AI world, this process is also known as Customer segmentation [40].

The RPA has been a game changer when the volume is very high because of the high Return on Investment (ROI). The AI does as Customer segmentation and segregates into coherent groups, and RPA process all these groups in parallel. Sometimes, the different strategy requires for each group. AI (strategy) and RPA (customised processing) work together in these scenarios [12].

10.5.2 Variety

The bank has almost all kinds of customers. Their requirement is different. That is why banks have different saving accounts, credit cards, loans, and other customised products. These are also known as personalised products by banks.

10.5.3 Diversification

Presently more than two-thirds of the population is covered by banking services. Depositors who are connected to the banks are ranging from different geographical boundaries. The banks face a hard challenge to meet the varied expectations of the customers.

10.5.4 Non-internet Access

Several "Jan Dhan" account holders are transacting on a date without access to the internet.

10.5.5 Physical Accessibility

There are numerous account holders staying residing away from branches and rarely visiting them.

10.5.6 Operational Costs

The reduction in operating expenses and risk has always been on the bank's improvement plan. Most banking operations are primarily digital and hardly a few are human-intervened processes. These are, by and large, repetitive steps. RPA plays a major role in automating these recurring processes.

10.5.7 Risk

Any business transaction has an inherent risk in it. Moreover, the risk must be managed or priced by accepting, mitigating, insuring, or postponing till the occurrence. The estimation of risk requires complex processing. The data with all possible situations (many of the situations may not have occurred) must be analyzed. The AI system uses the well-known "Monte Carlo simulations" (MCMC) technique to get a high-level estimate of risk exposure. This processing requires vast amounts of data and high computational power.

The main points of above are "complex processing", "huge data", "MCMC" and "high computational power". All these are motivator for RPA and build huge ROI.

10.5.8 Customer Experience

The customer experience stands to be a critical area that demands tactical management. Customers often expect certain products and services, but occasionally banks go with those products and services. Transactions are executed when the expectations match. Below provided some of the scenario analyses of customer experience.

What the customer is saying. In today's scenario, there are many channels through which customers talk about their experience with the bank's products and convey their satisfaction/dissatisfaction. The banks extract these comments from social media, customer care emails, and product review sites. The analysis helps the banks to make faster responses to the comments or redress the grievances if any.

Who is my customer: The Bank need to know their customer all-pervasively, be it- through social media profiles, customer care emails, phone calls to call centres, complaints and discussion blogs. It helps to build customer profiles and map the right products and services. Once the history of logs is created, then consistent optimization gets momentum.

Customer's needs and wants: The customer may come to the bank for their needs. Once the bank understands the customer profile, the bank can proactively go to the customer for their need and wants. It helps in customer retention. It also reduces churn. Nowadays, AI helps to predict what customers require in the near future. It also helps in building customer trust.

The RPA's part in the section "Literature review" explains multiple times that the RPA has played a significant role in customer care experience. It has enhanced customer experience by providing all necessary information to the customer executive before the customer care executive starts talking to the customer.

10.5.9 24 × 7 Banking

The banking hours are limited in most parts of the world, although the banking need (aka financial need) is 24 × 7. Many of us experienced long waiting queues at call centres. Even after such delayed waiting times, the customer care executive often could not resolve the customers' issues and advised to visit the branch in the future for the accomplishment of the task. The chatbots (conversational systems also one of the important applications of RPA) are available 24 × 7 and help to answer the questions based on the system's maturity. Its usefulness was been witnessed during COVID-19 lockdowns. The internet and mobile banking have also proven their usefulness for a long time.

10.5.10 Regulatory Compliance

The need for compliance reporting has been increasing almost every quarter. It has reached a state where manual reporting is not possible. The bank needs to record the various transaction and provide the report within the stipulated time. In this situation, the RPA helps. A suspicious activity report (SAR) needs to be filed if any suspicious activities are found, and the RPA helps to prepare such SAR. The Central Bank (RBI in India, Fed in the USA) closely watches the SAR. It also allows banks to avoid high risk and hence can avoid large-scale defaults. It also prevents banking customers from participating in financial crimes. In these scenarios, the AI and RPA help the bank to comply with regulations and at the same time uphold customer privacy. The AI and RPA combination help to prevent financial crimes (money laundering, and so on) [8–15]. The violation has high-cost factors and can also damage the brand. It also impacts business. The two recent times violations by Yes bank (a private bank in India) (business-standard, 25-Jun-2020) and PNB (a govt bank in India) (business-standard, 29-Jan-2018) have been in the news for financial irregularities.

The RPA has automated many reporting systems. As ordinary people, we have seen reports from banking systems (whether saving accounts, loan accounts, or credit cards…) come at the right time and frequency. The government authorities need various compliance reports (SAR, high-value transactions, …), and all these reports are automatically sent to their respective authority at the right time and frequency.

10.5.11 Fraud

The banking systems are still evolving and will continue as all technologies became more sophisticated day by day. The AI helps to prevent fraud by segregating "normal" behaviour from "non-normal" behaviour by analysing histories of consumer behaviour. There are various scenarios of fraud, and a few are mentioned here. The naiveness of users (predominantly non-tech-savvy bank customers) falls prey to fraudsters by sharing KYC/OTP details. The AI helps to detect an anomaly, in real-time, as per the user's past behaviour or similar users. When a user is a giant tech-savvy, they try to take advantage of technical gaps in the system, primarily rule-based. In these scenarios, the AI analyse the current transaction against the group's transaction histories and decides whether any anomaly sign is present. Based on the discussion (during consultancy areas) with bankers, machines are indispensable in fraud detection. The machine can analyse a massive amount of data and can apply different algorithms in almost real time [41, 42].

Fraud Detection is a hydra head problem and needs a multi-dimensional solution. Alone, AI or alone RPA or alone any other technology may not be sufficient. The introduction of digital systems has made banks efficient and count fraud-related concerns. Humanly, tracking all the transactions and scoring them for possible fraud is impossible. The RPA can help monitor the transactions in 24×7 mode. As soon as anomalous behaviours are suspected, the RPA raises the flag. It helps identify possible fraud patterns in real time (or a few minutes' delay). Where suspicion (based on score) is high, then the RPA blocks transactions.

10.5.12 Loan and Credit

The loans have been the primary source of income for banks and the foremost reason for NPA. It is a double edge sword and needs to play very cautiously. The world has seen many financial crises, and non-payment of loans has been the root cause of most financial crises. Banks now determine a person's credit risk based on several reliable (most derived from other factors) factors. The AI uses almost all information within the bank and communication with a third party (maybe a regulator or social media data). It provides a holistic view and helps in determining whether a loan can be provided or not. Nowadays, most banks have a risk score for each customer, and this score gets updated monthly or after every significant transaction. The bank has been analysing (with the help of AI) most of the parameters impacting creditworthiness to create a final score to determine creditworthiness. These parameters are quality banking transactions, credit scores by various authorities, credit history, and customer references. If the above points are unavailable with respect to a few customers, AI analyses behaviours and patterns to derive temporary credit scores.

The RPA's contribution to instant loans and overdraft are visible to an ordinary person. There have been many mobile applications which provide instant loans. The RPA helps get the loan amount to the bank account instead of a visiting branch.

10.5.13 Investment

Customers, in general, are having expectations of high rate of return from deposits. The financial instruments (fixed deposits, shares, precious metals, currency, real estate) are changing in rapid space. There is a different return rate for each alone and a combination of these assets. Portfolio planning has been a cumbersome process and needs a dedicated portfolio manager. In these scenarios, the AI systems are helping to build a customised portfolio based on risk and return expectations.

10.5.14 Ethic

The data security guidelines (PDPA, 2018) [43] mandate that data should be used for the purpose mentioned during the collection, which means they cannot be used for any other purpose. If someone is providing data to open a bank account, it should not be used for other purposes. It will continue to be an open question till comes a matured solution.

10.6 How RPA Can Expedite Repetitive Works

The RPA has been synonym for quick turnaround, and a combination of AI and RPA is meeting the expectations. The AI makes predictions in real time, and the RPA transforms those predictions into better decisions. It means machines are sourcing data, processing various information and making intelligent decisions on their own in real time [8–15]. Let us understand the value of the combination of AI and RPA using the following graph trend. The following Fig. 10.5 has been created from publicly available data from sites mentioned in [4, 5, 44] in appendix, hence permission to publish is not required.

The above graph shows that 600 million ATM transactions per month are happening in India. It raises the following questions (known as use cases) [44].

- How (frequency, amount, denomination) refill ATM so that ATM does not run dry?
- What should be the location of the new ATM installation?

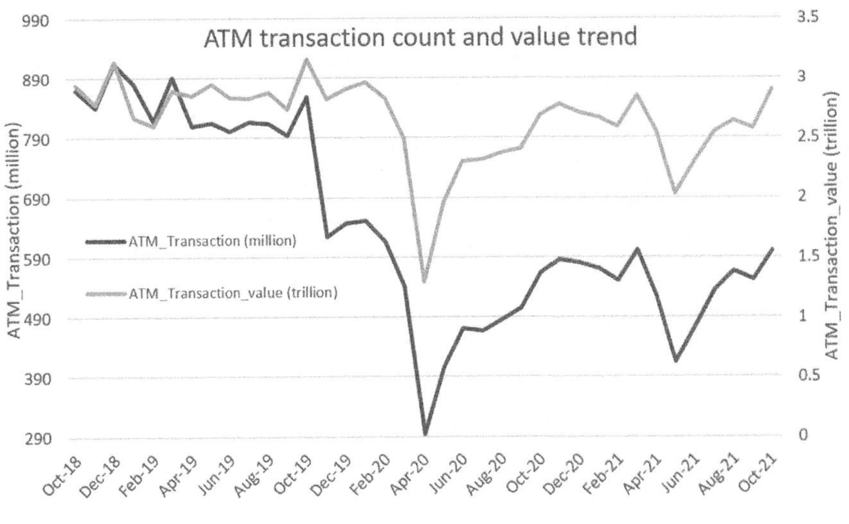

Fig. 10.5 ATM transaction count and value trend

- Which all ATM will go out service and when?
- What non-financial transactions are happening at the ATM site?

The answer to each above question (known as use cases, too) comes under the purview of AI. The project runs in the month to answer the above questions. The AI takes care of the solution above, and then the RPA comes into the picture to implement the above solution in the day-to-day scenario [8–15].

In summary, AI brings the knowledge and the RPA [8–15] brings smooth automation.

10.7 What Are the Use Cases Getting Implemented in Banks?

Stakeholders will always ask the usages to plan and communicate ROI (return on investment) to the investor. Here are a few use cases being implemented [45] in various banks (Table 10.1).

10.8 Details of a Few Use Cases Implemented in the Banking Industries

10.8.1 Fraud Analytics

The willful act of committing an illegal activity is fraud. The fraud in the context of banking sector is an act of procuring the money from the banks and financial institutions by some illegal means. Fraud analytics aim to detect such improper transactions by a combined use of analytic technology and human-intervened intervention (Table 10.2).

In Fig. 10.8, the prediction of "fraud" when actual is "Normal" is highly unacceptable. Hence this error is also known as type 1 error, should be avoided. The prediction of "Normal" when actual is "fraud" is again unacceptable because fraudsters are escaping. Hence this error is known as type 2 error, should be avoided. There lies a question as how to resolve the issue.

Fraud detection is the art of managing various balances. The AI uses different analytical techniques [52–55] to identify potential fraudsters. A few of those techniques are.

- Trend identification by unsupervised machine learning Association Analysis and Clustering [56].
- Deep learning techniques to identify various unknown patterns and complex feature engineering [57].

Table 10.1 The AI and RPA Use cases being implemented in various Banks

S. N	AI and RPA use case	Description
1	Customer segmentation for optimized offers 	Naturally, customers have various segments (groups), and the requirements for each group are different from others. That is why distinct categories or segments are necessary for relevant customer services. The segmentations start with basic demographic information, income, and other available information. The segments keep on changing or strengthening based on customer transaction behaviours. These segments help the banks to decide on appropriate banking products and services. These also help plan their sales, promotion, and marketing campaigns accordingly
2	Risk analytics	Risk is an inherent part of a human's life. It occurs in day-to-day activity in personal or professional life. Hence, being prepared is one of the ways to fight the risk. The bank leverage risk analytics in identifying the risk where transactions happen in a million per minute. It further helps to predict the occurrence of risks in advance to start preparing for the same. The outcome for this activity is a risk score for all entities within the bank (customer, transactions) [46, 47]
4	Fraud detection 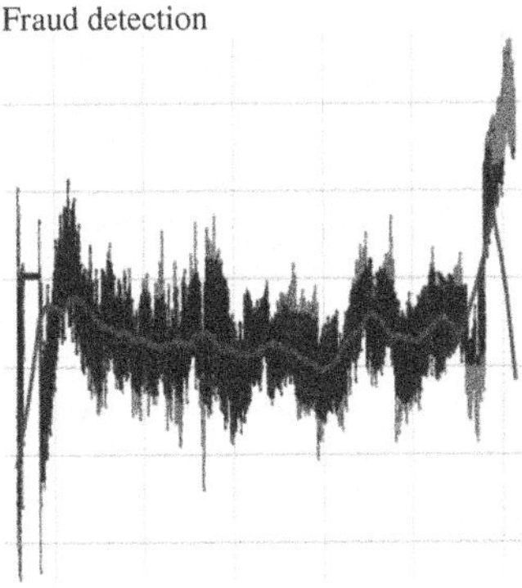	The banks have been traditionally known for white-collar activities and are accountable for every rupee. Historically, bank's businesses are built on trust. Fraud has been the primary cause of brand loss in the banking Industry; hence fraud identification and prevention has been part of all banking activity. The detection and prevention of suspicious and fraudulent activities have been part of most banks. It started with rule-based and then moved to real-time fraud analytics, happening nowadays. Fraud analytics enables banks to segregate normal behaviour from fraudulent behaviour. As the system matures and more data is ingested, more precise models are built to detect accurately [48, 49]

(continued)

- Social network analysis (SNA) to identify the relationships among various entities [58].
- Min–Max boundary conditions [59].
- Ratio technique [59].
- Finally, fraud Score prediction [60].
- It is strengthened by out of box techniques—Benford distributions [61].

Table 10.1 (continued)

S. N	AI and RPA use case	Description
5	Compliance reporting	All banks need to comply with the regulatory authority. They need to document all the information related to each prescribed trade and give regulators access to it within a stipulated period. To achieve this, banks implement a system to monitor transactions, collect all the necessary data and report to the compliance authority. The report is known as SAR (suspicious activity report)
6	Credit risk assessment	The memory of the subprime mortgage crisis, happened during 2008, is still used as a lesson learnt for most loan-related mishaps in financial institutions. Since then, the creditworthiness of individuals has been the primary criteria for granting the loan. The credit scores are being calculated externally (CIBIL in India) and internally (by various parameters to be detailed later) [50]
7	Personalised product offerings	The segmentation, as explained above, provides cohesive groups with similar financial behaviours. The marketing used to happen based on it, although lead conversion has been pathetic through this route. To pinpoint what my customer wants and their needs are, the planning for each customer need to be done. The banks (especially retail units) have almost all financial touchpoints customer data. These have been gathered from various customer transactions data, the internet, social media, and various monitoring agency. These have been churned out and provided more insights to customise new products and services for each customer. These insights help the bank to play a helpful partner in the customer journey and increase lifetime value [51]
8	Customer sentiment analysis	Human being is known for their diversified attitude and emotion most of the time. It is often impacted by the surroundings and the jobs they do. The same is valid for the banking experience as well. Every time the customer interacts or uses any banking channel, there gets created impressions (or experience either good or bad). Since there is a considerable number of customers and many banking channel experience them and hence analysis of sentiment is enabled by AI to bring real-time experience. It helps to resolve customer issues before the issue is blown up and paves the way for customer success

(continued)

Table 10.1 (continued)

S. N	AI and RPA use case	Description
9	360-Degree customer service	Most big banks provide almost all financial products ranging from a savings account, insurance, loans, and investments. To understand the customer, from both banks' and customer expectation perspectives, the touchpoints data need to be present at one location to provide a holistic view. It is the concept behind 360-degree customer service
10	Contact centre efficiency and problem resolution	There is a group 'Software reliability engineering (also known as SRE)' in most of the banks. They predict the issue before the issue gets bigger and resolve it immediately. If issue resolution takes time, information has to go to the customer support team to enable them aware of ongoing problems. It helps to maintain the high efficiency of the customer support team and smiling face to both banks and customers
11	Optimise offers and cross-sell 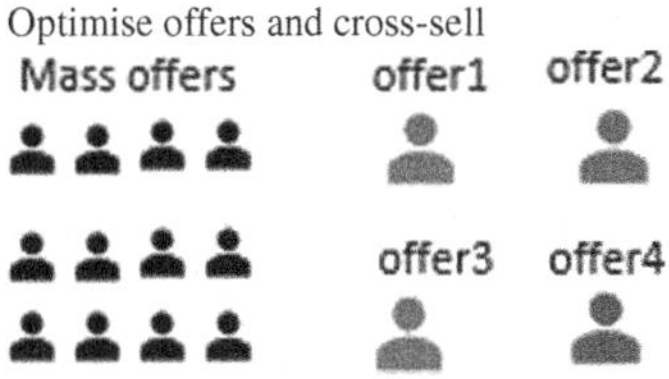	All of us are different in terms of banking needs at other times. When customer count is in a million, the combination becomes infinite when need and time dimensions are added. The AI helps to build offers for each customer from time to time based on their transaction behaviours
12	Customer churn analysis	When the customer goes away or does no transaction for a long time, the customer conveys the reason behind that. The AI helps to get various probable causes of the customer going away or maintaining a state of inertia. The prediction helps to keep the customer engaged by resolving their pain points. The AI also helps in mining the causes of churn so that existing customers may not face similar issues
13	Customer experience analytics	As customers start using more services from banks then his or her touch points (saving account, insurance, mutual fund, investments) increase. This generates enormous data and gets fitted in the AI engine, which creates customer experience analytics. It helps the bank with actionable insights for better customer understanding and offerings

Table 10.2 Confusion matrix for fraud detection

Actual	Predicted	
	Fraud	Normal
Fraud	True positive	False-negative (2)
Normal	False-positive (1)	True negative

Nowadays, the latest technologies and infrastructures are readily available. Fraudsters use these techniques to commit fraud at read time in a very short period. To detect and mitigate similar (or better) technologies, infrastructures are required. Besides these are a lot of other challenges which are as follows:

- Silos (with high ceiling) business units
- Data Volume
- Poor data quality
- Ever-changing data
- Limited or partial view
- Manual analysis of rule-based systems
- High False positives.

The solution has been multi-dimensional (a combination of many solutions) because of the hydra-headed nature of the problem. Here is a high-level architecture (Fig. 10.6), created after analysing open-source fraud data set mentioned from [22–35] in appendix.

Traditional (rule-based or gut feeling) approach. Consciously or unconsciously, most of the fraud-finding approaches start in an ad-hoc way based on the experience of various people responsible for preventing frauds.

Analysis of past trends or patterns or behaviours. The unsupervised techniques (for example—Association Analysis) helps in mining patterns and here comes the predictive analytics to predict the fraud scores.

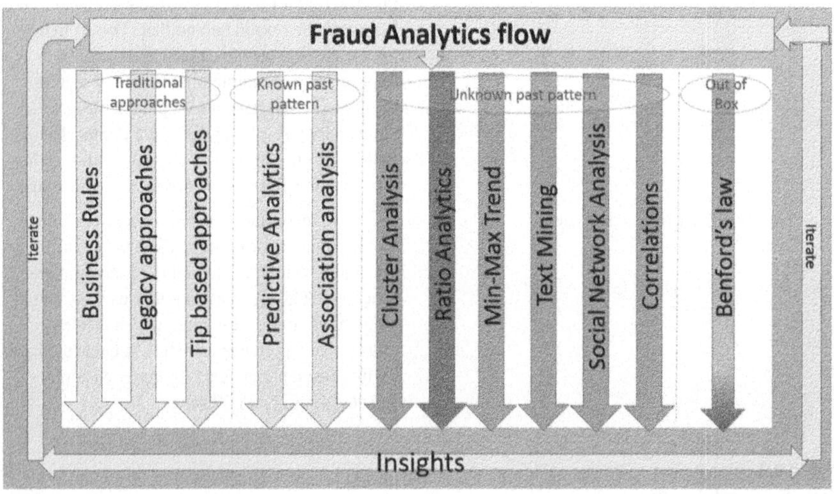

Fig. 10.6 Workflow of Fraud detection methodology

Unknown trends or patterns or behaviours. When we do not know what all information and insights are present in the data, then we leverage unsupervised patterns detailed below:

- Clustering
- Ratio Analysis
- Min–Max Analysis
- Text Mining
- Social Network Analysis
- Correlations.

Out of Box. Fraud is the art of manipulation of numbers. And, hence natural distribution and manipulated distribution will differ. The Benford law, also known as the first digit law (invented by physicist Frank Benford in 1938), helps to identify non-natural distribution [61].

As mentioned earlier, the high-level view of the techniques is that fraud happens in groups supporting or felicitating fraudulent behaviours. They indulge in forgery, identity manipulation and counterfeiting checks or currencies.

Network Link Analysis. Based on various activities (banking transactions, buy-sell of real states, buy-sell of golds), the link is established, and close communities are identified (Fig. 10.7) happens [62], created after analysing open-source fraud data set mentioned from [22–35] in appendix. These graphs are from huge data source and just for readability, few samples have been shown here. For analysis, the graph's trend matters and not the clarity of each point over here.

Association Analysis. It refers to the activities which are associated. Here, trends of activities going together or one after the other get detected.

Outlier and Anomaly Detection. Refers to activities which are deviated from normal behaviour and trends.

Clustering. Refers to the formation of the group and identifies the data which are non-similar with group (Fig. 10.8). They are created after analysing open-source fraud data set mentioned from [22–35] in appendix. These graphs are from huge data source and just for readability, few samples have been shown here. For analysis, the graph's trend matters and not the clarity of each point over here.

Predict the Fraud by Combining All Above

Ratio Approach. The absolute value cannot be compared across multiple years or domains or from different situations. That is why the ratio, with base reference, is analysed for identifying outliers and anomalies. Once the probable links are placed with fraudsters, they become the first potential fraudsters. A few more iterations are done, and the intersection of the various approaches indicates the first set of potential fraudsters. The following Fig. 10.9 is created after analysing open-source fraud data set mentioned from [22–35] in appendix. These graphs are from huge data source.

Like Ratio Analysis, the same steps were followed in Cluster Dimension and Min–Max dimensions.

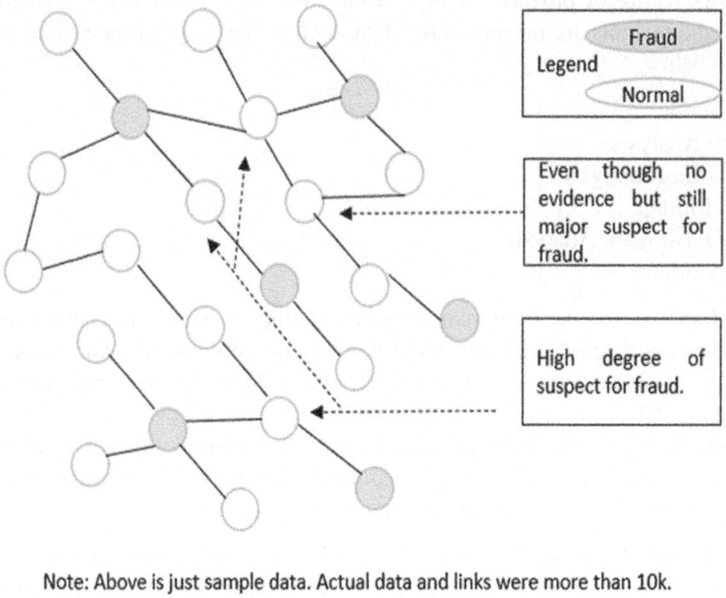

Fig. 10.7 How Fraud data are connected

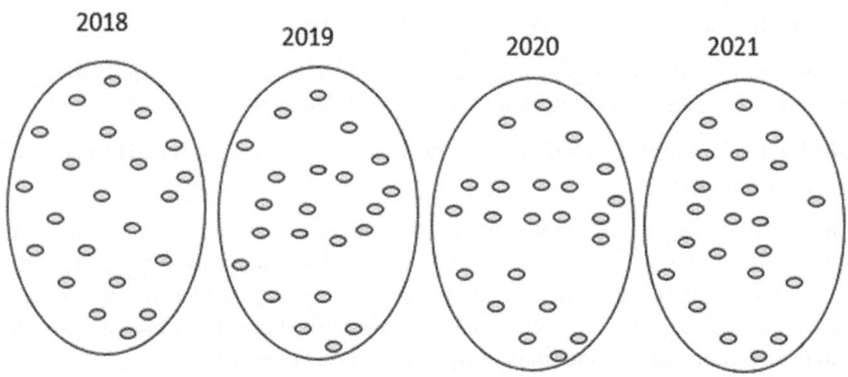

Fig. 10.8 How various Features are forming a group

The result of fraud analytics from all three dimensions was put together in the following figure. After taking a union of all three methods, unique probable fraudsters were identified. Link analysis further strengthened the findings. The following Fig. 10.10 is created after analysing open-source fraud data set mentioned from [22–35] in appendix. These graphs are from huge data source.

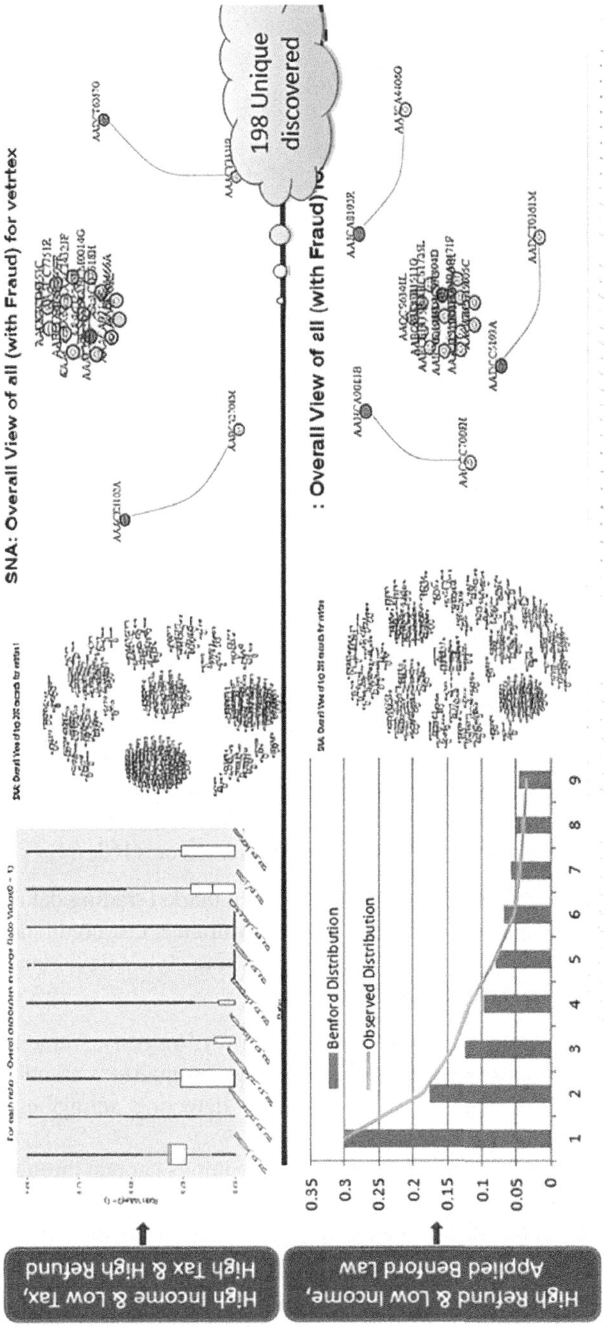

Fig. 10.9 From ratio dimension

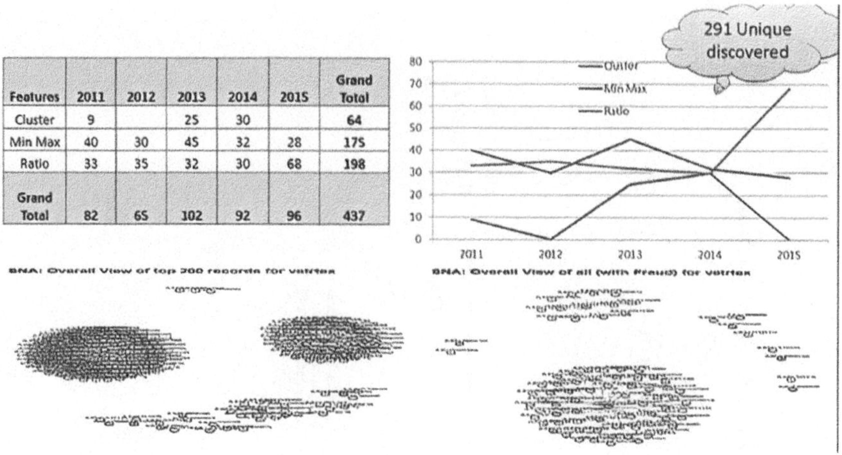

Features	2011	2012	2013	2014	2015	Grand Total
Cluster	9		25	30		64
Min Max	40	30	45	32	28	175
Ratio	33	35	32	30	68	198
Grand Total	82	65	102	92	96	437

Fig. 10.10 Combining all three dimensions

10.8.2 Anomaly Detection

A hypothetical situation is imagined where everything is normal. In these scenarios, nothing is suspicious and therefore no need for fraud detection arises. But the reality bites. Therefore, we find an anomaly—a behaviour different from usual in any direction.

Nowadays, data are often streaming, not static. Hence time aspect of anomaly is essential. Besides the detection of anomaly, the critical part is time too. The detection should happen as early as possible and in real time. The time aspect of detecting anomalies as early as possible or in real time can be precious. Here are a few scenarios as per the discussions made in various papers (numenta.com) [63, 64].

- Detect significant deviation or anomalies in stock market trading data [65].
- Fraud in various businesses such as in case of insurance, tax, credit card).
- Industrial processes or Machine (like ATM) working style's deviation.
- Insurance claims fraud detection [66, 67].

In today's world, the four attributes (Volume, Variety, Velocity, Veracity) of data are growing in all dimensions and hence detecting anomalies methodologies are exploding along with it. That is why accurately detecting anomalies has been challenging for a long time [68–70].

There is no magic in the analytics world, and top things happen through data and its quality. Similar opinion is there for various algorithms as well. Any one technique is insufficient to detect the fraud. The fraud itself is more complex and it is difficult also for simple algorithm to detect. The following high-level architecture documents show that many algorithms are used in fraud detection and many iterations take place before reaching the final score. As more and more data (both in terms of rows and columns) are ingested, the solution becomes more mature and trustworthy [71–74].

Figure 10.11 is created after analysing the open-source fraud data set mentioned from [22–35] in appendix.

Unsupervised Anomaly score identification

Time Series Analysis. It takes care of univariate anomalies concerning time. The data are streaming, non-stationary and with a clear trend. The data consist of high seasonality as well as low seasonality. Hence a combination of 3 algorithms (Anomaly detection, TS Outliers and Trend Analysis) was used to predict anomalies in the series. One of the detected anomalies was shown in the following Fig. 10.12. The following figure is created after analysing open-source fraud data set mentioned from [22–35] in appendix. These graphs are from huge data source and just for readability, few samples have been shown here. For analysis, the graph's trend matters and not the clarity of each point over here.

Clustering. To understand the logical grouping of data, anomaly can be detected at the group level, the clustering was done, and logical clusters (groups) were identified, as shown in the following Fig. 10.13. The following figure is created after analysing open-source fraud data set mentioned from [22–35] in appendix. These graphs are from huge data source and just for readability, few samples have been shown here. For analysis, the graph's trend matters and not the clarity of each point over here.

It looks like there are two colossal groups and many small groups are inside those big groups.

Multivariable Anomaly. The various combination of independent variables may show anomaly of different nature. The technique "Isolation Forest" has been used to identify the anomaly in the multivariable situation. The anomaly has been detected in comprehensive data and clusters of data. The total observation was 1.5 million,

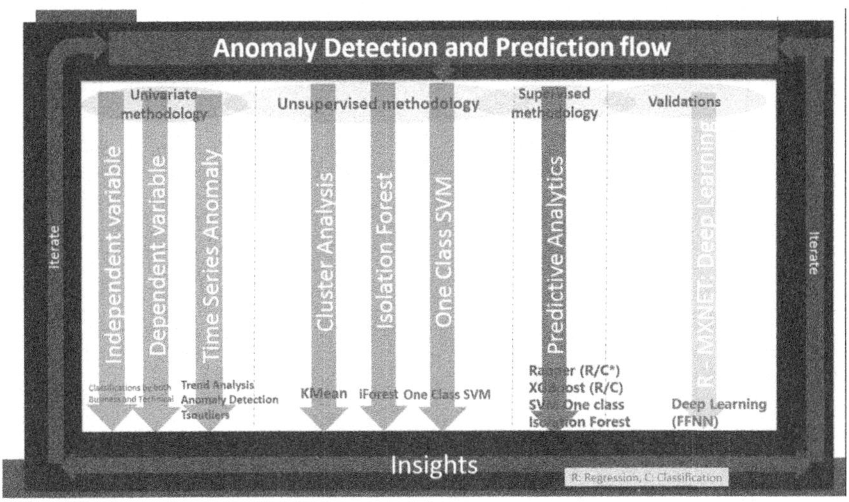

Fig. 10.11 Overview of advance analytics approach for anomaly detection

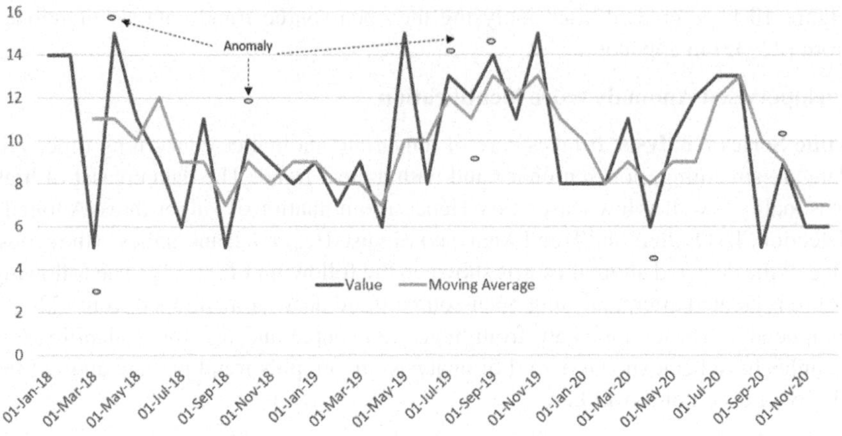

Fig. 10.12 Time series anomaly detection for Univariate data

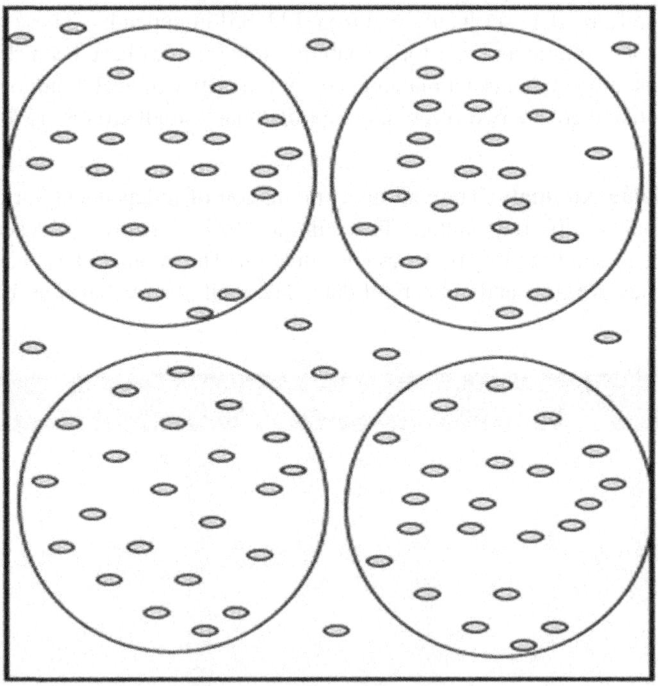

Fig. 10.13 View of clusters of overall data

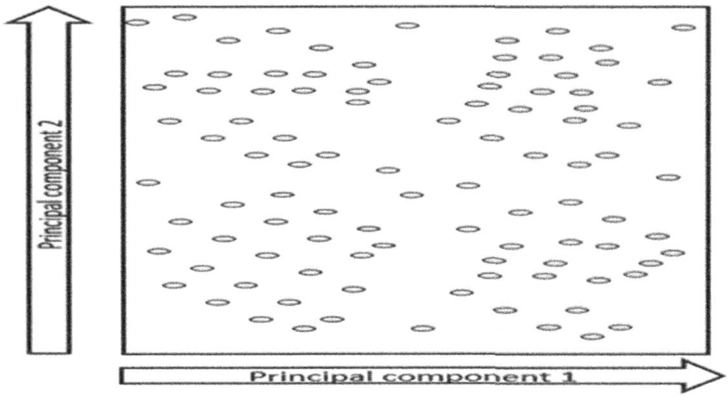

Fig. 10.14 View of anomalies by Isolation forest

and anomalies identified were highlighted in the following Fig. 10.14. The following figure is created after analysing open-source fraud data set mentioned from [22–35] in appendix. These graphs are from huge data source and just for readability, few samples have been shown here. For analysis, the graph's trend matters and not the clarity of each point over here.

A few anomalies are highlighted (in red text) at both the Global and local (for each group) levels.

The above three approaches (Time Series, Cluster and Isolation Forest) have been suggested probably anomaly as per the expertise of those algorithms. All 3 are unsupervised methods, and the calculation of the Anomaly score is done as follows:

$$\begin{aligned}\text{Anomaly Score} = {} & \text{f(time Series Score)} + \text{f(Isolation Forest Global Score)} \\ & + \text{f(Isolation Forest local Score)} \end{aligned} \tag{10.1}$$

Supervised (prediction) Anomaly score

The Anomaly detection should happen as early as possible and preferably in a few minutes if not in seconds. This kind of response will happen only when all other works (Model building) are already done and Compute System is ready to predict anomaly. That is why the supervised method of anomaly score is done.

The data has been highly complex and have all types (Nominal, Ordinal, Numeric) of data. The nature of data includes Dense, Sparse, Nonlinear separation, Unbalanced, Balanced (few subsets), pure Normal (absence of Anomaly records). Considering all the above, the ensemble of many methodologies has been used for Supervised Predictive Analytics.

The unsupervised anomaly score (from Eq. 10.1) has been used as the response variable for the supervised method. The supervised anomaly score has been calculated as follows.

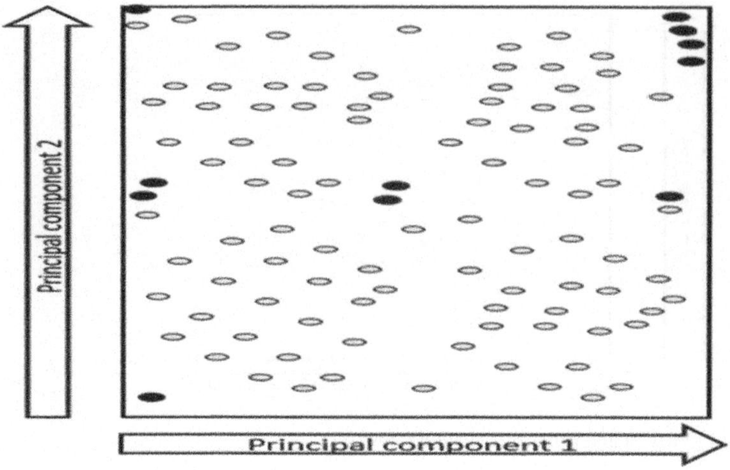

Fig. 10.15 View of final anomaly suggested by an ensemble of algorithms

$$\textbf{Supervised Anomaly score} = \text{Ensemble}\{\text{f(xgboost)} + \text{f(ranger)} + f\text{(One calss SVM)} \\ + \text{f(Isolation Forest)} + \text{f(Deep Learning using FFNN)}\} \quad (10.2)$$

The supervised anomaly score with the threshold (will vary from industry to industry and nature of anomaly being detected) was used for taking further action on the Anomaly score. The outcome of Global and Local anomaly is shown below in Fig. 10.15. The following figure is created after analysing open-source fraud data set mentioned from [22–35] in appendix. These graphs are from huge data source and just for readability, few samples have been shown here. For analysis, the graph's trend matters and not the clarity of each point over here.

The red points are probable Anomaly records with rank (decided by threshold).

10.8.3 Stock Market's Growth from New Users' Point of View

Across the world, one thing is expected from all of us is to invest surplus money to start earning money. Equity has been part of the portfolio for many investors irrespective of age, profession and geography. To understand the trend, the data from Nifty (India's largest stock exchange, www.nseindia.com) and Zerodha (India's largest stockbroker, www.zerodha.com) have been explored.

User growth has been a phenomenon. The next question comes: Do we need to hire the staff appropriate to several customers. No, not at all—why—because the RPA plays a significant role here. The RPA takes care of most activities, from onboarding to enabling the trading [8–15].

Now, the next question comes—how we can leverage this growth. The answer comes from the AI world. The AI helps cross-sell, up-sell, investment suggestions and future trends to all customers. There are various use cases for that, and a few of them are mentioned in previous sections.

10.8.4 Customer 360: Understand the Customer from All Dimensions

Nowadays, a customer uses many products (often from the same bank). The 360-degree view (Fig. 10.16) provides the actual view of customers' usages of various products and services (not necessarily by the same bank). The following Fig. 10.16 has been created from summary of above sections, hence permission to publish is not required.

The 360° helps to view the lifetime value and hence augment the lifetime value by offering various products and services.

Fig. 10.16 Datamart to build a 360 customer analytics view

10.8.5 Credit Card Customer Life Cycle

The journey to building a credit card business has several steps. Here are a few points and each of the issues is a use case for AI.

Acquire

- Prospect segmentation
- Prospect value modelling
- Prospect profiling
- Prospect selection & Targeting
- Manage
- Measuring current profitability
- Card activation trends
- Usage monitoring and analysis
- Balance build campaign
- In-activity modelling.

Grow

- Estimate the potential of a customer
- Share of wallet (SoW) analysis
- Cross-sell and Cross category spend analysis
- Spend modelling.

Retain

- Customer lifetime value (CLTV) modelling
- Loyalty analytics
- Balance retention analytics
- Attrition and dormancy models.

10.9 Ongoing Trends and May Reach to Maturity in a Few Years

10.9.1 Conversational AI (Also One of the Major Solutions by RPA)

The Conversational AI started a few years ago and most of the banks have implemented it. As per user experience of websites of a few top banks (name not mentioned deliberately), the conversational AI is still rule based and can do only a few steps only. As soon as customer query starts getting complex, the conversational AI is unable to proceed and hand over the request to live customer service executive. In coming years, the conversational AI will see paradigm shift by solving most of the customer's query and taking actions (pre-approved) on the spur of moment.

10.9.2 Personalised Service with High Accuracy

Many of us have received call from tele marketers for credit card or loan or overdraft. Did we accept those offers? Not necessarily, always. This is the major gap that will be addressed in coming years where personalised marketing will be done with high accuracy. As consumer, we will get what we need most of the time. Indirectly, improving customer satisfaction as well.

10.9.3 Data Collection Points

The data is being collected at most of the touch points. Are these really used (within guidelines of respective authority) by Banks? The answer was not convincing. In coming years, the data collection will become matured and it will handle with ease the customer satisfaction too. It will also pave the way for most of the banks to act before customer reaches them for banking needs.

10.9.4 Almost Real-Time Fraud Detection

The fraud detection is already happening, and credibility is increasing. It has been found that few genuine transactions are marked fraud and can be taken as machine error within tolerance. In coming years, the fraud detection will continue real time (with time lag) and prevent recurring fraud transactions from various dubious sources.

10.9.5 Ethical AI

The ethic will continue to focus as Banks have been known for doing ethical business.

10.9.6 Big Spend on AI and RPA

As per various survey reports by Accenture [75], the 75% of CXO believes the spending on AI projects will go up. It will pave the way for mature AI systems with visible outputs.

10.10 Limitations: Is AI and RPA Panacea?

The banking sector has been a leader in adopting new technology to make the banking process up to date and smooth. Moreover, still, there is much improvement that needs to be done even today. That is why AI and RPA are not a panacea. Here are a few of the challenges faced by banks.

- Based on the interactions with various MNCs, the implementations of AI and RPA cannot be a panacea, and expectation mismatch happens for lack of correct data, skills and infrastructure.
- The attachment with existing legacy tools is high on the mind of people. Even though banks have invested in new systems, the adaptability is slow. Except for sales and marketing, most units believe that the new system will not help them much.
- The automated process may or may not resemble the existing manual process. Hence validity takes time. The answer by a chatbot, AI replacing human decision-making, also differs a little bit.
- The cost of having a mature AI system is very high. The time to have a mature approach is around five years.
- The RPA has some limitations too. It can help detect Fraud in the offline system but not in real-time—ATM transactions, Online banking transactions.
- The RPA works like a double edge sword. If a decision is wrong (or flawed), then those are also delivered instantly without review.
- The RPA is just a technology, and by nature, any technology may have errors; hence RPA can also have errors.

10.11 Conclusion

The recent addition of many bank account holders has added business opportunities and challenges to banking systems. The opportunities are enormous in numbers of account holders and depositors. These vast numbers have caused a load on existing infrastructure, which needs to be revamped. The latest technology helps the user execute the transaction quickly of their choice. The same has also created a problem for non-tech savvy account holders. In this way, opportunity and problem have become like two sides of a coin.

The combination of AI and RPA has been acting as a boon and empowering banks' workforce by automating its repetitive tasks and enhancing knowledge by predicting the growth rate of various financial instruments. A few examples are—financial fraud and money laundering activities are getting monitored runtime. The regulatory reporting is automated, and hence compliance is increased. The AI understands the customer's transaction behaviours and sentiment in advance. The AI can predict the need and want of customer, and hence banks can customise products for an individual customer. This clears the path for banks to produce financial products and services for each customer and build positive relationships with their customers. The RPA helps deliver these services in real time. The presence of AI and RPA has been found in almost all business units of banks, and the business unit "Accounting" is a leader in implementing RPA.

Most importantly, the willingness is visible too. These have come from the investment by banks and overcoming the mindset of using a legacy application. These are not the panacea for all problems. A lot has been done, and a lot is yet to come.

References

1. Turning, A.: I.—computing machinery and intelligence. Mind 236, 433–460 (1950). https://doi.org/10.1093/mind/lix.236.433
2. Test Doubles, Fields of study, Abstract, Principal terms, https://science.jrank.org/programming/Test_Doubles.html, Accessed 15 October 2021
3. History of RPA, https://www.javatpoint.com/history-of-rpa/. Accessed 1 October 2021
4. Reserve Bank of India. (Rbi.org.in, 2021), https://www.rbi.org.in/commonman/English/Scripts/BanksInIndia.aspx. Accessed 1 December 2021
5. Reserve Bank of India—Reports, M.rbi.org.in (2021), https://m.rbi.org.in/Scripts/PublicationReportDetails.aspx?UrlPage=&ID=931. Accessed 11 December 2021
6. Marr, B.: Artificial Intelligence (AI) And Smart Process Automation (2021), https://www.linkedin.com/pulse/artificial-intelligence-ai-smart-process-automation-bernard-marr/. Accessed 4 February 2022
7. The Rise of AI in Financial Services, https://narrativescience.com/wp-content/uploads/2018/11/Research-Report_The-Rise-of-AI-in-Financial-Services_2018.pdf. Accessed 9 December 2021
8. RPA in Banking, https://marutitech.com/rpa-in-banking-and-finance/. Accessed 1 October 2021
9. RPA in Banking, https://automationedge.com/blogs/robotic-process-automation-rpa-in-banking-industry/. Accessed 1 October 2021
10. RPA in Fianance, https://itrexgroup.com/blog/rpa-in-finance-banking-use-cases-implementation-expert-advice/. Accessed 1 October 2021
11. RPA in Banking, https://www.comtecinfo.com/rpa/use-cases-of-rpa-in-banking-industry/. Accessed 1 October 2021
12. RPA in Banking, https://www.productiveedge.com/2020/06/18/banking-robotic-process-automation-intelligent-automation/. Accessed 1 October 2021
13. RPA in Banking, https://nividous.com/blogs/rpa-use-cases-in-banking-industry. Accessed 1 October 2021
14. RPA in Banking, https://www.uipath.com/solutions/industry/banking-automation. Accessed 1 October 2021
15. RPA in Banking, https://research.aimultiple.com/banking-rpa/. Accessed 1 October 2021
16. Pruitt, R.: Addressing-fraudulent-payment-activity-106347 (2013)
17. Standish, J.: Leveraging Best Practices with Advanced Analytics—Making the Right Decisions in Fraud Investigations (2012)
18. Aberdeen, G.: Fighting fraud with big data visibility and intelligence. Sept. **24**, 2012 (2012)
19. Gill, W.: Fighting Fraud with Advanced Analytics. Canadian Underwriter **76**, 9 (2009)
20. Lormel, D., Barta, D.: Can You Think Like a Fraudster?. SAS Conclusion Paper (2013).
21. Asaro, J.: Chris Poisson (2016), https://www.businesswire.com/news/home/20160628005456/en/ID-Analytics-Secures-Patent-for-Detection-of-Fraud-Rings. Accessed 16 December 2021
22. Fraud Data—Credit card fraud, https://www.kaggle.com/datasets/mlg-ulb/creditcardfraud. Accessed 16 December 2021

23. Fraud Data—Ecommerce fraud data, https://www.kaggle.com/datasets/aryanrastogi7767/ecommerce-fraud-data. Accessed 16 December 2021
24. Fraud Data—The Enron Email Fraud Dataset, https://www.kaggle.com/datasets/wcukierski/enron-email-dataset. Accessed 16 December 2021
25. Fraud Data—Ethereum fraud detection dataset, https://www.kaggle.com/datasets/vagifa/ethereum-frauddetection-dataset. Accessed 16 December 2021
26. Fraud Data—Fraud with Tensorflow, https://www.kaggle.com/code/currie32/predicting-fraud-with-tensorflow. Accessed 16 December 2021
27. Fraud Data—EDA, https://www.kaggle.com/code/jesucristo/fraud-complete-eda. Accessed 16 December 2021
28. Fraud Data—IEEE, https://www.kaggle.com/competitions/ieee-fraud-detection. Accessed 16 December 2021
29. Fraud Data—genneric, https://www.kaggle.com/code/viswajithkn/fraud-detection. Accessed 16 December 2021
30. Fraud Data—Credit card fraud, https://data.world/vlad/credit-card-fraud-detection. Accessed 16 December 2021
31. Fraud Data—Bank fraud, https://www.statista.com/statistics/1012762/india-value-of-bank-fraud/. Accessed 16 December 2021
32. Fraud Data—Amazon fraud, https://paperswithcode.com/dataset/amazon-fraud. Accessed 16 December 2021
33. Fraud Data—Genric, https://datasetguide.com/dataset/b8c96f8e9fca90a451a8337bdcadc4a1/. Accessed 16 December 2021
34. Fraud Data—India number of bank fraud cases, https://www.statista.com/statistics/1012729/india-number-of-bank-fraud-cases/. Accessed 16 December 2021
35. Fraud Data—Genric, https://data.gov.au/dataset/ds-sa-a51d5494-ece2-4c7e-a139-229046f0d9cc/details. Accessed 16 December 2021
36. World Retail Banking Report (World Retail Banking Report, 2021), https://worldretailbankingreport.com. Accessed 25 December 2021
37. IDFC AMC introduces Artificial Intelligence powered PMS (The Economic Times, 2017), https://economictimes.indiatimes.com/mf/mf-news/idfc-amc-introduces-artificial-intelligence-powered-pms/articleshow/60958477.cms. Accessed 9 October 2021
38. AI Powered Equity ETF (AIEQ) Launches on NYSE Arca, First Exchange Traded Fund to apply artificial intelligence (AI) and machine learning throughout investment process Businesswire.com (2017), https://www.businesswire.com/news/home/20171018005986/en/AI-Powered-Equity-ETF-AIEQ-Launches-NYSE. Accessed 5 January 2022
39. Doe, J.: The World's First AI Exchange-Traded Fund is Set to Run in Canada: AI investment managers are free from human emotions and investor bias (2017), https://futurism.com/the-worlds-first-ai-exchange-traded-fund-is-set-to-run-in-canada. Accessed 5 December 2021
40. Sandeep Uttamchandani: PD—2022 technology landscape to Democratize Data, https://www.linkedin.com/pulse/2022-technology-landscape-democratize-data-uttamchandani-ph-d-/. Accessed 16 January 2021
41. YES Bank loan fraud—CBI charges Rana Kapoor with criminal conspiracy (Business Standard News, 2020), https://www.business-standard.com/article/current-affairs/yes-bank-loan-fraud-cbi-charges-rana-kapoor-with-criminal-conspiracy-120062501509_1.html. Accessed 6 December 2021
42. What is PNB Scam—(Business Standard India, 2021), https://www.business-standard.com/about/what-is-pnb-scam. Accessed 7 November 2021
43. Data protection in India (State IT Secretaries Conf, 2018), https://digitalindia.gov.in/writeread data/files/6.Data%20Protection%20in%20India.pdf. Accessed 7 December 2021
44. Reserve Bank of India, (Bank wise ATM/POS/Card Statistics, 2022), https://www.rbi.org.in/Scripts/ATMView.aspx. Accessed 7 November 2021
45. Cross-industry standard process for Data Mining, https://en.wikipedia.org/w/index.php?title=Cross-industry_standard_process_for_data_mining&oldid=1071424927. Accessed 15 February 2022

46. ID Analytics Secures Patent for Detection of Fraud Rings, https://www.businesswire.com/news/home/20160628005456/en/ID-Analytics-Secures-Patent-for-Detection-of-Fraud-Rings. Accessed 7 December 2021
47. Barta, D., Stewart David: A Layered Approach to Fraud Detection and Prevention (p.15).SAS (2013), https://www.sas.com/content/dam/SAS/en_us/doc/whitepaper1/layered-approach-fraud-detection-prevention-106072.pdf. Accessed 11 December 2021
48. Industry insight: Fighting cyber fraud with analytics: Fraud analytics, https://www.analyticsvidhya.com/blog/2016/08/industry-insight-fighting-cyber-fraud-with-analytics/. Accessed 7 November 2021
49. An Inside Perspective on Automobile Insurance Fraud, Part 1 | Insurance Thought Leadership, https://www.insurancethoughtleadership.com/personal-lines/inside-perspective-automobile-insurance-fraud-part-1. Accessed 7 November 2021
50. Leveraging best practices with advanced analytics: Making the right decisions in fraud investigation, https://www.insurancethoughtleadership.com/data-analytics/leveraging-best-practices-advanced-analytics-making-right-decisions-fraud. Accessed 7 November 2021
51. Lee, B.K., Lessler, J., Stuart, E.A.: Improving propensity score weighting using machine learning. Stat. Med. **29**(3), 337–346 (2010). https://doi.org/10.1002/sim.3782. Accessed 7 December 2021
52. Singh, A.: Fighting frauds using Big Data Visibility and intelligence, https://www.algoworks.com/blog/fighting-frauds-using-big-data/. Accessed 7 November 2021
53. Huggett, J.: Is Big Digital Data Different? towards a new archaeological paradigm. J. Field Archaeol. **45** (2020)
54. Fighting fraud with advanced analytics, https://www.canadianunderwriter.ca/features/fighting-fraud-with-advanced-analytics/. Accessed 16 December 2021
55. Leveraging Best Practices With Advanced Analytics: Making the Right Decisions in Fraud Investigation | Insurance Thought Leadership, https://www.insurancethoughtleadership.com/data-analytics/leveraging-best-practices-advanced-analytics-making-right-decisions-fraud. Accessed 7 December 2021
56. Pruitt, R.: Addressing fraudulent payment activity with advanced decission management analytics. SAS (2013). https://www.mwsug.org/proceedings/2013/FS/MWSUG-2013-FS09.pdf. Accessed 11 December 2021
57. Routolo, R: Detect and Prevent More Fraiud improper payments. SAS (2013). https://www.mwsug.org/proceedings/2013/FS/MWSUG-2013-FS09.pdf. Accessed 11 December 2021
58. Neuman, E.J., Mizruchi, M.S.: Structure and bias in the network autocorrelation model. Soc. Netw. **32**(4), 290–300. https://doi.org/10.1016/j.socnet.2010.04.003. Accessed 11 December 2021
59. Standish, J.: Leveraging Best Practices With Advanced Analytics; Making the Right Decisions in Fraud Investigations (2012). https://fraudtaskforce.org/2012/06/leveraging-best-practices-with-advanced-analytics-making-the-right-decisions-in-fraud-investigations-2/. Accessed 11 December 2021
60. Breiman, L.: Mach. Learn. **45**, 5–32 (2001)
61. Benford's law, https://en.wikipedia.org/w/index.php?title=Benford%27s_law&oldid=1072079124. Accessed 15 February 2022
62. ID analytics secures patent for detection of fraud rings [Online]. Business Wire: 2016, https://www.businesswire.com/news/home/20160628005456/en/ID-Analytics-Secures-Patent-for-Detection-of-Fraud-Rings. Accessed 10 November 2021
63. Abe, N., Zadrozny, B., Langford, J. Outlier detection by active learning. In: Proceedings of the 12th ACM SIGKDD International Conference on Knowledge Discovery and Data Mining, pp. 504–509 (2006). Accessed 11 December 2021
64. Agarwal, D.: An empirical bayes approach to detect anomalies in dynamic multidimensional arrays. In: Fifth IEEE International Conference on Data Mining (ICDM'05), pp. 8-pp (2005). IEEE
65. Basu, S., Meckesheimer, M.: Automatic outlier detection for time series: an application to sensor data. Knowl. Inf. Syst. **11**(2), 137–154 (2007)

66. Agarwal, D.: Detecting anomalies in cross-classified streams: a bayesian approach. Knowl. Inf. Syst. **11**(1), 29–44 (2007)
67. Aggarwal, C.C., Yu, P.S.: Outlier detection with uncertain data. In: Proceedings of the 2008 SIAM International Conference on Data Mining, pp. 483–493. Society for Industrial and Applied Mathematics (2008)
68. Das, K., Schneider, J.: Detecting anomalous records in categorical datasets. In: Proceedings of the 13th ACM SIGKDD International Conference on Knowledge Discovery and Data Mining, pp. 220–229 (2007)
69. Anomaly detection, In Wikipedia. https://en.wikipedia.org/w/index.php?title=Anomaly_detection&oldid=1070297230. Accessed 11 December 2021
70. Anomaly (2022), https://dictionary.cambridge.org/dictionary/english/anomaly. Accessed 22 February 2022
71. World Map (2022), https://irlen.com/world-map/. Accessed 22 February 2022
72. Graves, E.: (2021), Gravesee/isofor [R], https://github.com/gravesee/isofor. Accessed 22 February 2022
73. The Science of Anomaly Detection [White Paper], https://numenta.com/assets/pdf/whitepapers/Numenta%20White%20Paper%20-%20Science%20of%20Anomaly%20Detection.pdf. Accessed 11 December 2021
74. Keogh, E., Lin, J., Lee, S.H., Herle, H.V.: Finding the most unusual time series subsequence: algorithms and applications. Knowl. Inf. Syst. **11**(1), 1–27 (2007)
75. Joyce, L.: Six AI and Big Data Trends in Banking for 2022.The Financial Brand—Banking Trends, Strategies & Insights (2021), https://thefinancialbrand.com/124980/six-big-data-and-ai-trends-in-banking-for-2022-cloud/. Accessed 11 December 2021

Chapter 11
Integration of RPA and AI in Industry 4.0

Ganeshayya Shidaganti, Karthik K. N., Anvith, and Neha A. Kantikar

Abstract The transition beginning from early Industry 1.0 to the current Industry 4.0 can be seen as movement from traditional methods to the present digital processes. These digital processes claim to be the foundation of Industry 4.0 and have led to the creation of vast volumes of data, collectively referred to as the Big data. As much as this proved to be a boon to the developing technology, the storage and effective utilization of this data, consuming unimaginable human efforts, was a crisis in itself. Here's where RPA (Robotic Process Automation) and AI (Artificial Intelligence) play their respective roles. While AI techniques mimic human thinking, RPA processes automate repetitive tasks. The Integration of RPA and AI leads to Intelligent Automation, thus streamlining human efforts along with increasing speed, productivity, and quality. Apart from these benefits, combining AI algorithms, strategies, and techniques with RPA tools improves RPA processes' effectiveness and precision in recognition, data extraction, forecasting, classification, and process optimization. In this regard, the main purpose of this chapter is to brief about the evolutions up to the present industry 4.0 and discuss the essence of integrating AI and RPA along with its benefits to mankind. It continues to explain the challenges faced in effective integration of the two booming technologies, elucidate a few misconceptions related to it and also illustrate the use cases of the same. In addition, the chapter throws light upon different RPA tools in great detail.

G. Shidaganti (✉) · Karthik K. N. · Anvith · N. A. Kantikar
Department of Computer Science and Engineering, M.S. Ramaiah Institute of Technology (Affiliated to VTU), Bangalore, Karnataka, India
e-mail: ganeshayyashidaganti@msrit.edu

Department of Electronics and Communication Engineering, M.S. Ramaiah Institute of Technology (Affiliated to VTU), Bangalore, Karnataka, India

S. Bhattacharyya et al. (eds.), *Confluence of Artificial Intelligence and Robotic Process Automation*, Smart Innovation, Systems and Technologies 335,
https://doi.org/10.1007/978-981-19-8296-5_11

11.1 Introduction

Artificial Intelligence (AI) and RPA (Robotic Process Automation) possess a firm connection. Both of them are becoming more popular, with enterprise deployments increasing on a yearly basis. They both have the potential to alter businesses pursuing digitalization. Until now, they've both been restricted to organizational boundaries, needing highly qualified—but scarce—professionals to deploy them successfully. The integration of RPA and AI technologies, commonly referred to as intelligent automation, allows the development of more efficient end-to-end business processes. Ever since the 1960s, the concept of AI has held its position in the industry. It basically refers to machinery, mostly computers, that execute jobs that formerly required human intervention and intelligence. A wide range of businesses and environments are infusing AI to change the way people work and live in the current day-to-day life.

In Robotic Process Automation, the term "robot" does not relate to AI robots or a physical robot [1]. It's a bot, which is essentially a software robot that automates repetitive operations and tedious procedures, reducing human blunders and significantly enhancing production, efficiency, and precision. All rule-based processes are automated by RPA bots. They can also cut/copy/paste data, transfer files and directories on a regular basis, perform web scraping, fill out applications/forms, and also perform data extraction within a document. Machines can either execute jobs according to rules provided by a Private equity employee or by utilizing built-in AI capabilities. Supply chain management and logistics, financial services, human resources, customer service, accounting, health sector, data entry, and other disciplines can all benefit from the integration of AI with RPA.

The chapter begins with giving insights about the industrial revolution, Robotic Process Automation (RPA), and Artificial Intelligence (AI). It further explains about the integration of the two developing technologies, i.e., RPA and AI, its benefits and challenges faced while doing so. The chapter also covers different use cases of intelligent automations and RPA tools used for the same.

11.2 Literature Survey

Numerous studies and researches have been carried out in the field of AI and RPA, and the summation of the two, Intelligent automation. Van der Aalst et al. [2] state that "RPA is an umbrella term for tools that operate on the user interface of other computer systems". An RPA tool maps the procedures for the software robot to follow, using its tool language. As a result, RPA solutions are designed to relieve employees of the strain of doing repetitive, uncomplicated activities.

Ng et al. [3] illustrate Intelligent automation (IA) to be a mix of RPA, AI, and soft computing that has the potential to go beyond standard DM to achieve new performance and efficiency, better decision-making, and reliability and safety. They further explain the key applications of intelligent automations such as aviation, supply chain, city logistics, and portfolio managements.

Ribeiro et al. [4] explain the contribution of RPA tools affiliated with AI, towards improving the business organizational processes of current Industry 4.0. The paper continues to give introduction to different RPA tools, such as Kofax, Automation Anywhere, to name a few. These tools help immensely in implementing AI algorithms, like classification, recognition. Ansari et al. [5] throughout their paper portray the practical application of RPA in any domain in a firm without causing any disruption to the existing system. As a result, RPA will be extensively used in different areas such as big data, manufacturing, analytics, and legal in the near future. Amit Kumar Tyagi et al. [6] in their conference paper discuss the underlying structure, evolution, and usefulness in business applications.

According to the IBM internal analyses, Williams and Allen [7] RPA automations offer a return of investment (ROI) ranging between 30 and 50%. Thus, IBM, in their research paper, addresses the methods by which the cognitive technologies could address the existing technical challenges.

11.3 From Industry 1.0 to Industry 4.0: The Evolution

Different people have different perspectives while defining industrial revolution. Some view it as an increase in results in the production of new hardwares, energy sources, technology, or a mixture of all the above; while others view it as a shift in social and economic organization as a result of replacing handheld tools with machine and capable power tools, as well as the advancement of production lines along with commercial production on a large scale. Formal definition of Industrial revolution by Schwab states "*new technologies and novel ways of perceiving the world [that] trigger a profound change in economic and social structures.*" Society for Industrial Management and Engineering [8] might be close to accurately defining the industrial revolution.

The Industrial Revolution was a decisive instance of the chronological past, which influenced almost every feature of day-to-day lifestyle. Particularly the population and its average income began to grow at an unheard-of rate. The local population's leading lifestyles witnessed a noteworthy never seen before improvement. While this was believed by one class of economists, the other class strongly claimed that improvements occurred only after the early twentieth century.

11.3.1 Industry 1.0 (1750–1830): The Beginning

The first industrial revolution was initially confined to Europe, particularly the United Kingdom around the 1750s. Britain's economical, governmental, and geographical benefits compounded to make it an effective contender for the commencement of the

Industrial Revolution [8]. During this period, the industries switched from handmade production to machine-based productions, powered mainly by coal, steam, and water. Many industries like the textile industry, coal industry, iron industry, mining industry were influenced by the adaptations.

Lombe's Silk factory built around 1721 was one of the first ever documented factories known. This factory mainly mass produced silk products with the assistance of water-powered machines. His motivation to open such a factory came when he saw silk throwing machines while touring Italy. Upon returning to the UK, he hired an architect, George Sorocold to sketch and construct his new "Factory". After the construction of the mill, he employed around 300 people and was believed to be the first known mechanized factory in the world [8]. Water- and steam-powered equipment were invented in the 1800s to help employees with their responsibilities. This greater efficiency and capability prompted expansion in other sectors; enterprises evolved from personal hobbies to business organizations composed of clients, managers, and owners. In particular, the textile and transport industry profited efficiently from the advent of industrialization. The application of machines in the process of manufacturing immensely grew by cascading it with coal as a fuel source. In fact, the first steam engine was also built in this era by Newcomen [9]. Economically, the first country to be remodeled was Belgium, despite the industrial revolution beginning in the UK. Thus, by the development of machines shops at Liege, the two Englishmen, John Cockerill and William were successful in bringing about the very first industrial revolution [10].

Industry 1.0 marked the beginning of the industrial culture focusing more on product efficiency, cost effectiveness, quality and scale changing the industrial standards forever.

11.3.2 Industry 2.0 (1850–1914): Building on the Bedrocks

The commencement of the nineteenth century began with the initiation of Industry 2.0, also referred to as the second Industrial revolution. The water and steam-based machinery which were developed during the first industrial revolution, proved to be excess resource consuming and ineffective. Thus, the introduction of electrically run machines confirmed to be both effort and cost effective, making them the primary factors of this industrial revolution [11]. Effective production processes occurred during this revolution were mapped from the industrial cultures of the previous revolution—Industry 1.0. Division of labor, manufacturing within the stipulated time, and several improvised manufacturing processes multiplied the underlying processes, which added to the outcomes of higher productivity and quality.

The Bessemer process by Henry Bessemer resulted in a low-cost industrial technology for mass-producing steel, enabling the construction of thousands of railroad lines, including tracks for steam engine trains with its unique designs. The improvisation of the steam engine of Newcomon was done by James Watt, with the engines attaining exceptional cargo speeds, thus bringing about tremendous revolution in the transportation sector. An American mechanical engineer, Fredrick Taylor studied resource allocation, optimization of labor and workplace performances and pioneered in them. Earlier, the construction of a complete vehicle occurred at only one station. Currently, the production happens at particular stages on the conveyor belt, which is significantly speedy and less costly. We witnessed world expositions for the first time, when traders assembled to converse about technology. Based on Benjamin Franklin's scientific work, Alessandro Volta invented the first electrical battery in 1800. Later on, crude oil was discovered in 1875 [9]. Hence, the introduction of new power sources to the reign of production and industry guided humanity into the present modern era.

Industry 2.0 marks the entrance of steel into the common market, as well as greater automation of production machinery and the design of new, more efficient engines.

11.3.3 Industry 3.0 (1940–2010): Automatons Taking Over Industries

The development in the field of electronics which happened in the past years of the last era during Industry 2.0 became the spark plug of the third industrial revolution—Industry 3.0. We can observe how manufacturing automation has progressed to where it is at present. Digital communications, software, robots, transistors, and other technologies all appeared or were developed during this time frame. During World War II, governments were enforced to investigate factory robotics due to the requirement of speedy output. Fighter planes, landing craft, warships, and tanks were all built with a lot of automation. Following World War II, these procedures were dutifully reproduced in commercial enterprises. Companies began automating parts of their production lines. This entailed the elimination of low-skilled occupations, as well as the steady reduction of current highly skilled posts. In the past, machinery introduced by industry aided the human workers; but, in this case, human workers were completely supplanted.

The development of the enigma machine during World War II resulted in the production of the first computers and computerized software. Electronic equipment such as integrated circuits and transistors proved to be more efficient in effort and time, resulting in greater speeds and precision, thus automating the machines [3]. A scale of such devices was manufactured aiding the effective replacement of a human medium, at certain situations.

Programmable Logic Controllers (PLC) [12] were developed in the 1960s in the automotive industries of the US. The Abstract view of PLC is as shown in Fig. 11.1. Their creation became an influential invention in the automation field of electronics. These provided well built, reliable, flexible, and easily programmable control systems thus replacing the old relay logic ones. The application of software programs in the electronically integrated production systems provoked the need for development of software business. Thus, techniques involving the application of electronics and

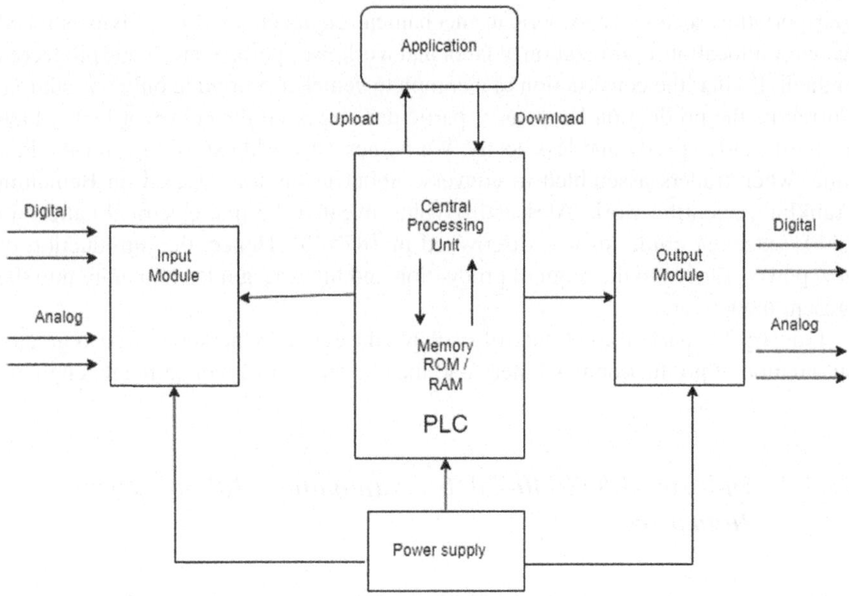

Fig. 11.1 Programmable logic controller

information automated a whole sector. Ever since, software systems and automation technologies have exponentially outgrown the sector of electronics and IT industries.

11.3.4 Industry 4.0 (Present): The Age of AI and RPA

We have briefly gone through the evolution of Industries beginning from Industry 1.0–3.0, now let us see what the hype on Industry 4.0 is all about. This revolution is the logical continuation of programs that were initially created during the Industry 3.0 phase. Product life cycle management, manufacturing executive systems, and other forethoughtful strategies created in the late 1900s lacked the necessary technology to be fully implemented. We are presently amidst the fourth industrial revolution, and its principal inventions are products that most modern customer base will be fully acquainted with: the smart machines. Industry 4.0 integrates the Internet of Things with construction practices, allowing systems to communicate, analyze data and use that data to guide intelligent actions. Industry executives and analysts are already claiming that Industry 4.0 has begun. The Unification of IoT with everything beginning from AI and analytics to sophisticated products to collective processes is the current competitive criterion that enterprises are provoked to meet [12]. Smart machines are becoming more self-sufficient; they can monitor on their own and interface without human interaction, relieving personnel for extra activities. A range of sectors, including electronic products, horticulture, agriculture, and farming practices are expected to reach out to the technical sector in order to increase manufacturing growth by converting their traditional factories into smart factories. These smart factories would combine conventional production with the wide-ranging digitization to user in the manufacturing processes of the fourth revolution, which would raise production and efficiency by more than a third.

Unlike previous revolutions, whose primary focus was to improve the economy, cultural and political developments played equally vital roles in revolutionizing the remaining world this time with the application of technology. The majority of these improvements were accelerated and subsequently expanded internationally with the assistance of the Internet of Things as discussed earlier. This resulted in an earlier unseen level of interconnections. But, the local people were not aware of events taking place on either side of the world, and interwoven economies ruined their lives. Despite its societal significance, Industry 4.0 is currently mainly centered on production developments. The present manufacturing methods are more comprehensive, connecting the digital and physical worlds. The cyber-physical environments would thus give rise to the current definition of industrial revolution. The key components of Industry 4.0 technologies include IoT comprising interconnected web of sensors, cognitive and cloud computing, the cyber-physical system (CPS) which is a system controlled by computer algorithms. Cyber-physical systems use sensors and algorithms which are integrated into most parts in the manufacturing process. Here the sensors capture the data which is then fed to the processors of the system which is responsible for making the decisions [13]. These decisions are then carried out by the physical entities of the machine. Ordinary systems work independently with just the algorithm with pre-recorded data. But with bringing IoT into the picture, the system gathers and exchanges the data with all the other resources and divisions of the manufacturing process. The entire process can be further optimized by analyzing the flow of data and the data itself with the help of cloud computing and AI. With integrating IoT, management and maintenance of the process becomes much easier as a lot of data will be involved. The decisions involving minor as well as major changes of the process can be inferred clearly and quickly by studying the huge data generated. Future production will be dominated by smart factories and early transition to the same will prove to be beneficial to the companies.

Advanced industrial robots will be designed to do sophisticated non-repetitive tasks in collaboration with humans. In 2016, just 8% of factory jobs were automated, and even so, the tasks automated were less difficult and more repetitive like the task of pumping a certain flavor of yogurt into a specific marked yogurt pot. By 2025, 25% of all our regular tasks will be automated, and the machines will depend on something we have in less abundance currently: artificial intelligence which will be discussed later on in this chapter. This means that the automaton may, in theory, be able to recognize which yogurt pot, the yogurt is being pumped into. Upon the pot being labeled inaccurately, the machine will be halted, potentially saving billions of dollars in food waste globally each year [9]. They would hire big data and analytics to take in from the past errors, by using a concept called machine learning which is one of the most popular trends today.

11.3.5 Comparison Between Industry 3.0 and Industry 4.0

As we have gone through each of the industrial revolutions, let us now see what constitutes the fourth industrial revolution differentiating it from the third one.

The third industrial revolution employed the use of information technologies and logic processors in the automation of operations. Such processes frequently run mostly without the intervention of humans despite a human component at work. Industry 4.0 takes part in the effective usage of massive amounts of data available in abundance on the manufacturing floor. When boiling down many aspects, the main distinction is that in Industry 4.0, machines operate independently without the assistance of people. In contrast, machines in Industry 3.0 are merely automated under human supervision. For example, consider the working of a CNC Milling machine. If the machine was operated in Industry 3.0, the changes of the tool would be automatic, but the spindle speed must be corrected and checked by the operator itself. But, if the machine was operated in Industry 4.0, along with tool changes being automatic, the spindle speeds and the other necessary parameters are registered by thousands of sensors available in the machine, thus optimizing the process by comparison with the huge amount of ideal data stored [14].

Let us consider an item designing firm and imagine what it might be like in industries 3.0 and 4.0. The manufacturing sector of a product designing firm of Industry 3.0 consists of an assembly-mounting workshop and a mechanical workshop. The mechanical division, a production division of Industry 3.0 processed metal, plastic, biological glass, and other materials. Parts made with various technological equipment are verified in the quality control department against the technological documentation and corresponding developments from the company's technical documentation library. Thus, the parts created in accordance served as the foundation for the beginning of production, and were transported to the storage of the mechanical workshop. Accordingly, the assembly workshop of Industry 3.0 was also a manufacturing division that complied with the electronic equipment on a circuit board and tasks of montage of radio and assembly finishing of an item production.

The production sector of an item designing firm of the fourth industrial revolution is a part of a corporation that is intended for the manufacture of items without the aid of humans and paper, by uniting a collection of technically operated automatic systems. The application of digital technologies such as IoT, cloud technologies guide in such manufacturing processes. This ensures a constant technical cycle of manufacture of item designing elements. Hence along with reducing human labor a flexible automatic (automated) manufacturing is carried out in the Industry 4.0 production division while compared to strenuous one of Industry 3.0 [15].

Thus Industry 4.0 has succeeded in successfully integrating human efforts into machine-based ones, thus enabling easy, efficient, time and cost-friendly techniques.

11.4 RPA: Catalyst of Industry 4.0

There are challenges which must be addressed in order to fully actualize Industry 4.0. The most typical one is a reluctance to alter legacy systems that are critical to an industry's business processes. Providing a single platform for all systems to interact in a business process will take some time. Furthermore, in this technologically developing world, the huge data generation is frequently unstructured, i.e., information that does not have a predefined data model, such as data from email, social media, or sensor data, and is difficult to analyze, interpret, and process using traditional methods. Here's where Robotic Process Automation (RPA) plays its role.

RPA is a technique that uses software robots or intelligent automation to automate commercial activities without the need for human participation. It is a new technology that is being used in a variety of industrial processes today, particularly in back office procedures where tasks are rule-based and do not require human judgment. RPA systems create an action list by capturing the touchpad (keyboard and mouse) controls used by the user to accomplish the given task in the graphical user interface (GUI) of the application and automating it by replicating the action in the GUI [16]. The Bots don't require a physical screen to function. Bots electronically read the screen display, and all of the automation operations take place in a virtual environment. RPA saves money on the acquisition, setup, and maintenance of IT infrastructure, in addition to regulating the investment of human resources in tasks that do not require the application of human brain processes. Reading and writing to a network, extracting structured and semi-structured large datasets, scraping websites, opening emails and attachments, and performing computations are just a few examples of frequent applications. Banking & Finance, Insurance, Healthcare, Energy & Utilities, and other industries are some of the major applications where RPA is widely required today. This is a non-invasive technology that does not necessitate any changes to the underlying systems, making it simple to incorporate into existing business operations. The technology's application range can be significantly broader than the ones listed above. RPA, when paired with other developing technologies like AI, will allow us to use advanced analytics to derive insightful and actionable insights from unstructured data.

Upon summarization, it can be rightfully said that RPA is a catalyst in Industry 4.0 enabling the automation of time consuming, monotonous tasks, round the clock, and hence has brought us one step closer to creating a fully-connected world.

11.5 AI in Industry 4.0

The next step of the enhancement of industrial technologies includes the connection of computers and robots to the Internet of things which would be further enhanced by machine learning algorithms. Thus, the fourth industrial revolution can be declared to have immense potential to be a significant engine in the growth of the industrial economy. Business analytics is polished by AI (artificial intelligence), which is a meaningful expansion of the global economy. In accordance with an IBM report designated, "The Global Race for AI," 82% of Spanish businesses have begun experimenting with AI [17].

Artificial intelligence (AI) includes a collection of technologies that enable machines to perform the tasks of humans such as sensing, comprehending, learning on their own, thus helping us attain efficiency in comparison to the effort consuming human labor. Thus with the application of artificial intelligence, robots can be created to execute difficult pieces of work which might seem beyond the capacity of humans, such as the management of microscopic components or managing hazardous raw materials.

Smart factories are composed of hyper-linked production methods and are built of various machines communicating with each other. They collect, analyze, and process all types of data with the help of AI automation platforms. Original equipment manufacturers (OEMs) are already successful in enhancing industry 4.0 by working in smart factories by integrating AI into their manufacturing operations [18]. Certain manufacturers are using AI and ML to refine quality, standards, and maintenance of the data sets of their organizations. They analyze the functionality of the equipment to achieve this, thus excessively reducing factory lines. Manufacturing is one of the world's most important industries, and implementing mechanisms in this area lies in releasing the real potential of systems and applications for consumers. In industry 4.0, Internet of Things and its analytics will play a pivotal part in spotting patterns and behaviors, as well as handing real-time data to the manufacturers.

When used correctly, artificial intelligence has a number of benefits for the manufacturing industry. A compiled list of three of the most important benefits for the manufacturing industry today are:

Reduction of Errors—After being educated, advanced algorithms can function very well in jobs that are prone to human errors. Algorithms should be immune to the effects of external factors because they are not vulnerable to them.

Cost-Cutting—Robots are being used by a number of e-commerce sites and banks to commence client assistance. If the problem is more complicated, the human employee is summoned. Companies can cut personnel costs or reassign people to more strategic responsibilities, allowing them to enhance profits and focus on more important business.

Sales Growth—With fewer errors and workers focusing on more vital operations, sufficient time to focus on the core company activities would be available and leave other AI chores to the robots.

Artificial intelligence has revolutionized the operation of businesses, resulting in a new network of human–machine interfaces. Smart factories, the spark plugs of Industry 4.0, are now indicated by cloud-based interconnections between cyber-physical systems and humans.

11.6 Integration of RPA and AI: Intelligent Automation

Despite the fact that more business processes are being digitized than ever before, employees including both large and small organizations are still required to complete boring, repetitive tasks such as tapping in customer details from bills, cutting and pasting data from one form to another, and answering frequent customer queries via chat. Employees may find it more difficult to interact with clients, invent new goods, and adjust to changing business conditions if these procedures cannot be automated. However, Intelligent automation, that is to say, Robotic process automation (RPA) together with AI is changing all that. Today, almost any digital business activity can and ought to be automated. Formally, the usage of automation technologies such as robotic process automation and artificial intelligence to help business simplify and balance decision-making process was known as Intelligent automation, also called Cognitive automation [19]. Intelligent Automation is basically RPA integrated with Artificial Intelligence with minimal human intervention. The Intelligent side is brought about by Robotics, artificial intelligence, and other new technologies that can carry out human operations and operate autonomously by making judgments or interpreting data without direct human intervention. Improvement in Machine learning techniques, sensor upgrades, and ever-increasing processing power have all contributed to the development of a new generation of hardware and software robots with real-world applications in practically every industry area. This advancement has piqued the interest of venture capitalists, technology enterprises, and an increasing number of clients who are integrating intelligent automation in both physical and information systems. Intelligent automation has a wide variety of uses, including streamlining processes, liberating up of resources, and increasing productivity.

Self-driving cars, also known as driverless cars, self-checkouts at grocery shops, smart home assistants, and gadgets are examples of common uses. Businesses may use data and machine learning to create predictive analytics that react to changes in customer behavior, or they can use RPA to optimize factory floor operations. Intelligent automation may be used in an automobile industry to speed up production or limit the possibility of human error, or it may be used in a pharmaceutical or life sciences firm to cut costs and get resource efficiencies when repeated operations occur. The technique has also been used to automate the distribution of Covid-19 vaccinations. Data from electronic health records in hospital systems may be used to identify and educate patients, as well as schedule immunizations. Businesses use artificial intelligence, or AI, to build a knowledge base and make data-driven predictions by using machine learning and complex algorithms to evaluate unstructured and structured data. This is IA's decision-making engine [20]. RPA uses software robots, or bots, to execute back-office operations including data extraction and form completion. Since RPA can leverage AI insights to solve increasingly complicated tasks and use cases, these bots are an excellent complement to AI.

11.7 Benefits of Integrating RPA and AI

In terms of human perception, Aristotle felt that "the whole is larger than the sum of its parts". This assertion is exemplified by the addition of embedded AI capabilities to RPA. AI gathers data from a variety of sources and nourishes its tools in order to improve the value of their intercommunications. RPA is useful for automating structured data-enabled activities that previously required manual intervention. Each one adds value on its own. However, combination of the two helps in mapping technically based knowledge to modernize methods to bring about effective application interactions thus adding immense amounts of value to the solutions. The following solutions are quicker and much precise [19], and they help you achieve the below-mentioned four efficiencies.

Increase in Productivity: Applications and practical processes that are automated run more quickly. Organizations gain increased efficiency and accuracy in their planning cycles by automating applications and processes, as well as decision-taking, predicting, and projections using many resources of structured and unstructured data in real time. For example, Deloitte, a financial services IBM customer, recently employed RPA to construct bots to automate the preparation of monthly performance information.

Decrease in Costs: "Executives think intelligent automation will yield an average cost reduction of 22%", according to Deloitte, however "organizations now growing intelligent automation indicate they have already achieved a 27% cost reduction on average from their deployments to date". Thus the above example clearly shows the advantage of cost reduction resulting from the integration of two booming technologies.

Improved Accuracy: Better decision-making is ensured by the utilization of both organized and unorganized data, as well as the repetitive procedural automations, and less human participation leads to more exact results. Deloitte's use of bots to create monthly performance reports revealed that the automation reduced errors produced into that particular process by physical data entry, boosting the precision and accuracy of reports. OCR also fastens processing of data thus bringing about data extraction automatically from various resources.

Enhance the Customer Experience: Technology allows companies to better analyze their customers' demands, communicate productively, and introduce higher-quality goods to the commercials. Customers are often more happy with their purchasing experience as a result. Bots were deployed by GAM, an IBM asset management customer, to deliver efficient customer service and pricing quotes. Thus, the optimization reduced the time it consumed to respond to consumer queries in half, enhancing the customer's overall experience and expediting the purchasing processes.

11.8 Challenges Faced While Integrating Intelligent Automation Solutions

Fragmented Processes: Adopting intelligent automation entails far more than simply automating what is currently being done. Companies nowadays have hundreds of diverse processes, the majority of which are split into functional departments or divisions. Nonetheless, many critical business operations cross organizational borders. Payroll, finance, and IT are all impacted by HR hiring and onboarding procedures. However, rather frequently, these procedures are broken down as records or data are pushed over the figurative wall to the person in the next department. Intelligent automation will play a significant role in reengineering these fragmented processes by allowing lean and cost-efficient operations at significantly greater speeds and 100% accuracy as opposed to the traditional ways [21]. Workers who struggle with several systems to execute a single task can be liberated from repetitious (and which are prone to errors) manual touch typing or button-pushing in order to focus on higher-value tasks. However, before this can be achieved, the workers who do the said tasks must be involved. Experts and advisors in process reengineering are useful, but so are those who are intimately familiar with the processes. The input from these workers is just as valuable.

Lack of IT Readiness: Intelligent automation needs a substantial amount of assistance from the IT sector. Considering the standard RPA, which can be executed by business units with little or no IT help, intelligent automation involves significantly more computation, storage, and other infrastructural resources—not only on-premises resources. Intelligent automation should, by essence, be located on the cloud for scalability and capacity reasons. And this will very probably entail the participation, if not collaboration, of a fully equipped IT staff that is familiar with, if not currently functioning in, the cloud as it deals with huge amounts of data.

Employee Resistance to Change: Intelligent automation consists of several components, one of which is technology. The human element is equally critical. Businesses must carefully evaluate how upcoming changes to roles, procedures, activities, and styles of working will affect people from the get go. Even among firms in the process of growing intelligent automation, 58% had not yet done this type of assessment, according to a Deloitte survey. This vision of intelligent automation, which excludes humans from the equation, is shortsighted and will not pay off in the long run. A comprehensive strategy that develops resilience and adaptability by concentrating primarily on the personnel is required. More than half of the businesses polled by Deloitte (59%) [21] are retraining employees in process skills like active listening and critical thinking, and an equal number are retraining employees in cognitive talents like creativity and problem-solving, so that they could adapt to changes.

Lack of a Clear Vision: Getting it correctly needs the integration of vision and strategy. Many businesses are introducing intelligent automation gradually, headed by either IT or business departments, with no overall plan. According to Deloitte, just

26% of organizations are testing automation out of which only 38% of those who are adopting and expanding have an enterprise-wide intelligent automation strategy [21].

Quality of Data: For the success of AI programs, accessibility to precise, usable, and high-quality data is critical but in the manufacturing business, this might be problematic. Manufacturing data is typically distorted, outdated, and full of mistakes, which can be related to a number of factors. The sensor data received on manufacturing floors in utmost conditions such as excessive temperature, vibrations, and noise are one kind of such examples. Typically, plants have been constructed using a variety of proprietary software that do not interface with each other. This appropriate data may be spread over several databases in diverse formats that are inappropriate for analytics, forcing extensive preprocessing each time. Access to the supervisory control and data acquisition system or process recorders, for example, is required by a predictive maintenance system [22]. It may also be necessary to acquire and modify data via connections or customized scripts.

Trust and Transparency: The intricacy of the technology, as well as manufacturers' lack of confidence in its capabilities, is a key hurdle to widespread AI implementation. People lacking a data science background fail to grasp the knowledge of operations of predictive modeling and data science. They would also lack trust in the AI technologies based on abstract methods. Greater openness would disclose information about the AI process, such as the raw data utilized, the algorithms chosen, and how the model produced predictions [22] which may not be understood by them.

11.9 Misconceptions About Intelligent Automation (IA) in Industry 4.0

A number of myths threaten to inhibit IA adoption, but they can be easily dismissed. The following are a few common misconceptions that people might have.

Intelligent Automation replaces a human workforce: IA does not replace human worklets; rather, it extends it by taking control of monotonous duties, freeing up the human, thought consuming worklets to focus on more complicated issues. It opens up new possibilities based on new abilities that can be learned through retraining. This is a terrific chance for the human workforce to update their skill sets and develop a better foundation for future growth.

Intelligent Automation is expensive: IA is only accessible to larger organizations since it is still in its early stages of development, and early stage created technologies are expensive. As time passes and IA becomes fully established, it will be available to small businesses as well.

Intelligent Automation always makes unbiased decisions: IA makes judgments based upon information obtained and collected, most of which is circumstantial or obtained by the people and business organizations who are in charge of that information. As a result, the choices selected are unavoidably biased.

Intelligent Automation is just another passing trend: This is one of the many misconceptions expressed by skeptics of this breakthrough technology who aren't convinced of its long-term viability. This is far from the truth, as evidenced by the global intelligent process automation market's spectacular expansion, which is expected to reach $13.75 billion by 2023 [23]. While IA is one of many tools in the automation toolbox, it is one that should not be overlooked.

Intelligent Automation is difficult to learn: Intelligent Automation covers a broad range of complex concepts, including coding, deep learning, machine learning, and others. These things are tough to learn, but they PPLare not impossible to learn. There are numerous current technologies that make it simple to study these topics and their applications. Some tools even have drag-and-drop features which can even be done by a person with no-coding experience. It abstracts the user from the lower level implementation and provides a nice interface for the user to work on.

Intelligent Automation can automate anything and everything: It is the most common misconception that IA can automate anything. But that is simply not true. Only manual jobs and repetitive tasks such as sending emails to a group of members, preparing aging reports, manufacturing the same items, and so on are supported [24]; nevertheless, it cannot execute tasks that need creativity and imagination that only humans possess.

11.10 Use Cases of Intelligent Automation

Data accounting for over 80% in a company include unstructured data handled by the use cases of intelligent process automation (IPA) which make use of techniques such as deep learning, NLP (natural language processing), and ML (machine learning) along with optical character recognition (OCR). By unison of RPA and IPA, it can be supplemented by the IPA dealing with the unstructured data [25] and the RPA managing structured data, allowing the company to automate a wider range of processes. Let us see some of the few use cases that are applicable.

Customer Support: In today's environment, no organization can afford to disregard Intelligent Process Automation (IPA) and customer services. The right implementation of current technology to delight clients builds trust and ensures prosperity and a higher rate of return on investment. Today's impatient consumers expect prompt responses and fair remuneration. Due to human limits in handling such a huge number of clients, it is impossible for a human to offer to such a large number of consumers, and the only solution is to facilitate the success of an automatic rule-based solution. Businesses can now answer consumers' frequent inquiries and FAQs [26] with quick first responses due to the automated customer support solutions, thus enabling their human side of interactions to be more reliable.

Invoice Automation: The automation of data entry, reconciling fault repairs, and some decision-making in invoice processing is possible through RPA. However, dealing with the various invoice formats used by different vendors is a hurdle that's where IPA comes in handy: by utilization of machine learning techniques and NPA to evaluate and extract information that are relevant from invoices, standardizing that to a well-structured format, and later return it to the RPA platform for automation of data entry, error corrections, and so on.

Recruitment Process: Another great option for IPA system deployment in large enterprises is recruitment. IPA could enhance the recruiting process by assisting the HR staff in sourcing resumes from a variety of online portals, analyzing specific skill sets, determining value, and sorting through spam and undesired implementations. It enables the HR staff to have access to the necessary capabilities for candidates with significantly less work and cost effectiveness. IPA not only clears out undesirable appeals, but it also assists HR in sorting the proper resumes and enabling access to every application that is specifically tailored to the job requirement. IPA assists the HR staff at every stage of the recruiting process, starting from the process of screening to evaluation to final integration and management.

Financial Document Analysis: Large amounts of data must be gathered by financial organizations for reports on a monthly and quarterly basis. RPA may help the automation of data collection from various structured resources. However, when unstructured PDF documents are included in the process, RPA reaches its limit. That being the case, you need an IA solution with OCR and NLP skills [25] to extract essential data and transform it to a well-structured format that the RPA tool can handle. Financial analysts in equities research groups—those who report on public businesses and provide investment advice—would benefit immensely from an RPA/IPA tag team.

Payroll Transactions: The processing of payrolls consumes a lot of time and is a repetitive operation for any company's HR personnel, also demanding extensive data entry efforts. This frequently leads to anomalies of data, which can create payment delays and employee dissatisfaction. IPA can evaluate timesheets and deductions as well as examine the consistency of employee data across several platforms. The automation of salary calculation operations and handling benefits and payments are possible too. The end-to-end payroll procedures are automated by IPA [26] to eliminate delays and inconsistencies.

11.11 RPA Tools with Intelligent Automation (IA) Support

In recent years, real-world settings like industrial processes, digital amenities, and commerce have successfully adapted the techniques of Machine Learning and AI algorithms. Machine learning "teaches" machines, the methods of dealing with excessive data efficiently, simulation of learning concepts of logical beings and their implementation with the usage of AI techniques, thus forecasting the logical address of attributes such as genetics, statistics, connectionism, etc. It is possible to categorize, associate, optimize, group, predict, find patterns, and so on, using AI algorithms and a machine learning method. RPA has been able to steadily add to its automation capabilities the implementations of AI techniques or algorithms used for classification, recognition, categorization of certain settings such as Accounting and Human Resources. The other researches have shown various potentials of AI algorithms, which have to be used in performing well defined, mature and stable tasks such as increasing job profiles, improvising the accuracy in routing processes, strategic areas focusing on customer tasks, enhancing the customer and employee experiences, upgrading the analytical data analysis, and limiting fraud and "fines" processes. With regard to this context, and based upon the current ongoing, on one hand, if there are obstacles to the ideas of automation using RPA, on the other, these can be enhanced further with the usage of appropriate algorithms and approaches. The following presents the economical and free open-source softwares that are typical of RPA's recent applicability.

UiPath (https://www.uipath.com/): The tool that facilitates the evolution of RPA performances in its particular structure to generate and run programming scripts is called UiPath [27, 28]. This tool allows the scripts to be programmed with a block-based interface and customizes business processes. The three components that constitute the RPA UiPath would include UiPath Orchestrator, UiPath Robot, and UiPath Studio with the former allowing the orchestration of robots. The UiPath studio module is a tool that enables us to construct, model, and showcase the algorithm of workflows, along with assisting in the inauguration and preservation of package transfers, queue conservations, and robot interconnections. As a result, the storage of records and providing their connection to SQL Server and Microsoft's Information Services Server, including the Apache License biased Elasticsearch which is an open source search engine, thus enabling a more potent view of RPA process operated analytical data. A few AI algorithms and approaches are made available through the UI Automation module and have been published on their official website, with recognition, optimization, categorization, summarization, and information extraction standing out among them. For the information sought, AI techniques include picture and character recognition, optimization, and classification.

Kofax (https://www.kofax.com/): The automation of companies' and organizational processes is possible through particular software and Kofax [27, 29] is a kind of a company that creates this software. The applications include document recognition using Optical Character Recognition processes, the resonation of business processes through prescribed flow of business performances, advanced data analysis, RPA, modules for establishment of AI-related algorithms and procedures. The tools can identify the context and content of the document, as well as classify the data in web portals or emails. The unison of ML techniques cascaded with the identification and classifications of OCR documents and studies of web pages or emails is considered as superintendent learning in the classifying and validating the contents of prior sets of information. On the other hand, NLP (Natural language processing) is adapted in supervised learning for segregation purposes or is employed in unsupervised learning for analysis of contents with the aid of information clustering and extraction of density, relying on algorithms used. In this context, some AI algorithms are currently available through the Cognitive Document Automation module and Kofax's Intelligent Automation platform [25].

Blue Prism (https://www.blueprism.com/): Blue Prism [30, 31] is a Process Automation tool that enables you to use software robots to build a virtual workforce. This allows enterprises to automate business activities in a cost-effective and flexible manner. The Java programming language is used to build this tool and it is based on drag-and-drop interface. The four main components of Blue Prism include Application Modeler, Object Studio, Project Studio, Process Diagram. Software tools like process diagrams are used to represent the corporate workflows. The place where process diagrams are developed is called Process Studio. To automate operations, almost all businesses require contact with external programs which is done by Object Studio. You may design application models using the Application Modeler functionality. In this manner, the Blue Prism program can be accessed by the UI Elements of the target application. Blue Prism is a popular RPA tool for creating automation scripts for a variety of departments and tasks. Blue Prism can encrypt or decrypt keys both internally and externally. Apart from that, it also supports Audit Logs, which makes debugging easier for users.

Automation Anywhere (https://www.automationanywhere.com/): Another platform geared toward RPA operations is Automation Anywhere [27, 32], which provides necessary collective information based upon the approaches of AI-driven techniques and algorithms. As an RPA tool for ERP, i.e., Enterprise Resource Planning settings, it encompasses numerous areas of relevancy, including Human resources, Customer interface administration and Supply patterns, and is prone to be integrated with SAP and Oracle ERPs, as well as other ERPs from other organizations. The RPA solution includes a cognitive automation module as well as analytical data analysis tools for RPA procedures. Internally, the Bot tool of Automation Anywhere is used to execute some Artificial Intelligence algorithms, like fuzzy logic, Natural language processing, and Artificial Neural Networks, to extract information from documents and increase efficiency of validation of documents. In this regard, it seems that various AI procedures or algorithms are presently available through the IQ Bot platform by the intelligent word processing application of Automation Anywhere.

Keysight's Eggplant (https://www.eggplantsoftware.com/): Eggplant [33, 34] is a major software test automation podium that automates the generation and execution of tests using artificial intelligence (AI) and analytics. Eggplant's Digital Automation Intelligence platform tests all types of technology on any given device, a browser at all layers or an operating system. It can perform tests from the user interface to the databases. The Keysight and Eggplant's merger unites two complementary firms to become a market disruptor in the automated software testing business, covering the protocol, practical, and application layers. The merger allows both companies to benefit from bidirectional measurement technology, which results in improvised solution distinction in the extended portfolio. It has a worldwide fusion engine that lets it test any system, from smartphones to mainframes. It is compatible with Linux, MAC OS X as well as windows. Its software interface in windows is as shown in Fig. 10.5. It provides end-to-end automation and can handle a variety of platforms to finish the task.

AssistEdge (https://www.edgeverve.com/assistedge/): AssistEdge [27, 35] is an enterprise-grade RPA solution that specializes in helping businesses react to market challenges that necessitate Scalability, Security, Intelligence, and Innovation. It covers the entire automation spectrum, from deterministic to intelligent to human-assisted automation. EdgeVerve Systems, a subsidiary of Infosys owns the AssistEdge tool, which is a private technology with a "opensource" version for the community. Its features include OCR reading and document processing based on the content attached with the document type. It takes in AI algorithms such as Artificial Neural Networks for autonomous data gathering and their inspections by the study of process changes which are based on monitoring of individual processes, and information classification for recommendation processes using data from automation processes (Table 11.1).

11.12 Conclusion and Future Work

Intelligent automation, i.e., the integration of RPA and AI has proved itself to be an effective technology in the current industry 4.0, thus reducing human efforts and time. We now find the majority of business organizations adopting this technology in optimizing, classifying, and automating the commercial processes. The set of exclusive (Kofax, Automation Anywhere, UiPath) as well as open source (AssistEdge) technologies, were identified to be handy tools in achieving intelligent automation in the most effective way possible. These tools use AI techniques and algorithms to implement computer vision, statistics, neural networking, decision trees, natural language processing, and text mining, thus improving the users' ability to optimize and organize the workflow. On the other hand, Industry 4.0 encompasses the merging of intelligent automation, cyber-physical systems, IoT, and intelligent devices. All of these concepts and technologies, when combined, result in a remarkable change in industrial operations thus altering the flow of digital methods throughout the organization. They are now embracing automation with a few steps using robots, (RPA) in order to optimize the operations. Furthermore, RPA now includes intelligent approaches (AI) in many technologies, as we've seen in this chapter, allowing for higher levels of intelligence in process automation.

Table 11.1 Merits and demerits of each tool

Tool	Pros	Cons
Uipath	Good value: contains rich features and cost of entry is low Availability: powerful training and community edition Simplicity: easy to use, quick deployment techniques	Product support must stay up with improvements Security features are limited Eg: Custom privileges Additional features necessitate add-on price
Kofax	Flexibility: usage of already-developed components is really good Pricing: of the solution is quite good Stability: fairly robust, reasonably easy to manage on a server-based deployment and is a stable platform	Documentation is not widely available on the web Difficulties of keeping up with the capabilities as they evolve The discovery of the procedure should be improved
Blue Prism	Technical maturity: durable and scalable product Consumerism: has big user base and penetrations in market Protection: good governance is effective	Expensive Customers do not directly receive the product Deficiency of innovations
Automation Anywhere	Automation: with its no-coding approach, it is quick Cost effectiveness: low entry cost for starters provided by bot stores Reliability: one of the earliest market penetrations	Requirement of programming experiences Better OCR, queue management, UI needed Compound licensing costs
Keysight's Eggplant	Stability: the solutions are stable and scalable Great features: developer interface and the ability to quickly develop are good Availability: the solutions are made available at the highest level	A few of the online aids and user enabled documentation are a little bit outdated and could/should be updated and revised on a more frequent basis
AssistEdge	Automations: good email automation and XL automation Usability:user interface is pretty good Scalability: stability and scalability are fine	Lacks integration with third-party tool Less availability of AI machine learning, natural language processing capabilities built into the solution The support could also be improved

References

1. Sreedhar, B.: How do AI and RPA work together?, 15 May 2020 https://www.automationanywhere.com/company/blog/rpa-thought-leadership/how-do-ai-and-rpa-work-together
2. Van der Aalst, W.M.P., Bichler, M., Heinzl, A.: Robotic process automation. https://link.springer.com/article/10.1007/s12599-018-0542-4
3. Ng, K.K.H., Chen, C.-H., Lee, C.K.M., Jiao, J.(R.), Yang, Z.-X.: A systematic literature review on intelligent automation: aligning concepts from theory, practice, and future perspectives. https://www.sciencedirect.com/science/article/pii/S147403462100001X
4. Ribeiro, J., Lima, R., Eckhardt, T., Paiva, S.: A systematic literature review on intelligent automation: Aligning concepts from theory, practice, and future perspectives. https://www.sciencedirect.com/science/article/pii/S1877050921001393
5. Ansari, W.A., Diya, P., Patil, S., Patil, S.: A review on robotic process automation—the future of business organizations. https://papers.ssrn.com/sol3/papers.cfm?abstract_id=3372171
6. Tyagi, A.K., Fernandez, T.F., Mishra, S., Kumari, S.: Intelligent automation systems at the core of Industry 4.0. https://link.springer.com/chapter/10.1007/978-3-030-71187-0_1
7. Williams, D., Allen, I.: Using artificial intelligence to optimize the value of robotic process automation (2017). https://www.ibm.com/downloads/cas/KDKAAK29
8. Society for Industrial Management and Engineering. Industrial Revolution 1.0—Era of mechanization, 23 May 2021. https://medium.com/spark-by-sime/industrial-revolution-1-0-9e6dc9c62c8c
9. Buckley, C.: Industry 1.0 to 4.0: How industry & factory automation have evolved, 16 March 2020. https://www.ledcontrols.co.uk/blog/industry-1-0-to-4-0-how-industry-factory-automation-have-evolved/
10. Lovell, J.C., Streeck, W., Brody, D., Levine, S.B., Koenker, D.P., Munck, R.: Organized labor. Encyclopedia Britannica, 2 April 2020. https://www.britannica.com/topic/organized-labor. Accessed 17 Feb 2022
11. Catwell: Industry 1.0 to 5.0—a brief history of progress, 20 November 2020. https://community.element14.com/technologies/industrial-automation-space/b/blog/posts/industry-1-0-to-5-0-a-brief-history-of-progress
12. Industrial Revolution—From Industry 1.0 to Industry 4.0. https://www.desouttertools.com/industry-4-0/news/503/industrial-revolution-fromindustry-1-0-to-industry-4-0
13. L2L: The role of IoT and Industry 4.0 in creating digital factories of tomorrow, 22 February 2022. https://www.iotforall.com/the-role-of-iot-and-industry-4-0-in-creating-digital-factories-of-tomorrow
14. Rajkarne, V.: What is the main difference between industry 3.0 and industry 4.0?, 29 January 2020. https://www.quora.com/What-is-the-main-difference-between-Industry-3-0-and-Industry-4-0/answer/Vaibhav-Rajkarne
15. Zakoldaev, D.A., et al.: IOP Conf. Ser.: Mater. Sci. Eng. **560**, 012206 (2019)
16. RPA and Industry 4.0—a glimpse, February 2020. http://ictconnect.in/Tech/Article.aspx?articletitle=RPA-and-Industry-4.0-A-Glimpse
17. Meaning of AI for industry 4.0. https://nexusintegra.io/artificial-intelligence-the-driving-force-behind-industry-4-0/
18. Ribeiro. J.: A (very) brief introduction to AI in the Industry 4.0, 2 February 2021. https://towardsdatascience.com/a-very-brief-introduction-to-ai-in-the-industry-4-0-14e6f4b46cd1
19. Williams, P.: Intelligent automation: how combining RPA and AI can digitally transform your organization, 7 September 2021. https://www.ibm.com/cloud/blog/intelligent-automation-how-combining-rpa-and-ai-can-digitally-transform-your-organization
20. Intelligent Process Automation. https://www.uipath.com/rpa/intelligent-process-automation
21. Automation Anywhere Staff. Top 4 challenges in implementing intelligent automation, 15 September 2021. Retrieved from https://www.automationanywhere.com/company/blog/rpa-thought-leadership/top-4-challenges-in-implementing-intelligent-automation

22. Fujimaki, R.: Challenges to implementing AI in manufacturing, 3 December 2020. https://www.americanmachinist.com/enterprise-data/article/21149328/the-6-challenges-of-implementing-ai-in-manufacturing-dotdata
23. Muhavani, M.: 3 Common misconceptions about intelligent automation, 7 December 2021. https://www.salixdata.com/3-common-misconceptions-about-intelligent-automation/
24. Manasa, G.: Top intelligent automation myths decoded in 2022, 25 January 2022. https://www.analyticsinsight.net/top-intelligent-automation-myths-decoded-in-2022/
25. Indico Data: 5 Intelligent automation use cases where intelligent process automation complements RPA, 14 November 2019. https://indicodata.ai/blog/5-use-cases-ipa-complements-rpa/
26. Tyagi, K.: 6 Real world use cases of intelligent process automation (IPA) implementation, 29 November 2019. https://www.espire.com/blog/posts/6-real-world-use-cases-of-ipa-implementation
27. Ribeiro, J., Lima, R., Eckhardt, T., Paiva, S.: Robotic process automation and artificial intelligence in Industry 4.0—a literature review (2020). https://www.sciencedirect.com/science/article/pii/S1877050921001393
28. UiPath: Prerequisites for installation (2020). https://docs.uipath.com/orchestrator/docs/prerequisites-forinstallation
29. Kofax: Developer's guide version: 11.0.0 (2020). https://docshield.kofax.com/RPA/en_US/11.0.0_qrvv5i5e1a/print/KofaxRPADevelopersGuide_EN.pdf
30. Blue Prism (2022). https://www.blueprism.com/resources/
31. Kappagantula, S.: What is RPA Blue Prism? A beginner's guide to Blue Prism, 6 January 2022. https://www.edureka.co/blog/rpa-blue-prism/
32. Automation Anywhere: Robotic process automation to ERP (2020). https://www.automationanywhere.com/solutions/robotic-process-automation-to-erp
33. KeySight's EggPlant (2022). https://www.keysight.com/in/en/about/newsroom/news-releases/2020/keysight-technologies-acquires-eggplant-.html
34. Teschler, L.: Keysight technologies acquires eggplant, 25 June 2020. https://www.eeworldonline.com/keysight-technologies-acquires-eggplant/
35. Assistant Edge (2020). https://www.edgeverve.com/assistedge/robotic-process-automation-rpa/

Chapter 12
A Comprehensive Review on Artificial Intelligence (AI) and Robotic Process Automation (RPA) for the Development of Smart Cities

Jayanta Kumar Ray, Rogina Sultana, Rabindranath Bera, Sanjib Sil, and Quazi Mohmmad Alfred

Abstract In the future, there will be the conversion from the real world to the smart world which will be expected around 2025. The increase in smartness is possible due to the improvement of technology from 4G LTE to 5G NR. As a result, there will be formation of a new smart city from the old city. The reason behind the formation of smart city involves the application of intelligence in various sectors. The application of intelligence having the phenomenon of automation will increase the smartness and gives rise to the formation of a new smart environment which include vehicles, home, government sectors, market complex, hospitals, institutions etc. The smartness of the device specifies the usage of the Internet in various devices which generally denote Internet of Things (IoT). IoT involves the communication between human and device. In the future, the usage of Artificial Intelligence (AI) and Robotic Process Automation (RPA) will give scope for realization. Around 2030, there will be the completion of fulfillment for various targets and hence the conversion from IoT to Internet of Everything (IoE) will be done. This period specifies the development from 5 to 6G. In 6G, the device will able to communicate with another device. As a result, huge facilities will be available by the application of the internet.

J. K. Ray (✉) · R. Bera
Sikkim Manipal Institute of Technology, Sikkim Manipal University, Sikkim, India
e-mail: jayantakumar.ray@gmail.com

R. Sultana · Q. M. Alfred
Aliah University, Kolkata, India

S. Sil
A.K. Choudhury School of Information Technology, University of Calcutta, Kolkata, India

S. Bhattacharyya et al. (eds.), *Confluence of Artificial Intelligence and Robotic Process Automation*, Smart Innovation, Systems and Technologies 335,
https://doi.org/10.1007/978-981-19-8296-5_12

12.1 Introduction

Artificial intelligence is a smart system technology that directs computers and machines to work smartly, for this reason it replaces human labor to work for human proving to be greater and effective and gives speedier result. AI combined with machines and computers supports human kind of solution to problems and increases the working processes. In recent years cybernetic technology is rapidly developing so AI has been seen in all our daily life cycle like SIRI (Speech Interpretation and Recognition Interface) of information searching tool on computer.

Now AI is a combination of different technologies that work like Human intelligence. AI technologies include ML (machine learning), DL (deep learning) and big data. ML maintains the efficiency of a computer to execute functions having no explicit instruction regarding solutions, used for different types of task like classification, clustering and making prediction about information. Deep Learning of ML concentrated on parameterizing neural networks consists of many layers and can learn representation of data [1].

Intelligence is a combination of three things—perceive, analyze and react. Artificial Intelligence (AI) defines as intelligence system developed by an artificial entity for solving intricate difficulties in a computer or machine system. On broad aspects of Artificial Intelligence can be known as Digital Intelligence, which is able to control computer-controlled robot to perform tasks with intelligent beings. Basically AI is a computing technique given in Fig. 12.1 that allows machines for replicating the cognitive aspects as similar to the human brain having the capability of perception, automated reasoning and interference, common sense reasoning, machine learning, knowledge representation, multi agent system, Natural language processing (speech to text), planning and action, human error correction, complete task faster in much less time than a human takes, logic and learning. It can discover patterns in large data that can execute like human being in a complete manner and the machines have intelligences as equivalent to the human level which has mostly attracted this level of corporate interest. AI also helps in healthcare like serving as a robotic nurse assistant. AI is applied in medical science as two branches like Virtual branch and Physical branch, which help in managing times and automatic scheduling [2].

The Virtual branch—The Component is represented by 'Machine learning' is to transcribe the human brain, the most powerful techniques that have gained a huge transformation in virtual world since over few decades and are now a result of whole economic conditions that also grow up day by day through experience. It includes mathematical algorithms. Machine learning algorithms are divided into three main categories.

I. *Unsupervised*—that has ability to find patterns
II. *Supervised*—that has the ability to classify and algorithms for prediction related to the preview examples.
III. *Reinforcement learning*—that uses only sequence of rewards and punishment to generate a strategy for operation cases in a specified problem platform.

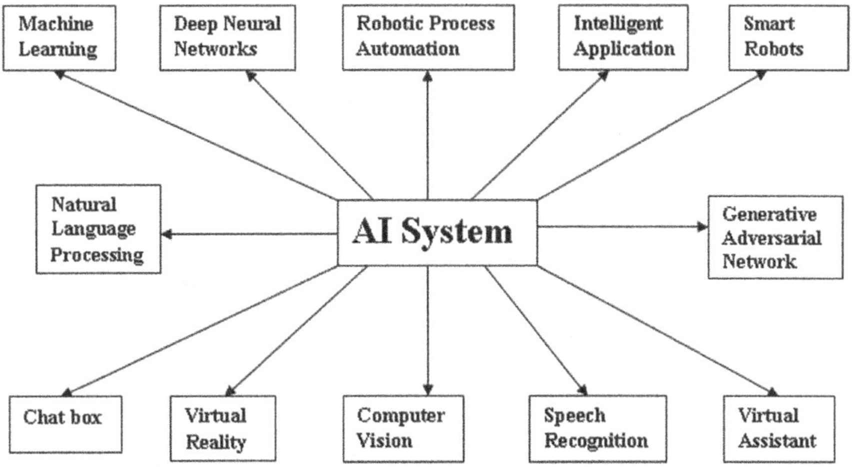

Fig. 12.1 A range AI applications considered in research

Weed Detection by image Processing—Now a days serious issues arise in agricultural research for wild plant growth, which is not in competition with cultivated plants. We can use pesticides to remove the unwanted plants which can also be affected. Recently various advanced weeds can be removed by autonomous developed systems. Variants of these plants are caught as images due to various conditions like moisture, wind and light. The caught images are directed with various algorithms using AI and ML [3].

Vertical Farming—due to lack of lands in city areas, indoor vertical farming could be successfully placed in terraces and balconies of existing buildings. Using recent advance technologies like Hydroponics that uses formulated solutions as well as Aquaponics that uses fish to provides nutrients where as Aeroponics uses nutrient mist to spray on the roots of the plants hanging in air. Sensor technology are used to measure, detect and control the temperature, light, humidity and carbon-dioxide levels in the environment 24/7. To get the estimated result nutrients are regularly inserted and calibrated according to stages of growth combined though AI and IoT [4].

The context of Industrial AI (Fig. 12.1) has references regarding the solutions in digital platform in which domain expertise along with AI techniques are merged for the improvement of business outcomes across the industrial and journey value chain and fulfill the target towards automation. AI emphasizes operational efficiency and performance through a unified data fabric that is straight forward for access, transformation and analysis. AI creates objective intelligence that is commonly used by the process expert. It is a user friendly analytical tool which changes data into easily perceivable guidelines that key in operational decision. AI in industry also identifies solving operational pain points through comparing customer words with the actual data that make sense for all users. AI helps to influence operational knowledge by connected factory nerve centers which perform as human nervous system that helps reducing down time prolonging equipment life cycles and formation of a digital storage of worksite wisdom acuity [5].

Robotic Process Automation (RPA)—RPA is the application of technology in which employees in an organization are allowed to construct software or a 'robot' for capturing and illustrating the existing applications to process various transactions, data manipulation, responses due to triggering and communication with other digital systems. The terms Robotic Process Automation (RPA) inspires to think about robots doing human tasks, and this solution is done by the software. For business processes, RPA means the calculation of a human worker by the utilization of technologies whose principle includes the tackling of ordered and constant tasks, profitably and quickly. The aim of RPA is to done in an outside-in manner by automation instead of people [6].

Now the future technology is based on Robotic Process Automation (RPA) which involves the experience and development of growth of interest in recent times. RPA is one of the most advanced technologies that a new technology has emerged that is mainly dedicated on the automotive features of repetitive work by the process of automation, routine, and human tasks based on rule. It provides good output to the sectors in which executions of solutions are based on software and hardware, and automation for doing very simple things [7] are decided. RPA is a technology of software to makes it easy to form, install and manage software robots that follow human actions linking with digital systems and software which can base on AI insights to done tasks with no lag time and enable us to accomplish digital transformation [8].

RPA contains hardware, software, automation and networking; this combination will perform the complete case turn into very simple cases. RPA is an arising form of business platform utilizing the automation phenomenon dependent on the software with robotic technologies or Artificial Intelligence (AI) workers [9]. RPA can be utilized in the set of approaches for processing the system. Mainly it can be specified as an optimization with process oriented and management approach having the resultant of a clear vision. These approaches include various stakeholders such as Subject Matter Expert (SME), Software Developer (SD), Business Analysts (BA) etc. at various situations which are arranged in a life circle with respect to the application of RPA techniques. The application of RPA techniques depends a life cycle based on Deming's cycle known as Plan-Do-Check-Act (PCDA) cycle. To optimize the process, *Fleshing* et al. combine RPA with Business Process Management (BPM). RPA life cycle has six phases like:

I. *Phase of Analysis*—It includes analysis and deciding the activity of achieving the automatic process.
II. *Design phase*—It is a branch of data flow, works etc. which must be the inclusion in RPA process.
III. *Phase of Construction*—It consists of executing every automatable parts of various processes in the previous phase (design phase).
IV. *Phase of Deployment*—The robot gets from the construction phase that needs an environment where it to be carried out, like humans need an environment for performing their work.

V. *Phase* of *Monitoring and Control*—When the robots are applied in their implemented environment, it supervises controlling and monitoring the action of a robot.
VI. *Phase of Evaluation and Performance*—End phase contains the evaluation and performance of robots action.

This chapter is organized as follows:

Section 12.2: **Literature Review** mentions the review of the Artificial Intelligence and Robotic Process Automation along with their contribution required for the development of smart cities.
Section 12.3: **Impacts of RPA and AI to Develop a Smart City** describe the importance of AI and RPA in various sectors such as industry, banking, energy sectors, 5G etc.
Section 12.4: **Applications** specify the role of RPA in intelligent auditing along with their steps for revenue audit, and its usage in Digital Forensics (DF).
Section 12.5: **Advantages** include the role of AI & RPA in business model design for digital platform representing various digital technologies, low cost outsourcing, decision support system, importance in pandemic situation regarding health care, education, travel and tourism, manufacturing and distribution, agriculture and food production, and facility of robot system for senior citizens.
Section 12.6: **Blessing or Curse** involves the importance of AI & RPA in human life regarding blessing or curse. The blessings involve various opportunities, increase in facilities, and advantageous than the previous stage. On the other hand, the curse include the disadvantages, unavailability of the facility due to network problem, absence of smart system in rural areas due to lack of infrastructure etc.
Section 12.7: **Smart Environment** specifies the usage of various sensors for the collection of data, execution and evaluation of the output. Both AI and RPA play an important role for the development of smart environment. This section includes various use cases such as customer support, fraud detection and prevention, smart city projects etc. The services in smart environment include the improvement of smartness in energy, buildings, irrigation, mobility, waste management, meters etc. Side by side, there includes the illustration of smart city services provided in cities with advanced technologies.
Section 12.8: **Internet of Things (IoT)**—A model having the massive number of heterogeneous devices are connected to the internet and the identification takes place through various IP address and protocols. The importance of IoT gives rise to the increase in smartness for various sectors such as banking, transportation, industries, homes etc.
Section 12.9: **Conclusion** provides the overall idea of AI and RPA and should be applicable for the development of smart cities. The application of AI and RPA gives rise to the development of 5G technologies. In 5G, the utilization of AI and RPA will be in a partial manner and gives rise to formation of smart environment. Day to day, technologies are developing which leads to the improvement in smartness. As a result, full smart environment will be created when AI and RPA will be fully utilized and give rise to the formation of 6G technology.

12.2 Literature Review

The literature review in this section spreads the important contribution of AI and RPA automation to develop smart cities. The first thing is to review about technologies that allow the Intelligent Automation as well as Robotic Process Automation. This review approaches the important topic since researchers and media articles eccentrically use different terminologies with regard to awareness and service work automation. Next, the actual impacts, applications and advantages of RPA and AI in industries, healthcare, energy sector, intelligent auditing, banking sector and digital forensic department to involve a complete understanding of the process of business value using automation was focused. Day by day digital intelligence levels are increasing incredibly, and human societies are also bound to upgrade themselves in this evolution.

The term 'Smart City' has been pre-owned for many years by so many technological companies. Now we are converting from 'Digital City' to 'Smart City' and serves as an explanation for the composite system to build up the urban framework like education, electrical, buildings, water supply issues and public safety [10].

Smart city idea incorporates a large number of services, inclusive of smart health care, smart attire, smart network system, smart transportation and smart services. Despite the fact these utilities are significantly disguisable according to their demands, all of them can be learned under an excellence operative architecture, where we have mainly five planes like:

1. Application Plane: It is the connection plane between a smart city and its citizens. To decrease the cities outlay its focus is to control assets misapply, assured future of the city though present everywhere and continual observation.
2. Sensing Plane: A sensing plane includes a large number of sensing devices and actuators to monitor the signals present in environment and interconnect with city lights. The sensing plane helps to execute with Wireless Sensor Networks (WSNs) in smart cities and is determined to the same restrictions.
3. Communication Plane: The communication plane provides the preliminary processed and the accumulated data is obtained from the sensing plane and brings them to the other layers, mostly associating the ground devices to the cloud.
4. Data Plane: It is the interaction end of the obtained data, where the expansion of apparently incoherent data is turned into sensible facts. These techniques need highly mathematically efficient host to elaborate further algorithms.
5. Security Plane: The field is now evaluated to be in its babyhoods. A number of multiple undeveloped services are present formally to conduct research with attainability of many more projects and favorable receptions by the citizens. In this circumstance privacy and security of the users are abandoned [11].

Global communications were established through Satellite systems in the last few years, whereever connections were within the limited bandwidth, and is costly but slow. The terrestrial and undersea fiber cables were to come into existence to overcome this problem and firmly fixed the networks and inverting communication in a faster and uncomplicated manner. It is contemplated that in the next few decades, more compatible networks and technologies will be established and will increase the carrying out of so many individual business organization.

In last decade, the IoTs, the innovation of Artificial Intelligence (AI) and Robotics Process Automation (RPA) its usages within so many respects of our lives have covered the digital age's read to association [12].

12.3 Impacts of RPA and AI to Develop a Smart City

12.3.1 RPA and AI in Industry 4.0

Now a days in any industries, digital services are available and seen as increasing fusions. Automation of services by RPA generate the work that also can humans do. The automatic process is done by AI robots that are skillful to execute, and perfectly, repetitive works.

In RPA some commercial and open source tools are highly applicable, like UiPath (RPA function developmental tool which form an execute programming script), Kofax (Process automation software are formed by this company), Automation anywhere, Win Automation (This tool incorporated in the RPA with automation Processes, that gives a flock of new features like automatic emails, PDF and Excel etc.), Assist Edge, Automagica (This tool is exclusively with an open source version, code available on GitHub, and formed in Python language) [10].

12.3.2 Importance of AI and RPA in the Banking Industry

The application of the robotic method automation software system like Blue Prism, UiPath is used to put in computer and end user device—Level software system robots that assist method banking work which is repetitive. The advantages of RPA in the banking sector is used for lower cost, faster speeds, better scalability, increased accuracy and improve compliance. The following processes are given below which for robotic process automation in banking industries is related to account (origination, receivable, payable), processing (Deposit, Mortgage, Loan, Investment, Cheque), duties of employee (on boarding and off-boarding), other Services (Surrenders, Lapse, Underwriter Support, Collections, Customer service, Billing, Service desk) etc.

AI and RPA has a huge contribution to develop banking industries like twelve times increase in speed to open accounts using computer on boarding and monitoring, three hundred hours per month save times to ready external reports for commercial loan organization, and two times ticket processing speed increase for customers service.

RPA and AI involvement to the financial world helps in banking sector to remove human environment to a huge extent in the implementation of prior human labor heavy works, so RPA improves both back office process and customer experience by saving money from labor expenditure while doing more work RPA. RPA also helps to detect fraudulent activities that are primary concerns of banking industries. RPA serve the customer information through web sites that are required for Financial Institutions [11].

With the help of RPA, banking sectors are resolved of many challenges like credit understanding by using software robots that reduced inefficiencies due to work load on the human labor, credit assessment by robots that check credit applications data include income/expense, then post it in a report for credit analysis, and fraud detection by RPA software robots that check internal and external data base for doubtful activity [6]. RPA is an important tool for the banking sector and its improved efficiency is required for significant growth for all banking sectors [11].

12.3.3 The Impact of Road to Intelligent Automation in the Energy Sector

AI with RPA closely at its high points in terms of awareness and capabilities, organizations is discovering what's beyond it. The road intelligent automation must involve a cognitive road map that each traveler's solution should consider it before introducing. There are three different types of technologies that involve behind which the systems are running:

- *Class I*—Basic process automation that consists of application, add-ons, macros which sit in the presentation layer and are not involved with an IT application.
- *Class II*—Enhanced/Intelligent process automation that contain technologies are using natural language processing and it can transfer unstructured data. This data are applied that understand process automation.
- *Class III*—Cognitive platforms are comprised which attempt to solution problems just like humans.

Some values with RPA solution are given below:

I. Transformed Cost & operating models—Enterprise wide application of IA (Intelligent Automation) across every business event involved RPA & AI tools and methods.
II. Enhanced and new insight—Intelligent process optimization solution permissive precise online forecasting of atmospheric distillates quality parameters involving IoT devices and Machine Learning (ML).
III. Improve ability and quality of judgment best processes—Mix of rules and cognitive automation leveraging to divide and review Disney products test records.

IV. More engaged and productive work forced—Automated IT service desk work to accurate IT team through chatbot along with RPA bots for resolution.

So Intelligent Automation (IA) will have a huge impact on the market that can be compared with the appearance of the internet back in the 90s [5].

12.3.4 Importance of AI in 5G

The new radio interface depends on Massive MIMO, where 5G has solved some of those challenges. The main problem is to provide facilities for citizens regarding markets, electronic health, Industry 4.0, smart grid, smart education etc. So these problems are furtherly solved by establishing more partnerships on multiple layers for sharing among various mobile operators, platform as a network service by providers [12]. The acceptance of AI and ML approaches as core part of AI is important to specify the evaluation of the environment and requirement of services to form a proactively and efficiently self-optimizing and self-updating network. As an example AI/ML has an important contribution for massive MIMO to recognize the user distribution dynamic change forecast by examining historical data, dynamically the better weights of antenna elements by the historical data to better cover in multiple scenarios on the whole inter-site interface bandwidth 5G, MIMO cell site etc. [13]. ML and AI are combined into the network edge that can be reached by utilizing 5G networks. At first, ML and AI are integrated, then with the help of 5G Multi access Edge Computation (MEC), the traffic steering across access better network can be controlled. The 5G networks are allowed by the combination of ML and AI that is essential for successful function [14].

12.4 Applications

12.4.1 Role of RPA in Intelligent Auditing

In this current situation RPA has been applied in business organizations from credit automatic calculation to customer accounts. According to the auditing respective, many audit tasks like reconciliations, internal control tasting, and detail tasting can be automatically done by RPA. In RPA, some automation tools are used in audit tasks like—Excel Macros, IDEA, and Python.

RPA vendor tools are UiPath and Blue Prism for Detail testing, Input—Collection of information, output—Compilation of audit test results.

The advantage of RPA in audit provides the deduction of time for processing, along with various facilities such as reliability, perfection in audit trails with expansion in service quality and better security etc. [15] (Table 12.1).

Table 12.1 RPA steps for revenue audit

Steps	Features
Reconciliation	• Audit evidence that is supplied by client is login into FTP for accessing • Enter query to search for sales listing and trial balance • Extract sale listing and trial balance • Important data of sales listing and trial balance to Excel or IDEA • Calculate total sale per listing • Compare total sale per listing to total sale per trial balance
Analytical procedure	• Login into audit work paper software to enter previous year audit work • Enter query to search for the audited revenue amount • Extract a report with previous year revenue balance • Import report to Excel or IDEA • Compare total revenue amount from recent year t o previous year • Forming alert if difference exceeds materiality threshold
Internal control testing & subtractive testing	• Login into FTP to acquire audit evidence provided by the client • Enter query to search for purchase order , invoice, shipping listing • Extract listing • Import listing • Compare price to quantity among the three listing

12.4.2 RPA in Digital Forensics (DF)

In DF, AI with RPA is the new command placed upon this sector which is really amazing. Presently the complex cases are easily solved by digital evidence that gives different types of investigation support in many criminal cases, where the specialist practitioners must be required for the extraction of various processes and evaluate the final report about this content having no delay in a well-thtrodden ground, budget limitations, multiplicity of devices and growing in the volume of digital information at needs to be processed.

The capability to increase the speed for the processing of digital data, with high accurateness and reliability has lasted with a long time target for research in this area, work done as having the potential to growing efficiency and keep steps with the demand placed upon expounders utilizing processes for criminal justice during

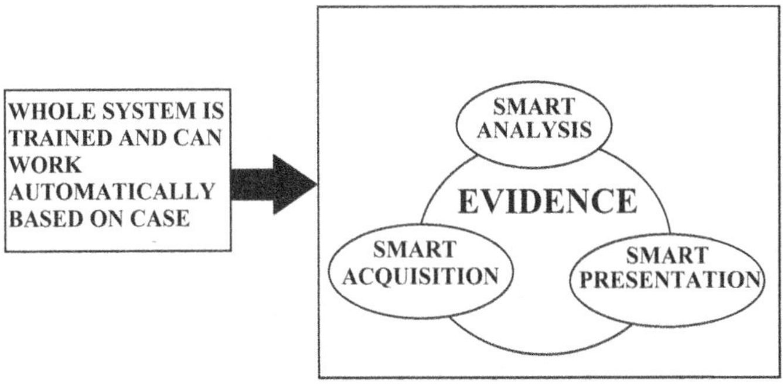

Fig. 12.2 Proposed AI based digital forensic framework (*Courtesy* Haidine et al. [16])

criminal cases. Figure 12.2 shows the proposed AI-based Digital Forensic Framework. To combat the challenges of handling huge numbers of caseloads, massive amount of smart data and management of the ability of examination, RPA is the only solution that improves efficiency. The efficiency of techniques that are involved include some digital forensic examination processes which develop the capability to do this work easily through RPA, a form of service task automation. The aim is offered by debating where and whether technology has an area in the growing efficiency in this field. This RPA is utilized within the DF environment and at the time of writing there are no needed academic lessons available documenting the use of RPA in this context. Works that automate DF processes are mainly joining with cynicism and concern. But they only depend upon automatic processes in criminal cases where the chance of risk remains high. Yet the given tasks are automated and are restricted to specific objective functions where results can be evaluated and the perfection of task completion can be judged, with the risk remaining low [17].

12.5 Advantages

RPA and AI automation is majorly involved in financial section to continue the ever changing and competitive industries.

12.5.1 Role of AI & RPA in Business Model Design for Digital Platform

See Table 12.2.

Table 12.2 New digital technologies improving the business model design for digital platform

Digital technology	Description
Banking, Financial, Services, Insurance (BFSI)	This sector is operating on automation business tasks ranging from data entry, compliance regulation and also speed and efficiency increased
Combination of light Detection and Ranging (LiDAR)	Lead computer vision solution for automated factory floor monitoring. It used AMRs to dynamically operate the map premises
An intelligent automation solution ML-driven text analytics and RPA components	It facilitates users to extract data from the huge corpus of legal documents for next time processing
Heat map generation using AI/ML	Broad Variety of applicability of this use across industries like retail, transportation, public service etc. that helps with guiding the movement of people with efficient space, improving layout design, traffic management etc.

12.5.2 RPA Emerges as a Threat to Traditional Low Cost Out Sourcing

Blue Prism, is a workforce operating system software that defined a robot as trained by a work flow of the procedure. This workflow is controlled and audited by a document procedure. Management information is collected automatically as the robot operates. All processes are developing a statistical data as a byproduct of doing the action. This permits tuning and expansion of a process in light of real data as fast and affordable. This technology introduces the best applied processes which are driven by rules and the needs for which it is planned for the justification of growth done by various organizations like Service Oriented Architecture (SOA) and Business Process Management (BPM) [18]. Some advanced technologies will further derive automation and insurance processes like Blue Prism version 6 which is a digital workforce operating system software, where technologies like AI, ML etc., that easily integrate with the business operation to deliver value are used [19].

To achieve the success in the business automation we need to apply the right technology and strategy using RPA, AI, Natural Language Processing (NLP), and Optical Character Recognition (OCR) [20].

12.5.3 A Decision-Support System in Selecting Process for RPA

A future source of success in business as a modernism is to drive not only the routine activities of a company but also in this digital period, needs companies to invest in value creation. With the help of artificial intelligence and BPM, RPA gives the assurance of robots with a virtual work ability that play these tasks in a self-determined way. By process mining techniques, we introduce an indicator system which is automatable as well as the current situation and judging decision support for many companies which seek to better their RPA activities and increase their return or investment [21].

12.5.4 Important Role of AI and RPA Due to Pandemic Situation

During the global Corona pandemic, the world has dramatically changed due to a rear disaster, has resulted in a huge number of human lives being lost. Due to this pandemic combat situation AI and RPA are one of the great contributions, such as Generative Adversarial Networks (GANs), Extreme Learning Machine (ELM), Long/Short term Memory (LSTM). AI describes a combined bioinformatics approach that has many aspects of data from researchers and doctors in a structured and unstructured form. Now the main purpose of AI is COVID-19 case screening. New Covid case improvements and prediction result in a better tomorrow, timely response and efficient outcomes etc. Due to the pandemic outbreak physical meetings are totally closed, even doctor's appointments were held remotely in order to decrease the contact with patients. In this situation all the sectors depend on both wire and wireless communication networks to continue lockdown situation is under control [22]. The key sectors which is affected by COVID-19 pandemic situation are—Healthcare, educational system, travel tourism, manufacturing and distribution, agriculture and food production etc. All those systems are more or less controlled by AI and RPA [23], which helps us to maintain to continue our comfort daily lives and helps us to grow in a smart human environment. In engineering, medicine, Psychology, economy fields, AI is the novel growing topic for researchers [22].

Role of AI with 5G in Pandemic Situations

Health care	
Application	5G Technology
Wearable	eMBB, mMTC
Telemedicine	eMBB, mMTC
Remote surgery	eMBB, mMTC, uRLLC

(continued)

(continued)

Contact tracing	eMBB, mMTC
Wearable	eMBB, mMTC
Education	
Remote teaching	eMBB
Remote assessment	eMBB
Remote conferences	eMBB
Immersive learning	eMBB
Travel and tourism	
Intelligent transport	eMBB
Autonomous vehicles	uRLLC, eMBB
Interactive maps	eMBB, mMTC
Immersive tours	eMBB, uRLLC, mMTC
Manufacturing and distribution	
Industrial Internet	eMBB, mMTC
Smart manufacturing	eMBB, mMTC, uRLLC
Robot control process	eMBB, mMTC, uRLLC
Remote supply delivery	eMBB, mMTC
Agriculture and food production	
Smart agriculture	eMBB, mMTC
Smart irrigation	eMBB, mMTC
Remote crop monitoring	eMBB, mMTC
Farm data acquisition	eMBB, mMTC

12.5.5 Communication Robot for Senior Citizen Based on RPA

Now a days an AI speaker namely a communication robot is popular among consumer services. Home applications that are connected with internet and operated by human voice are latest robot technologies of this communication but our elder community cannot control it properly. Although many elder persons can't use smart phone and they don't know about the latest technology they are not connected with the latest IT devices. Due to this condition, different watching services for seniors are offered. The formation of services gives rise to grasp and examine condition or identification of abnormality for the senior citizens that is done by a sensor and informed to their families.

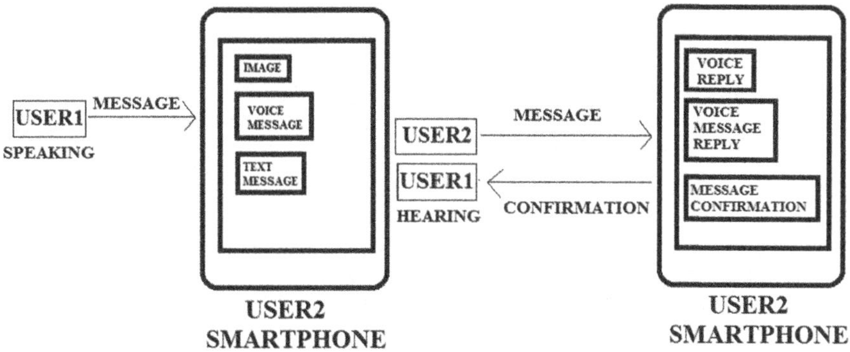

Fig. 12.3 SNS agency robot [24]

As an example, 'MIMAMORI Hotline' services by ZOJIRUSHI that transmit the information of elders by mail to the elder's families who stay in separate places when he or she utilizes this electric device. Other services like the SNS agency robot (Fig. 12.3) which authorizes the elders regarding communication in an interactive manner done by Twitter even though they are inadequate to use smart phones.

The principle of this robot is that the proper address is previously stored and executed on the said depending on its context and it responds just by the elderly talking to the robot, since the AI and IOT are not interconnected following the process of automation.

PapeRo is a SNS agency robot that was designed by service oriented architecture. It can be functioned by a voice. The elderly can transmit the video message through this robot.

A designed communication robot broadly judges dementia depending on normal conversation with the senior citizens who are living alone and detect suspected dementia to their families via social media. In SNS agency robot, dementia symptoms are previously loaded by which it can easily detect the disease.

Currently, the elderly's speech is changed to text by Google cloud speech API and it is processed further by IBM Watson/Assistant. Some suitable questions on the topic are chosen based on the replies of the elderly, then the answers are analyzed by Google cloud speech API that converts text to voice by the machining learning process. Google cloud speech API can detect more than 110 languages and dialects [25].

12.6 Blessing or Curse

AI & RPA is Blessing or Curse for our Smart Life!

Currently, our society is changing continually and at a growing speed, in this situation the corona pandemic does not change that but it also helps with the growing technological speed. From morning to night without technology and artificial intelligence our life is difficult to imagine and those who refuse to do so may miss the boat. AI and ML is not only a red signed robot but it also can destroy our next generation. In our daily lives we are searching for anything from google by using AI and ML. For example, the Californian tech uses Machine Learning to upgrade the accuracy of its search result to make data finding faster and easier. The main goal of AI is to attain more control and produce greater profits, be through medical or military technology, more efficient production, cheaper and better knowledge of the customer. Comparing human beings with technology, technology can makes our lives more comfortable, but they have not developed culture and morality in the world. Human intelligence is permeated by emotion and intuition but AI doesn't provide it.

For our next generation AI creates some opportunities by using this technology execution. Self-driving a car is the most famous example where human knowledge and expansion are growing at a rapid speed.

Another sector in life where AI will have a huge impact is the health care area According to BGV in 2018 the invested amount will be $6, 6 Billion. In the next generation, robots help in surgeries that will be designed by artificial intelligence for comparative better sureness and objectivity required for the treatment of patients.

AI conquers radiologists in recognizing the cancer of lung and that is a great blessing of our lives. But these technologies replacing workers by machines might be seen as a big curse for us [26].

Democratization of artificial intelligence:

> …the one who becomes the leader in this sphere [AI] will be the ruler of the world.
>
> —Vladimir Putin.

AI describes by National Institution for Transforming India (NITI) Aayog which refers to the ability of logical operations like learning problem solving, controlling, recognizing and sensing. NITI Aayog figured the basic 5 sectors which have a large positive on Indian residents and the world at a huge level.

- Enlargement of access and reasonable rate for standard healthcare.
- Increase in fertility of the soil suitable for agriculture and provides profit for the farmers.
- Development of the standard of education by enhancing access.
- Expansion of the urban community due to AI makes the city more advanced and smart than the previous stage.
- Smart transportation solves the problems for traffic and cloud gives rise to a smart and secured transport system.

Some barriers are identified at large-scale acceptance of AI:

- Lack of board-based facility create problems in analysis and execution of AI.
- Absence of allowing data ecosystems create problems in accessing smart data.
- Massive valuable assets and cheap recognition for verification.

- Security and protection, as well as a problems in following legal order around secured data.
- lack of collective attitude to appropriation and appeal of AI [27].

So, AI curse or blessing in our daily life—there are no right or wrong options, when discussing about AI. For our improving and growing future at an exponential rate there is mainly dependence on technologies rather than human performance [26].

12.7 Smart Environment

A smart city is a technologically modified modern urban area that uses various types of electronic method sensors to collect specific data, voice activation method etc. The information which is obtained from various resources is used to improve operations across the cities. Data collection in the city from citizens, devices which are processed and analyzed by monitor and control traffic with transportation systems, waste management, water supply chain, detection of crime, information systems, school, hospitals, libraries and other community services and hence it is defined as smart management both in the ways in which their governments applied technology as well as in how they perform surveillance, analyses agenda and rules or govern the city.

AI has the immense possible catchment basin. In this case for RPA, AI develops the functions of RPA permissively to manage unstructured information and enlarge its awareness utilizing Voice Recognition which was enabled by AI. As a result, RPA will be more reliable. Where RPA stops then AI takes over and helps to achieve a continual growth. In the market some AI technologies help to make RPA extraordinarily intelligent. In a simply unstructured information, the aim of AI is to the extract the information present in unstructured sources like emails, chat logs, social media feeds, free text word documents, pdfs, images, etc. As a result, the unstructured information becomes structured by utilizing RPA into a new structure level. The function of AI involves the decoding of human voice and it performs the actions according to the voice requests. The aim is that it utilizes NLP to digest the language in the manner that we talk. The effect of automation is taken by ML, and further enables the algorithms to learn on a daily basis, the solution becomes more easier and performance will be reliable due to the development of the intelligent system. Enabled by AI, RPA manage the requirement of human interference for judgment based actions chosen between various options to hold the smarter way. With the enhancement of AI, RPA includes the ability for the identification of text documents, where classification can be done, avoiding unstructured data and its conversion to structured, thus indicating it to make it explorable. RPA carries the explored information between systems and then it is converted into voice and text by AI.

When RPA look into doubtful blocks in face of varying arrangement, various documents, different content can take the responsibility of the next procedure level with AI. Various processes like invoice, claim process, on-boarding, etc. are now seamlessly handled with human interference.

12.7.1 Use Cases Where AI Can Develop RPA

Customer support: It includes the identification of different sectors in email and transfer it to the admissible workflow. The ability involves the identification of the chat request and carry the request to the right workflow to recover the information. The performance of all activities seems better due to higher accuracy and faster TAT in real-time resulting in customer satisfaction.

Fraud detection & prevention: The RPA and AI technology is a compact sitting on to a massive Data that are produced during various transactions and have the ability regarding the identification of different patterns, however, few of them may be fraud in nature. RPA analyzes these type of transactions following the procedure of scrutiny by differentiating them against the identified patterns and thus getting rid of frauds and other negligence.

Smart city projects: RPA collects data from various sources such as sensors and connected devices. These are taken by AI and the patterns are detected to generate the comparison via the desired values. This event takes place in real-time while the different conditions of a smart city are monitored. RPA then specifies the red flags for an announcement system. AI develops RPA to not only take the technological scene to a new high but also enables the facilitation of new business use cases, which were quite incomprehensible in the past [28].

Digital cities obtained from digitized presentation of cities that digitized image of cities, smart cities got from new intelligence of cities that present allied and distributed intelligence. The other name of smart cities is 'cyber city', obtained from cybernetics, cyber space, city surveillance, and control depends on data feedback from the cities which are collected from the city data bank.

For improving the quality of life in urban cities, municipalities have a planning to implement an efficient service through technologies that improve their daily life. So paper workout is replaced by electronic data by an e-office. As an example, e-participation was obtained via direct communication through social media. Public authorities consequently distribue a group of services and infrastructure, depending upon ICTs. Now web page design, management of content permit and opportunity to form its own web site are achieved for services that municipalities and other providers which deliver to the needed groups.

12.7.2 Services Applied in Smart Cities

The providing of real time information about urban environments is the most important thing for ongoing various helpful applications and services. In urban areas smart cities concept based on Information and Communication Technologies (ICT) that are worked through some data bases and collected data from buildings, human beings, cars, machines, devices in the city systems namely transportation, water, energy, material cycle, communication, surveillance etc., therefore formed an infrastructure to remote monitoring, diagnostics, fault detection, energy grid and power meters etc. So data, city systems and their connection to the sensing and control city subsystems are given.

Sectors of smart city applications are:

I. Smart energy and smart grid—energy saving and reduced cost
II. Smart buildings—cutting energy used
III. Smart water—decrease of water losses
IV. Smart public service—faster, more cheap
V. Smart lightening—energy efficiency
VI. Smart mobility—less traffic, reduction of travel time
VII. Smart waste management—improved cleanliness
VIII. Smart meters—accurate bill, control of energy usage [29].

12.7.3 Main Objective of Smart City Services

The aim to reach the destination is to be more active in receiving the quality of service and to stream line business processes and including their increasing requirements for the abilities needed to form smart city services. The next aim included needs for the formation and generation of electronic services than other competitors and improved performance level for good decision-making and processes. Figure 12.4 illustrates the services that are provided in cities with advanced technologies.

The target is to position monitoring and formation of a dispersed network of sensor nodes having intelligences that include the measurement of many parameters for a better operations in the city. Recent advance technology in wireless sensor networks have been examined utilizing the range of underlying technological advances, first being the improvement in MEMS sensor technology, and different ways to power consumption.

12.8 Internet of Things (IoT)

IoT is a model that consist of a large amount of heterogeneous devices that are connected to the internet and it can identify them via IP addresses and protocols [30]. Due to IoT, the transformation of Smart city, smart homes, pollution control, energy saving, smart transportation, Process of smart banking and smart industries etc. are increasing. A lot of investigation and research studies have been done to enhance the technology through IoT [31]. In this system all devices with sensors and actuators are wirelessly connected to the network. Radio Frequency Identification (RFID) is considered to be an essential of IoT since it is trusted that everything in our daily life could be identifiable with the use of radio tag. In cloud computing, the task of dividing computational resources and providing services to the device

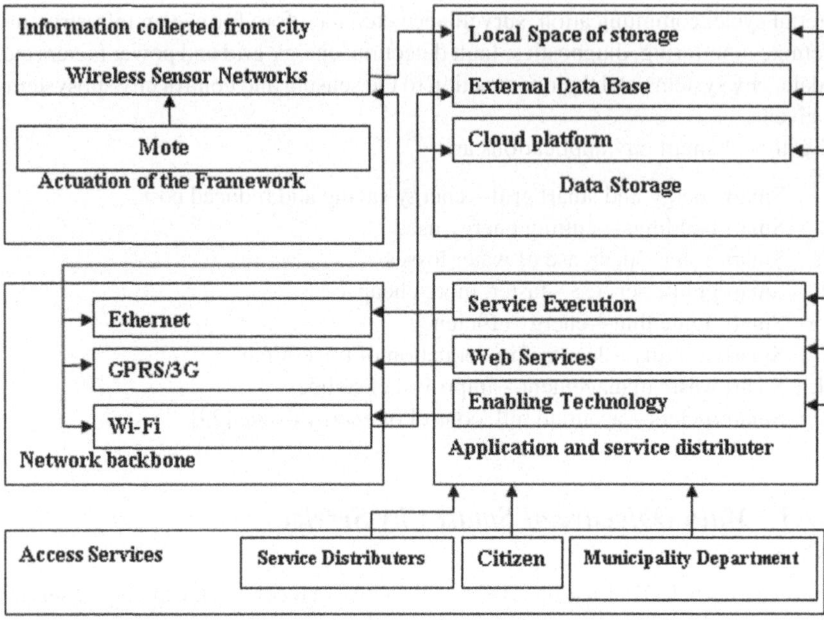

Fig. 12.4 Services that are provided in cities with advanced technologies

is achieved through internet. All these technologies are mainly used together and helped researcher whose works are related to Source Code (SC) [31].

In the remote areas, due to the lack of infrastructure in the hospital, patients are not getting proper treatment. Side by side, the unavailability of doctor is the another main issue. Hence the patients in the remote areas are misguided and have to come to the urban hospital for proper treatment bearing huge expenses. In the future, this problem will not faced by the patients staying in the rural areas. Now IoT with RPA will able to develop the Tele surgery system which will assist the patient and doctors to undergo proper treatment.

Figure 12.5 shows the tele surgery system where the signals and parameters will be exchanged between a remote surgeon and a robot present in the local hospital. The wireless connectivity is done by 5G communication systems. The remote surgeon sends audio and position/velocity signals to the robot. The robot consists of arm and finger. The robotic finger sends accurate haptic feedback i.e. kinesthetic, as force/motion or tactile as vibration/heat along with audio/video data. On the other hand two high definition video feeds (one for each eye), two way audio, one way haptic feedback, and robotic control signals (position/velocity) are received by the remote surgeon. As result, the Tele surgery phenomenon will be totally fulfilled.

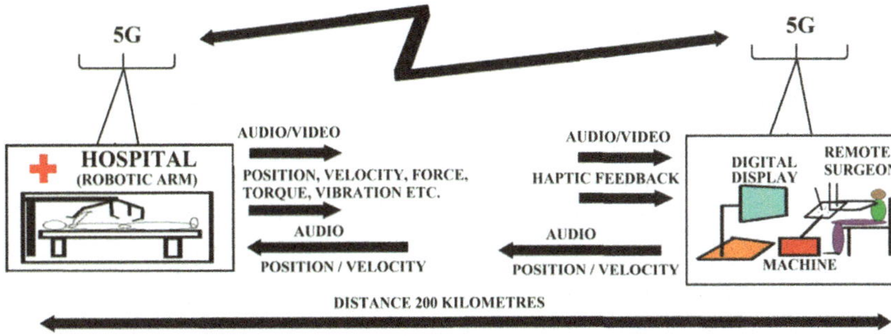

Fig. 12.5 Exchange of signals and parameters between remote surgeon and a local hospital having robotic systems

12.9 Conclusion

In the development of a smart city, the automation process is unavoidable as technology continues to develop faster. There is a significant opportunity for provincial and cross-country collaboration and intercity researching with concern to the design, development and management of smart cities and their basic structure. To develop a smart city, at first we need to develop network technology for faster upgradation. In this paper we analyzed that what a huge contribution of AI is needed to build up the cellular network. The intelligence requires to develop practically in all areas of cellular networks which operated by upgrade generation (4G, 5G and in future 6G) to explore the upgradation. In new generation with the help of RPA, the world of mobile wireless communication is quickly developing. The parts of the article presented ***Role of AI & RPA in business model design in digital era,*** RPA emerges as a threat to traditional low cost out sourcing, Importance of AI and RPA in the Banking Industry, Intelligent auditing, Impact of RPA and AI in industry 4.0, The road to Intelligent Automation in the Energy Sector, Important role of AI due to Pandemic Situation, Utilization of RPA in healthcare, Application of RPA in Digital Forensics (DF)—these sections processed through new automation technologies with 4G AND 5G that provides a Smart environment that we needed. But the future of this huge field is still unknown, these developments might be just the beginning of smart city formation. So we conclude that "SMART CITIES are those who manage their resources efficiently. Traffic, public services and disaster response should be operated intelligently in order to minimize costs, reduce carbon emission and increase performance."—Eduardo Paes.

References

1. Jiang, J., Naumov, P.: Data-informed knowledge and strategies. Artif. Intell., April 2022. https://doi.org/10.1016/j.artint.2022.103727
2. Kumar, K., Thakur, G.S.M.: Advanced applications of neural networks and artificial intelligence: a review. I.J. Inform. Technol. Comput. Sci. **6**, 57–68 (2012). Published Online June 2012 in MECS (https://www.mecs-press.org/ijitcs/ijitcs-v4-n6/IJITCS-V4-N6-8.pdf)
3. Prince Samuel, S., Malarvizhi, K., Karthik, S., Mangala Gowri, S.G.: Machine learning and Internet of Things based smart agriculture. In: 2020 6th International Conference on Advanced Computing and Communication Systems (ICACCS), pp. 1101–1106 (2020). https://doi.org/10.1109/ICACCS48705.2020.9074472
4. Krishnan, A., Swarna, S., Balasubramanya, H.S.: Robotics, IoT, and AI in the automation of agricultural industry: a review. In: 2020 IEEE Bangalore Humanitarian Technology Conference (B-HTC), pp. 1–6 (2020). https://doi.org/10.1109/B-HTC50970.2020.9297856
5. Anagnoste, S.: The road to intelligent automation in the energy sector. Manag. Dyn. Knowl. Econ. **6**(3), 489–502 (2018). https://www.managementdynamics.ro/index.php/journal/article/view/279
6. Ivančić, L., Suša Vugec, D., Bosilj Vukšić, V.: Robotic process automation: systematic literature review. In: Business process management: blockchain and central and Eastern Europe Forum. BPM 2019. Lecture Notes in Business Information Processing, vol. 361. Springer, Cham (2019). https://doi.org/10.1007/978-3-030-30429-4_19
7. Madakam, S., Holmukhe, R.M., Jaiswal, D.K.: The future digital work force: robotic process automation (RPA). J. Inform. Syst. Technol. Manag.—Jistem USP **16**, e201916001 (2019). https://doi.org/10.4301/S1807-1775201916001
8. Enríquez, J.G., Jiménez-Ramírez, A., Domínguez-Mayo, F.J., García-García, J.A.: Robotic process automation: a scientific and industrial systematic mapping study. Electronic ISSN: 2169-3536, https://doi.org/10.1109/ACCESS.2020.2974934, www.ieeexplore.ieee.org/stamp/stamp.jsp?arnumber=9001110
9. Axmann, B., Harmoko, H.: Robotic process automation: an overview and comparison to other technology in Industry 4.0. In: 2020 10th International Conference on Advanced Computer Information Technologies (ACIT), pp. 559–562 (2020). https://doi.org/10.1109/ACIT49673.2020.9208907
10. Novotný, R., Kuchta, R., Kadlec, J.: Smart city concept, applications and services (2014)
11. Smart city system design: a comprehensive study of the application and data planes. Hadi Habibzadeh, University at Albany, Suny, USA. Kaptan, C.: University of Ottawa, Canada. Soyata, T.: University at Albany, Suny, USA. Kantarci, B., Boukerche, A.: University of Ottawa, Canada. ACM Comput. Surv. **52**(2), Article 41, Publication date: May 2019. https://doi.org/10.1145/3309545
12. Yarali, A.: Intelligent connectivity IoT AI, IoT AND 5G ('Connectivity of Tomorrow' and 'Cybersecurity, Privacy, and Risk Management')
13. Ribeiro, J., Lima, R., Eckhardt, T., Paiva, S.: Robotic process automation and artificial intelligence in Industry 4.0—a literature review. Procedia Comput. Sci. **181**, 51–58 (2021). ISSN 1877-0509. https://doi.org/10.1016/j.procs.2021.01.104
14. Vijai, C., Elayaraja, M., Suriyalakshmi, S.M.: The future of robotic process automation (RPA) in the banking sector for better customer experience. Shanlax Int. J. Commerce, April 2020. https://doi.org/10.34293/commerce.v8i2.1709, http://www.shanlaxjournals.in/journals/index.php/commerce/article/view/1709/1997
15. Rahman, M.M., Manavalan, M., Neogy, T.K.: Artificial Intelligence in 5G technology: overview of system models. Asia Pac. J. Energy Environ. **8**(1), 7–16 (2021). https://doi.org/10.18034/apjee.v8i1.595
16. Haidine, A., Salmam, F.Z., Aqqal, A., Dahbi, A.: Artificial Intelligence and Machine Learning in 5G and beyond: a survey and perspectives. In: Moving broadband mobile communications forward—Intelligent Technologies for 5G and beyond. IntechOpen, London, UK (2021). https://www.intechopen.com/chapters/77411, https://doi.org/10.5772/intechopen.98517

17. https://www.deepsig.ai/how-artificial-intelligence-improves-5g-wireless-capabilities
18. Moffitt, K.C., Rozario, A.M., Vasarhelyi, M.A.: Robotic process automation for auditing. J. Emerg. Technol. Account. **15**(1), 1–10, July 2018. https://doi.org/10.2308/jeta-10589
19. https://www.researchgate.net/figure/Proposed-AI-based-Digital-Forensic-Framework_fig2_320758716
20. Asquith, A., Horsman, G.: Let the robots do it!—taking a look at robotic process automation and its potential application in digital forensics. Forensic Sci. Int.: Rep., June 2019. https://doi.org/10.1016/j.fsir.2019.100007
21. Fersht, P., Slaby, J.R.: Robotic automation process emerges as a threat to traditional low cost outsourcing. https://www.horsesforsources.com/wp-content/uploads/2016/06/RS-1210_Robotic-automation-emerges-as-a-threat-060516.pdf
22. Vajgel, B., et al.: Development of intelligent robotic process automation: a utility case study in Brazil. IEEE Access **9**, 71222–71235 (2021). https://doi.org/10.1109/ACCESS.2021.3075693
23. Shidaganti, G., Salil, S., Anand, P., Jadhav, V.: Robotic process automation with AI and OCR to improve business process: review. In: 2021 Second International Conference on Electronics and Sustainable Communication Systems (ICESC), pp. 1612–1618 (2021). https://doi.org/10.1109/ICESC51422.2021.9532902
24. Wanner, J., Hofmann, A., Fischer, M., Imgrund, F., Janiesch, C., Geyer-Klingeberg, J.: Process selection in RPA projects—towards a quantifiable method of decision making. In: Fortieth International Conference on Information Systems, Munich (2019). https://www.researchgate.net/publication/337289984
25. Jamshidi, M., et al.: Artificial Intelligence and COVID-19: deep learning approaches for diagnosis and treatment. IEEE Access **8**, 109581–109595 (2020). https://doi.org/10.1109/ACCESS.2020.3001973
26. Jain R, Bhatnagar R.: Robotic process automation in healthcare—a review. Int. Rob. Auto. J. **5**(1), 12–14 (2019). https://doi.org/10.15406/iratj.2019.05.00164
27. Kobayashi, T., Arai, K., Imai, T., Tanimoto, S., Sato, H., Kanai, A.: Communication robot for elderly based on robotic process automation. In: 2019 IEEE 43rd Annual Computer Software and Applications Conference (COMPSAC), pp. 251–256 (2019). https://doi.org/10.1109/COMPSAC.2019.10215
28. Niekler, C.: Artificial Intelligence—curse or blessing? Crystalloids News, July 2019. https://www.crystalloids.com/news/artificial-intelligence-curse-or-blessing
29. Kulkarni, U.: How Artificial intelligence transforms the robotic process automation landscape to make it more productive. Datamatics, December 2019. https://blog.datamatics.com/how-artificial-intelligence-transforms-the-robotic-process-automation-landscape-to-make-it-more-productive
30. Democratization of Artificial Intelligence for the Future of Humanity by Chandrasekar Vuppalapati Copyright year 2021
31. Novotný, R., Kuchta, R., Kadlec, J.: Smart city concept, applications and services. J. Telecommun. Syst. Manag. (2014). https://doi.org/10.4172/2167-0919.1000117, https://www.hilarispublisher.com/open-access/smart-city-concept-applications-and-services-2167-0919-117.pdf

Chapter 13
The Existing IT Functions and Robotic Process Automation

K. Devaki, V. Murali Bhaskaran, and S. Anjana

Abstract This article provides an overall view of Robotic Process Automation (RPA) evolution and its use cases in the IT Industrial sector. In recent times RPA has evolved and democratized all sectors of the economy for better and efficient production and usage of products, processes, and services. The chapter shows the reasons for the flourishment of the technology. The evolution of industrial sectors from steam engines to unattended automatic robots taking business intelligence decisions are driven by one major quotient i.e. automation. The chapter comprises the entire journey that RPA has taken to be now one of the affordable and efficient solutions out of all. The ultimate aim of every corporation is to generate revenue and to achieve it through the different processes that happen around it. Each of the processes involves different user personas and applications and thus humans act as the binding chains and deliver the end product or service to the consumer. As humans cannot handle complex computations and tedious monotonous tasks, automation minimizes these tasks and makes their work easier. This article also covers the sectors where automation is being a critical savior for industries. It also talks about the pros and cons of automation over a period and how it could be a disruptive technology in the upcoming years.

K. Devaki (✉) · V. Murali Bhaskaran
Department of Computer Science and Engineering, Rajalakshmi Engineering College, Chennai, India
e-mail: devaki.k@rajalakshmi.edu.in

V. Murali Bhaskaran
e-mail: muralibhaskaran.v@rajalakshmi.edu.in

S. Anjana
Department of Computer Science and Business Systems, Rajalakshmi Engineering College, Chennai, India
e-mail: anjana.s.2019.csbs@rajalakshmi.edu.in

S. Bhattacharyya et al. (eds.), *Confluence of Artificial Intelligence and Robotic Process Automation*, Smart Innovation, Systems and Technologies 335,
https://doi.org/10.1007/978-981-19-8296-5_13

13.1 Introduction

Industry 4.0 represents the fourth industrial revolution which has created a huge disruption in the manufacturing sector [1]. The first industrial revolution started with the introduction of power by converting water into steam slowly moved into the second industrial revolution where mass production and supply chains came into play along with the induction of electricity into the world. The third industrial revolution started with the adoption of computers and automation to improve the efficiency of production. The Fourth Industrial revolution took on where the third industrial revolution started and enhanced it with smart and autonomous systems that are motored through with the mammoth amount of data generated by the humans and next-gen technologies like Machine Learning, Cloud Computing, etc. (Fig. 13.1).

Industry 4.0 optimizes Industry 3.0's computerization. The introduction of modern computers with high computational capabilities changed the way in which the sector was working. As Industry 4.0 unfolds, computers have started communicating with each other and make decisions without human involvement with precision. Combination of the Internet of Things, Cyber-physical systems, Big Data, etc., has made the systems to be smarter and intelligent in learning human behavior and thoughts. As a result, these factories and other industries have become more efficient and productive with fewer errors.

Industry 4.0 paves a way to take business intelligence decisions. Big Data Analytics is one of the sub domains of evolution and it helps companies to analyze their data and give better insights into it. Bosch has integrated machinery with their computers and these computers store the data of hyper parameters specifying cycle time and machine conditions. With the help of advanced Data Analytics, the collected data is analyzed in real-time for any anomalies. Hence, failure of machines and maintenance can be easily found and scheduled accordingly.

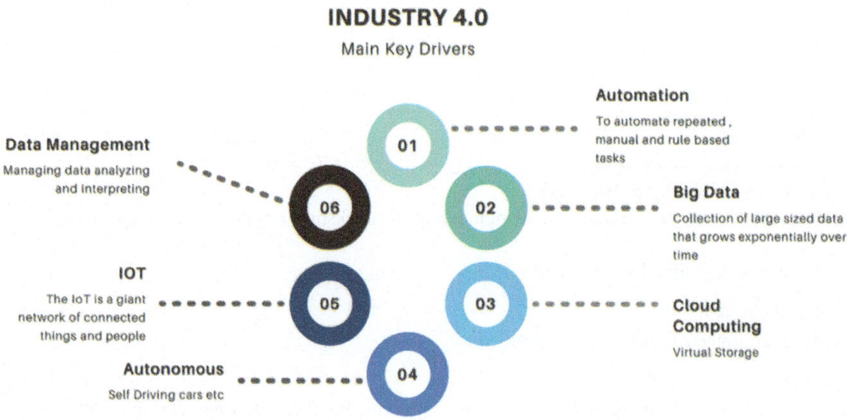

Fig. 13.1 Fourth industrial revolution life cycle

Ultimately the machines share and create the information thus resulting in the power of Industry 4.0.

Roadmap of the Chapter

This chapter comprises of how automation started, different tools available to build a bot, its features and the future. Section 13.2 provides a short history of automation and its foothold in all the sectors. Section 13.3 gives a brief on the RPA technology, different tools available, and the phases of automation, relationship with AI and the process of end-to-end automation. Section 13.3.7 provides insights into different capabilities that could be added to the bot to improve the efficiency at the maximum for better throughput using cutting-edge technologies like AI, ML, IOT and much more which leads to the applications of it in the sectors of the IT industry along with its pain points and impact. Finally we will see the growth in the RPA industry and the near future reverberations in the IT sector creating a disruptive revolution.

13.2 History of Automation

As we entered the Fourth Industrial Revolution, the strong driver acting is Automation. Robotic Process Automation (RPA) is the driver of revolution; it inherits the features of Cloud Computing and Machine learning and tries to mimic cognitive human actions with unattended automation mechanics. Lakhs of companies have already transformed them into fully digital and automated companies. To understand better, we need to go back the lane and understand RPA development [2].

13.2.1 The 1990s: Automation for User Interface Testing

User Interface is what the users see and try to invoke the functionality and that is where the journey of RPA started. UI testing is testing of the frontend elements and checking their triggers and making sure that it does not have any issues working with the app [3]. During earlier days there were very few computers and that too occupying a huge space. Slowly it has changed from big companies to governmental organizations to home-based users. The main reason for such a shift is the introduction of Operating systems like Windows, the recognized OS at that time.

As a result, the UI testing improved, as the requirements and the screen sizes became more diverse. During the late 90s and early 2000s software development processes changed and new methodologies came into the industry for delivering the projects better and more friendly to the end customer. Spiral, V models are few examples of the methodologies. One such methodology is agile development, where the Agile Manifesto puts people over processes and tools [4]. Firms have also seen the need to speed up their development so that they remain competitive in the market they are in. Thus Quality Assurance Testing, Software UI testing scripts and tools like Selenium were born.

13.2.2 The 2000s: Banking Sector Automation

The key elements and tools for RPA emerged in the early 2000s. A major revolution was the introduction of Screen Scraping technology which basically means fetching data from the computer screen [3]. This resulted in a significant improvement for businesses in terms of handling large volumes of data of all sorts and as Banking sectors were booming and becoming more structured and regularized, RPA helped to reduce the amount of legal paper works and compliance.

Although RPA has more pros it has few drawbacks and one of them is Set Up structure. If an organization wants to automate their data entry process for example then it would almost be requiring a complex IT environment setup. In order to set this up we need expert engineering skills and complex integrations which would be time consuming.

13.2.3 Timeline 2010–2020: Business Process Automation

The major breakthrough for RPA came in the year 2012 where many large scale corporations recognized the need for automation and started adopting those due combinations of factors such as financial crisis, easy-to-go solutions for users leading to a better product.

Managing the incurring costs for a company is very important as it affects their profit margin and selling price and when they were searching for a way to transform the processes digitally, RPA was the finest, easy, and most affordable solution. Considering these factors companies started using RPA for critical activities in their companies like Assembly line of their products, manufacturing them etc.

The demand for RPA slowly increased over the years and 2020 has been the time of RPA democratization in all segments of the economy. The reason behind this is the decrease in royalty and licensing fees and hence the companies started building partnerships with RPA platforms and service providers like Uipath, Automation Anywhere, Blue Prism, Power Automation etc. This enabled them to start a new model called RPA as a service thus allowing firms to expand their calibers and achieve more productive results.

13.3 A Brief of RPA

Robotic Process Automation (RPA) is defined as the use of software bots to handle repetitive and procedural based tasks [5]. These bots are used for interactive applications handling as we do on traditional-based systems. RPA will help us to replace traditional legacy systems which will in turn help in saving costs for the organization. It also improves the employees efficiency making them focused on more valuable business priorities.

RPA deals with several tasks which includes changing from application to application, entering data into various fields, real-time data recording and manipulation of those data. Almost every task is led by a set of rules and regulations. The robot works on software which does jobs such as fetching customer profiles, support and arranging data from multiple enterprise systems.

As process automation takes over multiple iterations, cyclic manual work, human error is annihilated, and there is no space for an expensive mistake. End users are expecting the services to be more efficient and personalized than before, and the only domain which makes use of virtual capabilities is RPA to keep up with more cultivated and changing customer demands and it will remain relevant in the long run.

RPA workforce has the following characteristics such as precise, accurate and also relived from working on mundane tasks. RPA can perform multi-staged tasks working with web applications, desktop applications or even cloud-based applications. It acts as a bridge to the world of interaction with computer user interface making human abled tasks doable by robots too.

13.3.1 RPA Tools

In order to build these bots, tools are to be used that have capabilities to capture the data that is available on the system. Predominant RPA Tools are given in Fig. 13.2.

Fig. 13.2 RPA platform delivery organizations

Uipath. Uipath was founded in the year 2005 based on Bucharest Romania and they are primarily based on creating Automation Product Lines for RPA [6]. In the year 2013–14, they first came up with the Desktop Automation Product to automate manual and repetitive back office tasks and now they have over 20+ services for different phases of automation. The Software combines a family of low-code visual integrated development (IDE) products called Studio for process creation, with client-side agents called Robots that execute those processes. The processes are deployed, monitored and managed remotely with a central management tool called Orchestrator.

Blue Prism. Blue Prism is a RPA Tool which holds the capability of a virtual workforce powered by software robots. This helps the enterprises to automate the business operations in an agile and cost-effective manner. The tool is based on Java Programming Language and offers a visual designer with drag and drop functionalities [7].

Automation Anywhere. Automation anywhere is a web-based management system. It has a Control Room that helps in managing automated tasks. It is mainly used at the enterprise level and changes the way the enterprises operate. The primary aim of Automation Anywhere is to offer scalable, secure, and resilient services to its users [8].

PEGA. PEGA is a popular Business Process Management (BPM) tool created by Java concepts that allow users to execute changes faster than Java-based applications. The primary use of PEGA is to reduce costs and improve business reasons. PEGA is created in Java and uses OOP and Java ideas [9].

Kryon Systems. Kryon delivers innovative, intelligent Robotic Process Automation (RPA) solutions enabling digital transformation for enterprises [10].

Jidoka. Appian RPA connects systems and applications to automate tasks and processes without the need for APIs or custom integrations. Intelligent document processing automates the classification and data extraction of structured and semi-structured business documents [11].

Work Fusion. Work Fusion integrates various technologies like pre-trained bots, proprietary artificial intelligence solutions, and advanced analytics to automate a wide range of business processes. Overall, Work Fusion can help organizations to reduce costs, improve productivity, and scale rapidly [12].

BeInformed. BeInformed platform lays the foundation for automated processes and decision making. Legacy integration or new build, the engine accelerates digital development. BeInformed helps to move forward in handling complex situations [13].

Gartner's Magic Quadrant is accumulation of research on a specific market giving wide spread angles on the company's relative position in the market [14]. By using this set of evaluation criteria on the graph, the firm would be able to know how well they are executing their vision and goals against Gartner's market View. It also

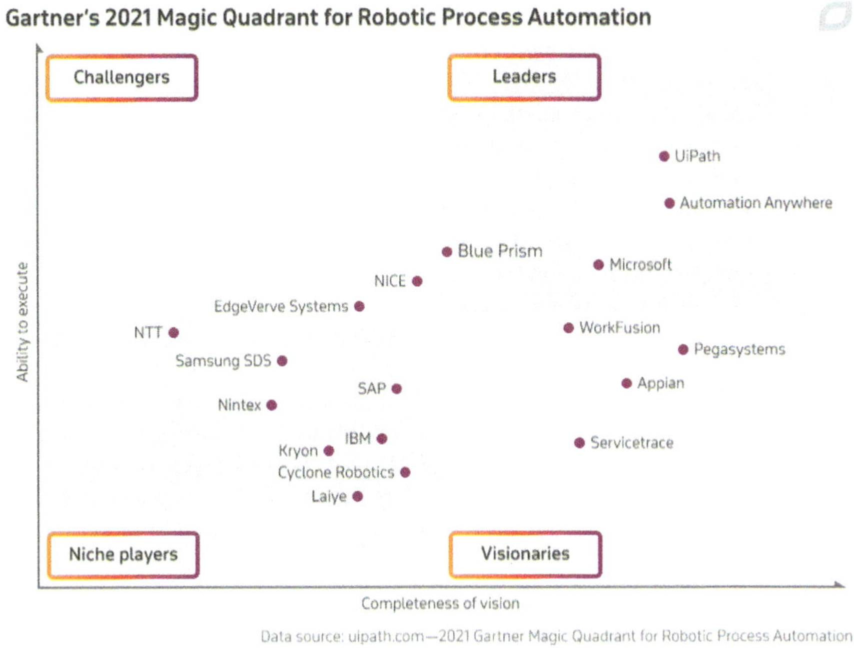

Fig. 13.3 Gartner's magic quadrant

allows customization where the firm can do it for their own business goals, perspectives, needs and priorities. Gartner also gives High Lens by providing high-impact additional perspectives on key industry, region and company size. This enhancement provides analyst commentary on the market and notable vendors from key client contexts. According to Gartner there are few divisions such as Challengers, Leaders, Niche Players and Visionaries against their vision and capabilities and Uipath is the leader of the RPA Industry (Fig. 13.3).

13.3.2 The Relationship Between AI and RPA

Artificial Intelligence (AI) and Robotic Process Automation (RPA) are key drivers of Industry 4.0 and in creating business intelligence systems for firms to achieve their goals in ease which in turn increases customer trust towards the company and also employee's remuneration and work productivity which reduces the operating costs of the company.

There has always been confusion between physical robots and automated robots. The major difference between them is that automated robots also called as "bots" work on computer-based software applications whereas physical robots use the software to do the computation or their results are fetched and displayed as a dashboard to the user which is implemented using actuators in the real environment. This environment can be partially or completely observable which we equate to bots as Exceptions and Application runtime difficulties. Thus Robotic Process Automation (RPA) uses specialized technology to standardize the workflow and automate repeated processes by analyzing the company's day-to-day interactions with the applications they use and create an automation pipeline. The bots complete the given tasks at the best possible time following the orders given while building the bot and hence its performance remains the same even if there are 1000 or 10,000 files to be analyzed. Therefore they do not learn from one iteration or neither learn from the exceptions; everything needs to be explicitly programmed. As such combining the Artificial intelligence and RPA gives an extra hand to handle exceptions, learn new things by itself and improve its intelligence to mimic actions higher.

Having 100 people to enter census data from the census survey increases the labor cost and the task happens to be a redundant and mundane task. Deploying virtual assistants like bots with the use of RPA weighs out the repetitive tasks of the employees and they can utilize their efficiency in other business entities that need human intelligence and interaction. RPA can be an advantageous solution for the following reasons.

Accuracy. It minimizes the erroneous or faulty errors made by the human during data entry or raising tickets. RPA can be an effective solution as it reduces the misapprehensions and the incurring disbursements associated with the same.

Compliance. When working with manual data entry, one has to keep track of the user's access and check whether the standards and compliances are followed. Bots have inventories, audit trails and regulatory rules more precisely.

Speed. With the help of RPA we can complete the tasks in half the time that the humans would do, improving the efficiency and perfection.

Reliable. It will work 24/7 without any tiredness, need to take breaks and off-days to complete the work. It can do the same task again and again timelessly.

Employee Spirit. It relieves employees by refraining from doing tedious, repeated and mundane work and helps to focus on productivity not on processing the data.

Let us take an example of a financial institution like a Bank that uses Robotic Process automation to detect the fraud happening in the bank regarding opening or closing of accounts. The process is like if a customer had been a victim of fraud, will raise a complaint to the specific bank with which it is related. Once the problem has been escalated, then the complaint will be taken and the bank registers a complaint stating the status as "Raised Complaint". Then the bank will interrogate the issue and send continuous updates to the customer regarding the status. This repetitive process can be automated where we can deploy virtual assistants to get the complaints forward to the mail and then move to the next customer concurrently. The main advantage of this is that the handling time is reduced, gives greater accuracy, better adhering to the compliance and regulations, more trustworthy employees and consumers.

Thus RPA is great for automating rule-based tasks and if the complexity increases then the use of new technologies like Cloud and Artificial Intelligence (AI) can be put into the use to take it to the next level.

Artificial Intelligence is a technique that came in to mimic human action and be intelligent as human by "computer systems" or "machines" [15]. While RPA works on automating repetitive and rule-based tasks, AI is a process to create end-end process automation that needs no human intervention (unattended robots). RPA as we said before requires explicit programming while AI has the capability to even derive insights from unstructured data and derive its own outputs through techniques like Unsupervised Learning which can be used to create autonomous systems to do complex computations. As an enterprise the data coming in/flows out is both structured and unstructured and thus features of both Robotic Process Automation and Artificial Intelligence are required to automate an end-end application processing or to make the existing process more efficient.

13.3.3 Creating End–End Automation with AI and RPA

End–End automation is used mainly in e-commerce sites where there is more requirements of personalized suggestions and the place order journey. In a larger view, any firm typically involves a lot of self-services options for the consumer. Considering the same example of Financial Institutions like bank, if a person wants to open a bank account, then with the help of RPA and AI it can do automatic KYC Verifications, Form Filling and Validating. Doing a background check using RPA results in less hassle for consumers to go to a bank and open an account, less tedious task for employees to verify validate and revert back if any errors making more efficient-money saving.

How does it work?

- The Chabot is activated and it redirects to a link where the data has to be filled.
- The customer fills the form and submits the legal documents required.
- Then it triggers the background running bot to verify and open a new account in the database of the bank.
- After that, using third-party software provided by Google and other social networks are activated automatically to complete the "know your customer" process online.

Document Verification. Using OCR technique the robots analyze the important details and identify any anomaly in the data and the legal documents that have been given and in case of any discrepancy the application is automatically rejected. These processes are done by the use of Text Summarization and analysis along with the use of NLP-Natural Language Processing to interpret the data fetched and also categorize them if needed from the data inputs fetched [16, 17].

Once these data are stored in a sequential manner with the statutes, then an account is opened and automatic "no-reply" mail is sent back to the customers with details and access credentials to their Net banking account.

13.3.4 The Role of AI and RPA in IT Processes

Organizations from all over the world have benefitted with the use of RPA as it multiplied the productivity after implementing RPA as a feature on their existing processes. The results signify that the companies can perform more innovative tasks allowing more profit to businesses.

With the automation of well-defined tasks, it has reduced the latency in the completion of business processes along with speed and cost-effectiveness. In the real time scenario, most of the data generated by the organization are unstructured like invoices, excel, email requests and all are different for various business clients. Thus performing the existing RPA methodology would not be effective.

Artificial Intelligence is one such technology where machines can understand the language in which we work i.e. ability to understand human language by acting like a human. With the combination of AI and RPA in automating tasks, it has expanded the capabilities to process such unstructured data, image content and understand natural language etc.

Cognitive Automation is a process where the system uses techniques that mimic human learning to help humans take decisive actions, complete tasks and reach other goals. When RPA uses AI algorithms to improve the experience, be it for the workspace that we are working on or the customer we term it as cognitive automation and this constantly adapting system does not need complex models and also it can be pushed into the production environment in few weeks of inception. The key abilities of cognitive processing include Natural Language Processing (NLP), Optical Character Recognition (OCR) and Machine Learning (ML).

Natural Language Processing (NLP)—Language Understanding makes it much easier to automate most customer service processes. In departments like IT Support resolving queries at the right time is important. Having NLP features involved in the automation of chatbots helps to resolve queries without much human intervention.

Optical Character Recognition (OCR)—It involves automating document formats like images, handwritten forms and scanned copies which will create a significant impact on the business processes in the document-intensive industries like invoices, handwritten applications, forms, cheques and so on. For example, production-based industries might receive over 1000 invoices a day and processing it might take huge number of man hours to process while, RPA along with OCR will reduce the time to process along with increased turnaround time and reduced cost.

Machine Learning (ML)—Decision-making can be done through this process. ML algorithms generate data patterns which are capable of learning from past data to understand the meaning and the context. With ML in hand we can automate processes by replacing human judgement with machine judgement. With past data now the bot can understand the support requests generated and automatically create a ticker in service desk systems.

Every workflow is unique but in most of the cases, once the automation process is done AI will take over from the end of an automation process. For example, in email processing automation, a bot might be used to open and route to the message to be viewed but that is not the end result. From there RPA will send it to an AI component which can open and read the contents of the attachment present on the email with OCR and NLP extracting critical information which is then given to ML algorithms to generate decisions. Moreover, it can also extract unstructured content like dates, invoice and dispatch numbers and then using bots we can reformat and use it for entering details in Content Management System (CMS) or Enterprise Resource Planning (ERP).

AI-Powered RPA thus becomes a powerful automation tool to perform successful RPA implementations. This can process the unstructured data types such as text, natural language, images and web content etc. In the rapidly evolving world of automation, firms will gain a significant competitive advantage with cognitive automation where they achieve increased efficiency and productivity.

13.3.5 Benefits of RPA

The RPA has a lot of benefits. The software robots can be used to achieve the significant results in the area of business that the company is in, thus improving customer satisfaction and employee engagement leading to more efficient cost reduction, speed, accuracy and precision. The benefits are as follows.

Personalized Journey. Giving services to customers that are personalized for them and suggestive in nature, makes the consumer think that the company or firm is more customer-centric and they want to avail the services to its best utilization as possible. It gives an improved loyalty and trust among customers thus providing greater ability to meet the requirements of value proposition made by the company.

Productivity. RPA can do the same task for n number of times without any tiredness retaining the same accuracy which implies that it works 24/7. The work done will be much quicker and creates more space to do other tasks and also a room for more rapid growth and expansion of the company.

Accuracy. These robots are 100% accurate, perfect, consistent and compliant to all the standards and policies framed. More the work to the robot, lesser the clerical work, more efficient throughput by the employees. It eliminates the errors made by human, cutting down the cost and more customer satisfaction.

Resource Efficient. Handling of tedious labor intensive tasks by the bots keeps the value increasing. These robots can be scaled tremendously enabling the organizations to respond to complex scenarios and also meet the demands faced by the company taking off the workload of the Human Resources Department to select, acquire, train and recruit fresher or experienced persons.

Features. The RPA does not work biased, it is gender-unbiased and creates a sense of equality. There is a buzzword that RPA will replace jobs. The answer is yes it will but not replace the full-time equivalents of the human. They will move into complex tasks according to the skill sets provided making them work for what they are capable of and thus not working on a task overqualified or under qualified. Since efficiency acts as a prime factor or quotient of advantage in RPA the ROI (Return of Investment) for the firm is more since the operating costs are low. The RPA can be scaled faster and also it is not that much cost-intensive as compared to recruiting a human. According to stats, a RPA developer can replace 6 software developer roles so the human workforce can be put into use at the right level at the right time to do a particular task [18].

13.3.6 Phases of Automation

Discover Phase. This phase involves identifying processes that could be automated. As a company has a lot of processes and applications used, it is important to know the sequence or workflow of how the task could be performed and which automation is most suitable for the identified process. The output of this phase is Process Definition (PDD) which will be used to produce automation solutions [19].

In order to identify this process the following steps have to be carried out.

Process Mining. Companies use a lot of software like ERP and CRM. Gathering the data from it and giving it as an input to the AI model will give us a detailed description of complex business processes, know what could be automated, how well it could be done etc.

Task Mining. This process tracks the everyday workflows of the company. With the empowerment of Artificial Intelligence, we can automatically identify the repetitive processes to add it to the company's automation pipeline along with a dashboard of how many times the same task is performed, actions done, users, applications accessed etc. and also quickly transform the activities into flowcharts for quick and easy scaling up of automation.

Task Capture. This phase facilitates the reporting of documents done by the company. This tool accelerates business process documentation by recording the regular workflows in detail. Once it is done the auto-generated process maps can be shared. It involves Diagram Builder or Task recorder [20]. It can be imported from either Uipath Studio or also from Process Definition Document designed. Once done, it can be integrated with the Task Mining, Automation Manager and Test Manager.

Build. This phase is where the bot is built for the task captured. There are different tools with which we can build the bot like Uipath Studio and Uipath StudioX, Automation Anywhere's AARI, Blue Prism, Pega, Microsoft Power Automate etc. Once the bot is built, the bot is sent to the production phase for running.

Manage. In this the bot is streamlined into an automation workflow and it is driven to the orchestrator for running, deployment, monitoring, measuring and tracking and ensuring the security of every robot in the organization with Orchestrator (on-premises) or Automation Cloud. Before running the bot, it has to be tested with the Test Manager for better analysis of the bots' workflow and using the capabilities of AI resulting in more returns for the firm.

Run. This phase as the name suggests is for running the bot which can be attended, unattended, semi-automated or hybrid bots spread across application workflows integrating APIs, UI's, Applications, Third party services etc. and get security endpoints for robots.

Engage. This phase brings the robot in contact with the people and empowers them to engage with automation to capture benefits across the business enterprise. Action Center is a critical element used for human validation or a correction station interface. Uipath offers an UI Action Center that is a part of the Uipath Orchestrator where robots create tasks for the humans to go through and verify. It is a part of Uipath Apps.

Measure. The measure phase is all about the baseline of the current process, data collection, validating the measurement system, and also determining the process capability. Uipath Insights, an RPA analytics solution, is a powerful embedded tool to measure, report, assign RPA operations and assess metrics that enables the business to track and manage the performance of the entire automation program—to scale the automation journey to the next level.

13.3.7 The IT Functions and RPA

As the technology is increasing day by day and companies or organizations concentrating on maximizing their growth, automation play a major role in decision making as well as improving their process. A company has different sectors like Manufacturing and Production, Customer Care, Quality Assurance etc. Let us try to understand the different use cases of RPA in detail [21] (Table 13.1).

Customer Service. In the recent study it is known that the companies make their sales between 5 and 20% of new customers while they jump to 60–70% for existing customers [22]. This shows that acquiring new customers is expensive for the companies. It is a double time investment that the companies have to spend money on

Table 13.1 Use cases of RPA

Sector	RPA applications	Disadvantages of the traditional systems
Human resources	• Employee monthly payroll • Employee data management • Employee on-boarding and offer Letters Generation • Tracking and monitoring the performance of employees • Travel and expenses management	• Manual tracking and monitoring • Tedious data entry • Biased analysis • Erroneous errors • Less structured and non-standardized
Sales and marketing	• Customer service • Customer feedback analysis • Lead tracking • Order booking and sales • Multilingual website • Report maintenance	• Tracking leads and complaints • Outage issues • Cumulative report collection and analysis • Requires a lot of manual working hours
Information technology	• Automation testing • Information • Configuration • Threat analysis	• Manual security configuration • Vulnerability scanning is difficult • Revamp after network connection failures • High maintenance and infrastructure cost • More time spent on analysis rather than building
Operations and logistics	• Shipment scheduling • Delivery notification • Tracking	• Multiple distribution channel tracking • Compliance and query processing • Deliverables maintenance

marketing as well on acquiring new customers and maintaining the customer retention rate. The Customer service departments not only handle complaints of customer and product queries but also maintain a constant interaction with them, feeding them about new product releases, maintenance guidelines, order tracking etc.

In order to create these personalized experiences manual human effort is impossible and hence bots have been introduced. For example, to answer your queries immediately, companies are using automated chat bots that give all possible replies to the query. Recommendations and constant offer mails are generated using bots which makes the companies to be quicker and faster on reaching customers thus improving the scalability (Fig. 13.4).

The use case for customer service would be.

Order Tracking Automation. In the recent study 97% of the customers want a customized experience and receive communication about the whole shipping process once they order online.

Fig. 13.4 Customer service

How does RPA help to do this process efficiently?

With the intelligence of AI, RPA software, automatic emails notifying the customer of the order that has been placed along with the tracking id and information of what they ordered could be sent. RPA can read and classify the emails that have been sent. For example the One Time Password (OTP) technology has the feature to send automatic verification codes with locks based on the intelligence of AI to check whether the phone is unlocked or not.

It can be used to update the customer details and their reviews directly on the website. Using this data we can send offers and incomplete order details making them ultimately buy the product.

Call Center. We always turn to customer service whenever a company's procedure, product, or service is unsatisfactory or we have some questions. Since they usually connect to several employees to answer the query, the average handle time (AHT) is longer and we frequently wait for hours.

This process could be automated by getting the details from the CRM system to gather data about the customer, his/her orders and any other information in order to provide a complete answer to the customer question. This would help the call agents to concentrate on the questions rather than going to multiple systems and getting the information needed [23].

Online Reviews. The reviews are the testimonials to the company about their product. They read all the reviews, reply to them if necessary and give those feedbacks to the product development team for further improvement.

How can RPA help to do this process efficiently?

- Reading the reviews, identify the sentiments and take only the negative reviews.
- Passing information about product defect to other departments.
- Automating the feedback survey and deriving insights from it automatically.
- Immediate reply to queries through Chabot.
- Automatic sending of notification to the employees working on it, in case of any anomaly. For e.g. based on the keywords present in the customer review the bot automatically classifies the messages and sends them to the respective authorities.

Human Resource. The study tells us that RPA would save 40% of the work time thus allowing employees to focus on more valuable and human centric tasks. The companies feel that they are working 10 times faster and get more ROI than expected. Deployment of RPA on these processes would save 10–20% cost incurred for the overall business process [24, 25].

The use case for Robotic Process Automation in HR includes.

Resume Screening and Candidate Short listing. The Company posts the open positions on the portal. The employees have to go through the applications and do the background checks. These bots can gather the files whenever a new application comes. Hence, a system can be built with predefined rules on how to do the selection procedure and background checks. Once the bots filter the employees through automation, it can automatically send emails to the new joiners.

Offer Letter Automation. Usually companies go both for on-campus and off-campus hiring with different sets of new joiners with different CTC and roles. Although these factors change the rules and regulations of the company, it is common. Thus these tasks can be automated by quickly feeding the bot with a template along with the employee details and sending the mail with the appropriate details.

Attendance Tracking. Big firms and organizations have 1000+ employees and tracking each person's work is difficult and moreover tracking their attendance and reviewing it is time consuming. The bots can cross check the reports submitted by the employees, the absenteeism and prevent workflow disruptions.

Employee Onboarding. When a new employee joins the firm, the data from several systems must be coordinated in order to create a new user account, email address, access rights for applications, IT equipment, etc. With RPA—Robotic Process Automation, the user account can automatically activate a particular template for the on-boarding workflow, and this streamlines the whole process. The bots can then make rule-guided decisions as to which credentials to be assigned to the new employee, which on boarding documents to send, etc. Bots make processes such as employee ID creation much faster.

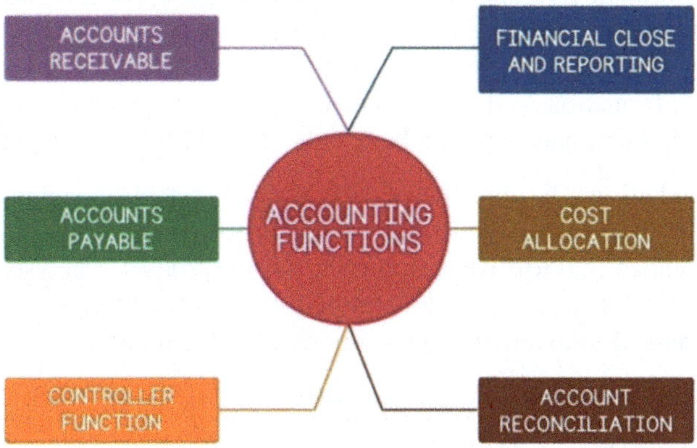

Fig. 13.5 Accounting functions

Payroll Management. Processing Payroll every month is an apt example of repetitive, monotonous tasks. It involves a lot of data entry and is often error related. The bots reduce the risk to zero while decreasing the waiting time. Avoiding delayed payments will have a positive impact on the employees increasing the job satisfaction and productivity of employees.

Accounting and Finance. The heart of the company lies in the accounting of finance. Over the last decade the demand for data has increased. The increase in transaction volume, repetitive work and those requirements needs to be managed (Fig. 13.5).

The use case for Accounting and Finance include.

Cash Format. A firm receives cash from stakeholders or customer orders or from some other source through different means like UPI's, NEFT transfers, cheques etc. RPA can be used to automatically read the required information or gather them from the applications and match them with the open invoices.

Distribution or Logistics Management. The firm has different distribution channels like Wholesale, Retail, White Labeling, etc. and based on the demand the company serves them with the products or services. With the use of RPA, we can reduce the amount of paperwork thus improving the management of inventory bills.

Invoice Processing. The Company uses ERP software like SAP, Oracle to store their details effectively. RPA can easily fetch details from the ERP system, match it with the seller invoice and do a 3 way matching of invoices. The same application can be integrated for payments automations, validating the payment done, notify the seller that the invoice amount has been paid etc. thus leading to no manual errors or space to store details physically [23, 25].

Financial Houses. Financial institutions only lend people the needed amount of money after thoroughly vetting them based on their background, credit history, and level of compliance with the law. RPA can greatly aid in the consistency of the names and papers provided, and if it matches, the bot can automatically create an account and transmit information to the company via mail. This procedure can be biased because it involves humans who can be bought off and influence the outcome.

Sales and Marketing. The objective of every business establishment is to ultimately get revenue. This revenue is generated by the customers. In order to generate the revenue the company has to drive customers to buy the product. Analysis and insights from the historical data would be a great help to the Marketing and sales team to invest on streams like Advertising, Promotions with Third Parties etc. (Fig. 13.6).

The use cases on sales and marketing includes.

Customer Details Storage. Medium and large-scale companies store their information in Customer Relationship Management software (CRM). The main stack behind CRM is RPA [23]. Whenever a new customer or a lead is given by the Sales Representative, bots detect the details and arrange them in an organized way so that the data representations are easy to understand and derive key areas where the representative or sales point gets lagged.

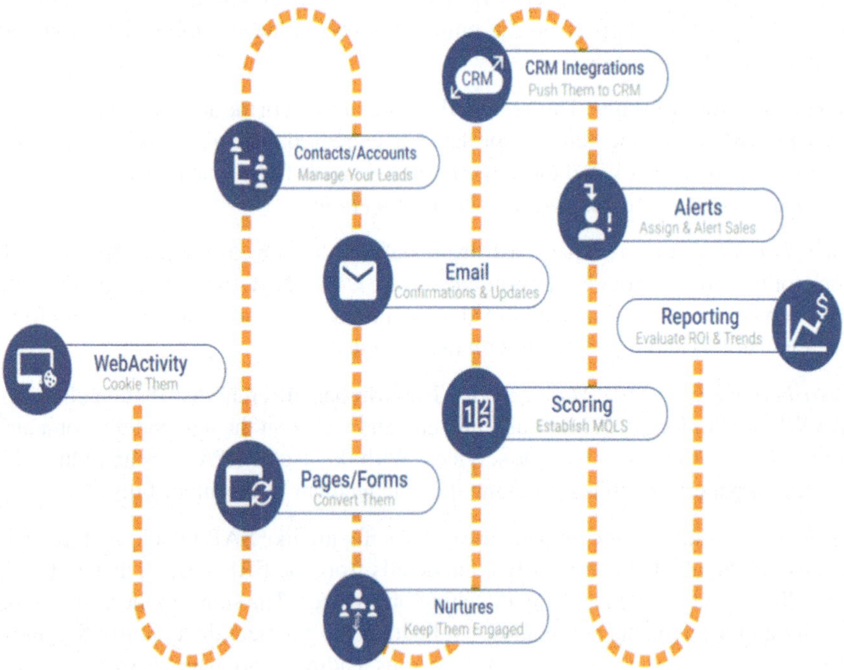

Fig. 13.6 Sales and marketing prime phases

Data Extraction and Visualization of Sales Data. Data appears in a variety of formats like JSON, XML, CSV etc. and it is difficult to read and process the bits and pieces of information for which RPA's OCR technology can be used to read the information from the unknown source formats and automatically push those details into the CRM database without losing the contextual essence and accuracy.

Sales Order Processing. Every data that a sales representative gets is a lead to the Sales Department and an indirect asset to the company. Therefore, these data are recorded by people into the CRM software in various sections that are distinct for finance on payments, separate for email processing, etc. Such things are always tedious and over the flow the accuracy and the value of information get lost. However these tasks can be automated through RPA where the bots enter the details, create automatic invoices and open payment cheques creating end-end automation. It also consolidates the data formats in a single database system.

Supply Chain Management. It is the flow of products and services between the businesses franchises and warehouses to furnished products supplied to the end user.

After-Sales Service. The main value of the product lies after it goes to the customer's hands. When a customer uses the product they may have different things to say to the company for improvement or an appreciation for good service which attracts other customers to buy the product. Their queries are important and must be answered instantly. We can create a bot that can score the text with a sentiment and escalate it accordingly to the employee levels and also check the service schedules and AMCs validity etc.

Intelligent Data Entry. E-Commerce services in the recent times have been the trend and customers buy tons of product on a regular basis and the order is automatically forwarded to the warehouse to check it and escalate it to different warehouses if not found. One customer will order multiple times and each time when we store the data it consumes a lot of space and effort of people. RPA can thus find out the redundancy and if it finds the same customer, then it can automatically forward the stored details to the warehouse.

A real time example would be franchised shops. Usually when a customer buys a product in a shop for the first time, all the customer details are collected and saved. After a few months, when the same customer comes back again, the system will verify for an existing user or not and points are automatically added for every purchase. The technology behind these tasks is RPA.

Adding RPA in supply chain processes means continuous, proactive, and preventive maintenance for network processes, as well as ongoing route optimization connected to other systems of record.

13.3.8 Hyper Automation

In simple terms hyper automation is the extension of traditional legacy business process automation by combining the efforts of improvements in AI as well as RPA. This will help us to automate any type of business process that involves tedious human efforts by dynamically analyzing the business processes and creating bots to automate the same. Hyper Automation is known as real digital transformation wherein the core technology behind is AI and ML. These technologies help in faster computation, flexibility and power. It has 4 components: Intelligent Process Discovery, Analytics and Insights, Robotic Process Automation and Intelligent Document Processing (Fig. 13.7).

Intelligent Process Discovery. This phase uses AI to reveal processes that could be automated and automatically create bot and increase the pace of automation upto 10 times. With organizations coming up with different types of platforms to facilitate the tasks like Automation Anywhere, Discovery Bot etc. the stakeholders can be involved in every business process. Managers, leaders, experts, domain knowledge experts on RPA can use this unprecedented technology to scale automation while maintaining regulations and compliance leading to secure data transaction and transfer [26].

Analytics and Insights. The Hyper Automation analytics will automatically derive the insights as the bots gather the work [26]. Before deploying the bot into the automation pipeline, the bots are attached to an embedded and a self-measuring monitoring system. Once it moves into the production the complete information about the bot is fetched and stored. The whole journey from production till retirement of the bot is captured and even future obstacles or hassles can be predicted from it. It uses vast level data detailing for processing of high-level detailed information fetched from the bot which gives meaningful perception and paves a way to predict what's next.

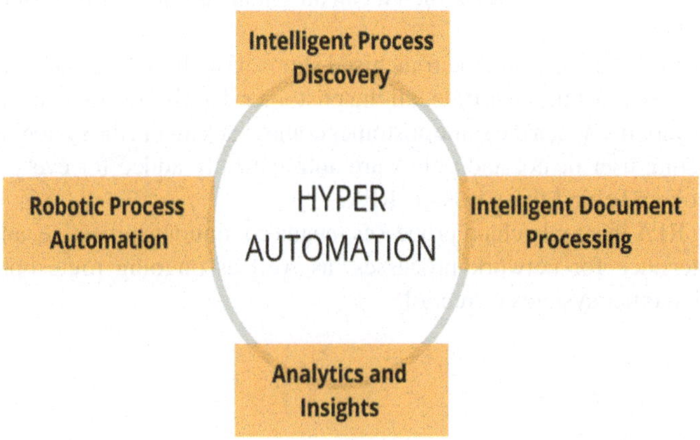

Fig. 13.7 Hyper automation phases

Intelligent Document Processing. It uses techniques like Optical Character Recognition (OCR) which has the ability to capture both structured and unstructured data. Raw Data from heterogeneous sources are fetched and processed for better research and analysis so that processed data can be unstructured. Techniques like Fuzzy Logic, Machine Learning are applied to capture and extract data. These intelligent cognitive powered bots are productive, efficient and intelligent to learn the job assigned to them [27–29].

Robotic Process Automation. RPA helps to execute processes in any environment and application which has the flexibility to work 24/7, highly scalable with great accuracy and precision.

13.3.9 Benefits of Hyper Automation

Speeding Up Complex Work. As Intelligent Process Discovery phase is part of Hyper Automation, all the stakeholders of the business can be involved to get knowledgeable inputs from people and thus create complex high-speed routes to automate tasks.

Digital Workers. The digital workers are the change agents of hyper automation. It will be able to connect different stakeholders of business, connect various enterprise applications, operate with different types of data and mine the data to analyze interesting patterns and measures, thus exploring new automation opportunities to transform the firm digitally (Fig. 13.8).

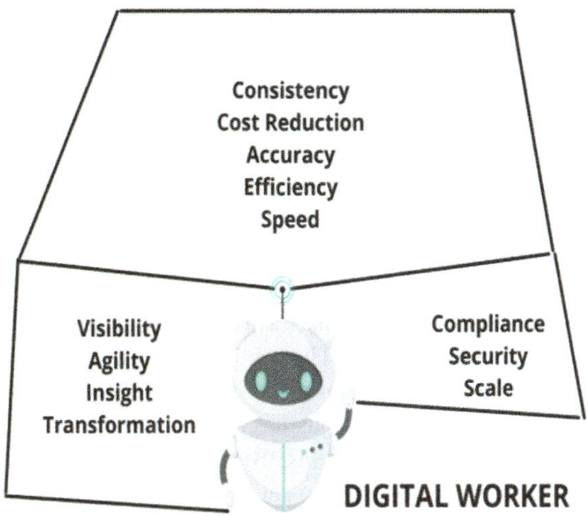

Fig. 13.8 Digital worker

AI+ RPA. With the above essential ingredients acting as a pioneer of hyper automation, it helps to uncover inaccessible data and processes. It also adds another feature named Digital Twin. The Digital twin of the enterprise explores unseen interaction between processes, data, functions, and KPIs framed by the company and thus the value creation of the business happens virtually incurring the intelligence to rapidly respond to identify new opportunities [28].

13.4 Scope and Future of RPA

It is said that automation will replace half of the global workforce in the industry. With the rapid development of Artificial Intelligence and Machine learning, the capabilities of understanding the human interaction with the computer systems has phenomenally improved leading to no or less human workforce to do the processes of an industry. Not only IT Industry, multiple industries like Manufacturing, Banking is adopting automation techniques to drive their company towards quality, profitability, safety and productivity. The famous example is that the Haldiram's factory in Nagpur is driven fully by automation from the start to finish.

The scope of automation in the IT industry is very high mostly in the fields of data entry and data validation. As the companies are concentrating on bringing out new gen technologies and innovations to facilitate the work of humans, data manipulation and pre-processing kinds of jobs make the workers sit for hours and finish them without erroneous errors. These manual, repetitive works are now automated with bots and it can work 24/7 monitoring and processing workflows without any trigger from humans improving scalability and efficiency.

The influence of RPA will not only lie in handling internal processes of an organization. It will go to range within the organizations, be it an internal process like checking any suspicious activities on the LAN network and immediately reporting it to the Chief Information Security Officer if any malicious things are found or for payroll processing of the employees at the end of every month. Not only these, now the companies move on to automation with customer oriented services like incoming email sorting on important promotions, updates in which the huge productivity can be satisfied by smart robots. These things show that RPA can improve organization structure in forthcoming years compared to the previous year's resulting in the company's succeeding levels.

The RPA tools and platforms are best performed when we integrate with other tools and technologies. Though RPA is more versatile and commanding, which can perform several functions alone, the final consequences would be gained by utilizing it with other tools. As business demands usage of diverse applications like CRM Software, IDS/IPS systems, Big Data centers, Cloud storages, different types of automation are required to satisfy the organization's process flow. The wholesome competitive advantage of a company lies in creating a total workforce consisting of people as well as machines that are intelligent to work on a vector of diversified applications.

The scope of RPA is huge and infinite. It is considered to be the future of the IT Industry. The career prospects for RPA practitioners and developers have increased by 52%. Tools like Blue Prism, Pega, Automation Anywhere, and Uipath have been adopted by businesses leading to over 65% of their processes automated. Still the appetite for RPA among companies is high. More innovations and inventions will make the business process smarter than today creating a better digital world.

RPA-as-a-service is the major equation for RPA market growth. The growth of RPA will involve more next-gen hyper stack technologies like Computer Vision, Optical Character Recognition (OCR), Machine Learning and Artificial Intelligence (AI). Companies usually get it from their Managed Service Providers MSPs and integrate it with RPA for broader automation [28, 29].

These integrations will outperform the automation market leading to automation of physical robots with a robot as-a-service model. The future of work is one and only automation and MSPs will continue to bring more innovative product, process and service with complex technologies bringing in a better service to the users.

13.5 Conclusion

Robotic Process Automation is one of the booming technologies in the current trend and knowing about its history and capabilities will help to understand the applications of it in a better way thus leading to more and more complex creations of bot in the future. The Fourth Industrial Revolution has opened up lots of interconnectivity between different domains due to which RPA can be applied anywhere anytime at any-level be it a small-scale industry or a larger multi-national company. With increase in data and customer-centric products and services RPA is a go-to solution for organizations to provide the data and get useful insights from it. Adding cutting-edge technologies like Cloud, Machine Learning and Artificial Intelligence the bot can now even be used to take crucial decisions for the firm instead of manual analysis. Thus the use-cases of RPA are widespread and the market continues to expand according to Gartner's Market Research on Robotic Process Automation [30].

References

1. Real Life Industry 4.0 examples. https://amfg.ai/2019/03/28/industry-4-0-7-real-world-examples-of-digital-manufacturing-in-action/. Accessed 31 Dec 2021
2. The Timeline of RPA. https://electroneek.com/rpa/history-of-rpa/. Accessed 14 Mar 2022
3. History of automation. blog.airslate.com/history-of-automation. Accessed 31 Dec 2021
4. Agile Methodology. https://www.wrike.com/project-management-guide/faq/what-is-agile-methodology-in-project-management/. Accessed 18 June 2022
5. Robotic Process Automation. https://enterprisersproject.com/article/2019/5/rpa-robotic-process-automation-how-explain. Accessed 18 June 2022
6. RPA finance and accounting automation|Uipath. www.uipath.com. Accessed 25 Feb 2022

7. Blue Prism. https://www.blueprism.com/. Accessed 18 June 2022
8. Automation Anywhere. https://www.automationanywhere.com/. Accessed 18 June 2022
9. Pega. https://www.pega.com/. Accessed 18 June 2022
10. Kryon Systems. https://www.kryonsystems.com/. Accessed 18 June 2022
11. Jidoka. https://www.lean.org/lexicon-terms/jidoka/. Accessed 18 June 2022
12. Work Fusion. https://www.workfusion.com/. Accessed 18 June 2022
13. Be Informed. https://www.beinformed.com/. Accessed 18 June 2022
14. Gartner Magic Quadrant. https://www.gartner.com/en/information-technology/glossary/magic-quadrant. Accessed 18 June 2022
15. Artificial Intelligence. https://www.techtarget.com/searchenterpriseai/definition/AI-Artificial-Intelligence. Accessed 18 June 2022
16. AI and RPA. https://www.nice.com/guide/rpa/rpa-ai-and-rpa-whats-the-difference-and-which-is-best-for-your-organizationlast. Accessed 18 June 2022
17. AI and RPA use cases. https://www.uipath.com/automation/ai-and-rpa. Accessed 19 June 2022
18. RPA Technology. https://www.blueprism.com/rpa-guide/will-rpa-technology-take-human-jobs/. Accessed 19 June 2022
19. PDD and SDD. https://www.probegroup.com.au/blog/what-are-pdd-and-sdd-in-rpa. Accessed 19 June 2022
20. Uipath Phases of RPA. https://www.uipath.com/product/task-capture. Accessed 19 June 2022
21. Real World Use cases in RPA. https://flobotics.io/blog/rpa-use-cases-across-industries/. Accessed 19 June 2022
22. RPA in Business. https://marutitech.com/benefits-of-rpa-in-business/. Accessed 18 June 2022
23. RPA's use in Marketing. https://electroneek.com/blog/rpa-in-sales-selling-that-pen-with-a-wave-of-the-hand/. Accessed 18 June 2022
24. Customer Relationship Management. https://focusonforce.com/crm/what-companies-use-crm/. Accessed 18 June 2022
25. Sales and Marketing RPA. https://www.automationanywhere.com/solutions/sales-and-marketing. Accessed 18 June 2022
26. Automation Anywhere. https://www.automationanywhere.com/rpa/hyperautomation. Accessed 30 Mar 2022
27. RPA and Machine Learning. https://www.anyrobot.com/rpa/machine-learning. Accessed 19 June 2022
28. RPA vs. AI vs. ML. https://www.crowdreason.com/blog/rpa-vs-ai. Accessed 19 June 2022
29. RPA and AI. https://www.capco.com/intelligence/capco-intelligence/rpa-and-ml-a-powerful-combination. Accessed 19 June 2022
30. Gartner Market RPA Review. https://www.gartner.com/reviews/market/robotic-process-automation-software. Accessed 18 June 2022

Chapter 14
RPA Adoption in Healthcare Application

K. Jayashree, R. Babu, A. Sathya, and S. P. Srinivasan

Abstract With the advent of Technology in the healthcare industry, there is a growth in revenue generation. Patients, doctors, insurance companies, and other entities are the key players in Healthcare Industry. Further effective and accurate back workplace method is urgently needed towards maintaining poise among the growing number of patients and the paperwork required for development plus insurance prerogatives, among other things. Hence, this concerns progressive mechanisation elucidations such as Robotic Process Automation (RPA) being able to assist healthcare organisations in increasing effective proficiency, lowering expenses, and reducing the risk of human error once handling data such as doctor credentialing, staffing, and patient fitness, as well as medical record maintenance, payables and denial recover, and patient scheduling.

14.1 Introduction

Changes in the overall thrift focused using the growth of innovative skills have the need for commerce towards grow into further nimble and towards rapidly reply towards the requirements, requests, and demands beginning with the consumers. Furthermore, economical plus commercial forces organisations towards being further capable, therefore continuously looking for novel tools as well as approaches which aid including further prolific, emancipate expenses plus improve cost towards their commerce. Specific elucidations that are developing as an innovative technology is Robotic Process Automation.

K. Jayashree (✉)
Panimalar Engineering College, Chennai, India
e-mail: k.jayashri@gmail.com

R. Babu
SRM Institute of Science and Technology, Chennai, India

A. Sathya · S. P. Srinivasan
Rajalakshmi Engineering College, Chennai, India

S. Bhattacharyya et al. (eds.), *Confluence of Artificial Intelligence and Robotic Process Automation*, Smart Innovation, Systems and Technologies 335,
https://doi.org/10.1007/978-981-19-8296-5_14

(RPA) that is able to substitute workers on monotonous jobs and systematise the aforementioned, and thus, facilitate workers towards being convoluted in further complex jobs that are able to take additional organisation importance [1]. RPA helps by way of a semi-intelligent handler including methods bycombining workflow, corporate rubrics, as well as "presentation layer" association by means of information methods. RPA is engaged within medical care aimed at monotonous processes including earlier authorisation, updating patient data, plus billing. It can be cast off towards extracting data as of when paired with other technologies such as recognition of an Image [2].

Two forms of RPA have been identified. They are as such Automated Assisting and Unassisted automation. Automated Assisting is a sort of automation that is appropriate for jobs that require interaction with a human system. The automaton remains worker-aided plus activated across different programs through the customer utilising robots placed on their computer. However, human involvement has advantages, including the ability to perform multi-faceted operations by means of a single mouse click. The aforementioned furthermore increases consumer satisfaction by lowering typical handling times. However, it is also reliant on the desktop environment. Unassisted Automation: This sort of automation uses machines that do not require human assistance. This type of automation does not necessitate a person logging on to the system, starting the process, monitoring its progress, and finally closing the mechanisation once the situation is finished. This may be done via the control room dashboard UI, which is used to execute operations like allocating tasks to machines, prioritisation, queues, bot performance monitoring, user administration, and so on. Because no human involvement is necessary, these bots can work through lineups. If something serves incorrect, the employee can be notified using the alert system. Unassisted automation achieves the desired extent of automation devoid of the need for human intervention. Of course, to reduce human involvement, digital information and precisely defined regulations are essential [3].

The following are two major causes pushing RPA adoption in finance and elsewhere [4]:

- Immediate Return on Investment (ROI): RPA initiatives consume small development phases and give fast results, such as FTE lessening plus increased amenability, centred on modest primary speculation and a great ROI in a shorter period of time.
- Increased Reliability: RPA-supported methods remain greatly auditable, devise low inaccuracy rates, are accessible, as well as deliver advanced analytics reporting to customer.

The automation of robotic developments diverges from the typical automation techniques [5]. The traditional automation and RPA process differentiation is shown in Table 14.1. Traditional Automation has certain drawbacks, hence it is considered as a motivation to include RPA in healthcare.

Some common steps in RPA [6] is Fig. 14.1. As per the author's opinion, some of the below-listed processes have to be deliberated as a distinct phase.

Table 14.1 RPA and Traditional Automation differentiation with various parameters

S. no	Parameters	RPA	Traditional automation
1.	Operating layer	Mimic the activities of application	Implement the set of instructions for automation
2.	Programming skills	No programming knowledge is required	Need expertise in a programming language to automate
3.	Complex system	It can automate any complex system	Automation depends on the constraints of Programming languages used
4.	Design time for scenario	Drag and drop is enough to design a complex scenario	Feasibility study and evaluation of test cases have to be involved in design
5.	Domain knowledge	Business Analyst should have deep knowledge of domain and process	Manual testers use a scenario-based approach, and automation testers develop the content relevant to each scenario
6.	Primary use	It is possible to automate RPA, sites, laptops, mobile apps, and mainframe-based applications	Traditional automated testing via a web-based or desktop-based software
7.	Execution time and scalability	Execution time is reduced by utilising more virtual machines	Parallel Programming can be simulated using programming methodology for improving the efficiency
8.	Maintenance of test scenario	Costlier to maintain	Maintenance is based on upgrading programming constructs

14.2 Related Work

In this work, [7] addressed RPA and reported on the evolution of a smartcard-based electronic health record monitoring system to improve the health record management system. The XAMPA platform, QR codes, HTML, and PHP are among the tools used to create software in this scenario.

The author [8] talks about the impact of automation on healthcare and how Robotic Process Automation and Artificial Intelligence may work together. In developing the digital twin of the future hospital, properly developing robust RPA solutions, and minimising the disruption from automation solutions, a new framework of simplification was introduced to address various impacts such as learning and understanding, emotions, natural interaction, judgement, complex problem solving, and creativity. The RPA solution creation, 3-D building information collecting, activities and flow simulation and optimisation, and scenario assessments are the primary components. The end-users confirm the details of the major components.

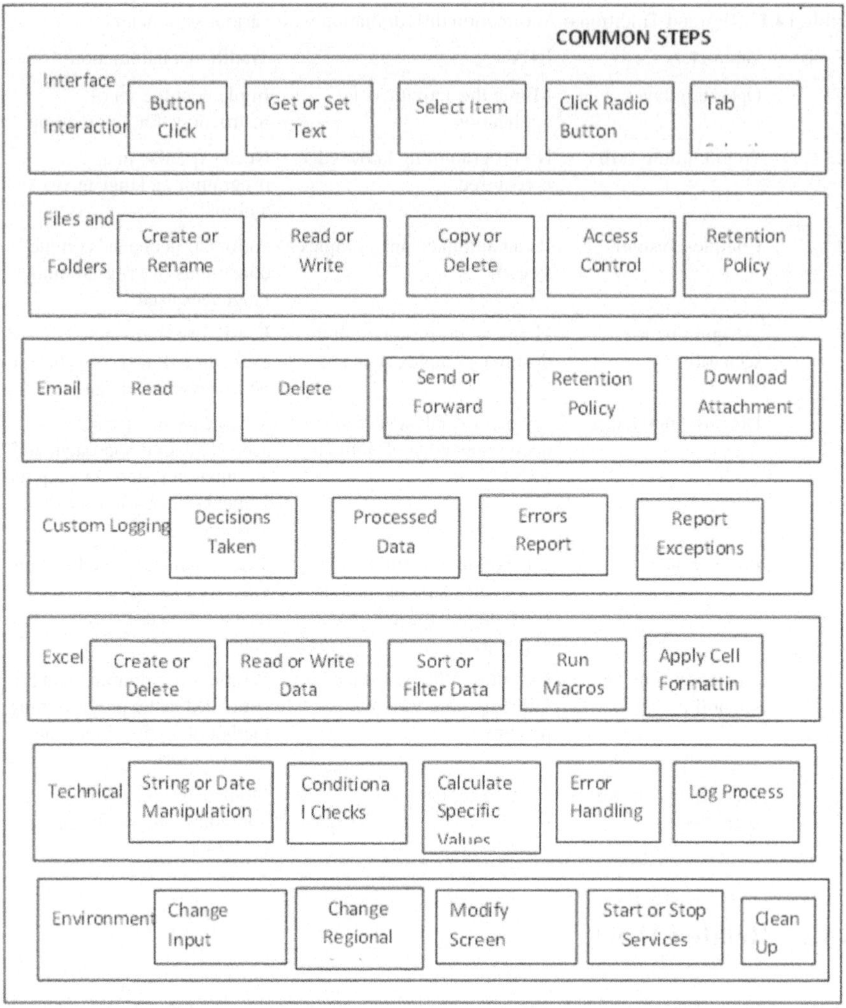

Fig. 14.1 Some common steps in RPA

We were able to learn about their perspectives on RPA installation by conducting eight interviews with employees whose tasks are automated by RPA [Zande/2018]. Employees viewed RPA implementation as good both before and after it was implemented. According to the survey, employees stated that RPA implementation reduced their workload due to the simplicity of the automated operations. Employees hoped that their occupations would become more varied and exciting. They remained favourable following installation, but expressed worries about process mistakes and how to handle them when technology knowledge is scarce. The author [9] proposed an eight-step RPA delivery model: 1. Identifying processes; 2. Assessing processes;

3. Reengineering processes; 4. Definition of user stories; 5. Automation; 6. Acceptance testing with users; 7. Excessive care; 8. Ongoing assistance. These steps can be outsourced or insourced, and alternative choices for some of them can be chosen. Vendors have suggested additional distribution mechanisms that are similar.

Validation is a major issue that RPA must overcome before it can be broadly deployed in pharmaceutical companies. Any system which accomplishes a decision-creating utility in a regulated environment requires validation and modification control. Any change, in particular, necessitates re-validation to demonstrate that the system either correctly understands the input and executes the expected result action, or flags the transaction for human intervention. There are extremely few situations when pharmaceutical companies use RPA to manage validation, and they still rely on people to do so.

Because RPA systems are precisely configurable and auditable, they have the ability to effectively meet the validation challenge, and may even be more dependable and accurate than humans. Because pharmaceutical innovation allows people to live longer and healthier lives, regulatory agencies oversee the process, imposing strict controls at every step. RPA allows pharma companies to focus on bringing safe and effective medications to market faster and at a reduced cost by automating some of the large volume, repetitive, rule-based, and error-prone operations. RPA is utilised to assist human workers and is not intended to be a replacement.

The following are examples of use cases of RPA implementation in the pharma industry [10]:

Review of Electronic Batch Records

Bots can inspect batch records to ensure that all essential documentation is there and that data obtained from multiple systems intricate in the process is consistent plus in working boundaries. The bot can close and log the reviews on its own if no abnormalities are discovered during robotic process automation inspections. If the bot detects an issue or a conflict which needs human intervention, the RPA system could be configured towards notifying one or more people before proceeding with the process's following steps.

Pharmacovigilance

Given the large volume of adversative event data for which organisations need procedures on a regular basis, pharmacovigilance is one of the primary sectors which can benefit from RPA. Because the data is inconsistent in eminence, building, plus format, and is difficult to integrate, companies are now managing case reports manually.

Financial Record-Keeping and Reporting

RPA elucidations are able to simplify data processing, data verification, and scheduling through account resolution plus transaction data reporting, removing manual job scheduling in the backend, maintaining GxP compliance, and making tax and legal activities easier to manage. It's vital to keep accurate financial records during drug trials in order to maximise tax deductions and avoid fines. Companies can minimise conflicts between taxable and non-taxable invoices by using RPA systems which make routine R&D drug sample work remits plus invoice abstraction. This automation could aid a corporation with R&D teams in several countries that have varying tax legislation to manage tax concerns more efficiently.

During the COVID outbreak, RPA was widely employed in testing and diagnosis. With the rapid spread of Coronavirus, testing facilities are overwhelmed, resulting in a long line of patients waiting for a test [11]. These data are then maintained automatically by RPA without the need for data entry professionals, resulting in a significant reduction in diagnosis time. RPA paired with AI can be used to assess COVID-19 from CT scans or chest X-ray pictures more efficiently and quickly, lowering patient waiting times by up to 70%. By automatically scheduling patient requests, RPA bots ensure that video consultation services are available 24 h a day, seven days a week. By computerising background checks, credential verification, and application data input into the HR system, RPA bots can help speed up the hiring process. After the processing is completed, robots notify the relevant department of the competent volunteers and assist them in assigning volunteers to task areas. RPA can help speed up the development and dissemination of the COVID-19 vaccine by coordinating and communicating essential clinical data among many teams or processes to shorten the time it takes to develop the vaccine [12].

RPA technologies that have been built for recovery and monitoring can significantly cut worker workloads. Screening bots keep an eye on employees' health and raise a red flag if they have any questionable symptoms, such as elevated body temperatures, allowing the appropriate authorities to take action. RPA bots aggregate COVID-19-related data from a variety of sources around the world, analyse it, provide insightful daily reports, and provide risk management instructions [13]. RPA has a lot of promise for use in back-office processes. RPA allows for real-time monitoring of pedestrian traffic flow and the triggering of risk alarms when traffic flow in public spaces becomes too dense.

14.3 Benefits of RPA

RPA is able to support companies in overwhelming repeated systematic mechanisation complications. The aforementioned stays a technique towards creating an effective labour force which lets organisations towards obtaining an economic benefit in terms of consumer satisfaction as well as enterprise swiftness [14]. The advantages of using RPA are shown in Fig. 14.2. Bots provide various degrees of advantages, including.

- Accuracy—Humans are prone to error, but RPA may dramatically increase the accuracy of business operations because it's resistant to errors made by humans.
- Consistency—Automation can be a repeatable and rule-based business process. It can carry out commercial tasks with incredible accuracy and speed that even the most intelligent humans are incapable of.

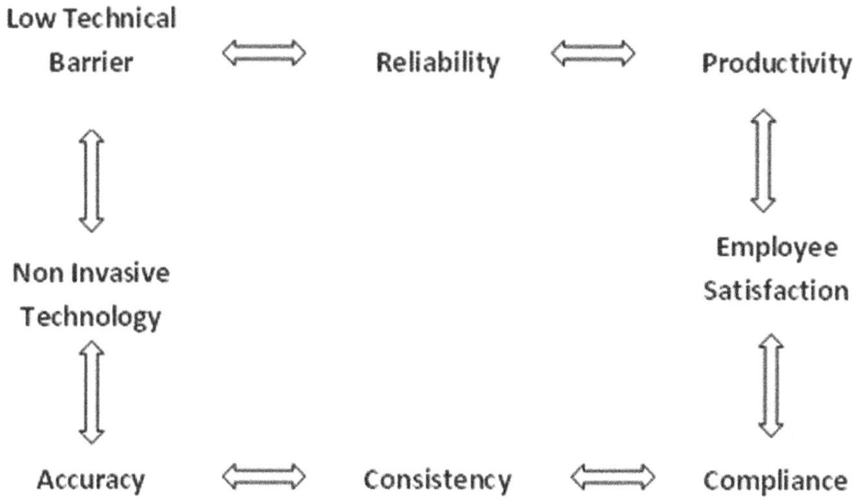

Fig. 14.2 Advantages of using RPA

- Reliability—RPA's built-in capabilities, such as Monitoring and Analytics, may provide extensive Audit logs and processes' health when unanticipated errors occur, employees can be programmed to respond appropriately using pre-defined protocols.
- Scalability—RPA methods offer centralised Bot management, permitting enterprises towards rapidly gauging up plus down. By means of a basic Bot manner and regular auditing of the bots after positioning, scaling up and down the RPA structure will remain straightforward.
- Lower Costs—Process Automation Bots are not as much as expensive as full-fledged bots. RPA can dramatically cut operation costs by deploying bots to automate business operations.
- Non-Invasive—RPA remains aimed towards imitating humanoid action, therefore it interrelates by means of data on the platform and application's display layer. As a result, organisations can employ RPA deprived of needing towards making significant deviations towards their current structures. The indicated extricate currency by obviating the requirement aimed at ongoing IT improvement as well as a methodical workforce.
- Improved Production—RPA's ability towards completing commerce processes in competition 365 days a year, 24 h a day, seven days a week, by means of extraordinary efficacy plus correctness, be able to lead to significant productivity gains for businesses. In this aspect, bots surpass humans in terms of reducing the amount of time it takes to accomplish tedious, repetitive tasks.
- Increased Employee Morale—RPA can liberate employees from tedious, repetitive jobs, concentrating on more engaging and difficult jobs. People's morale improves when they spend their phase and energy on jobs that are more engaging and less repetitive.

- Greater Visibility—Automation bots can discover and regulate poor data integrity problems. In every industry, this leads to transparency by revealing important flaws that impede management choices and operational effectiveness.
- Increased Compliance—RPA Bots perform processes according to the instructions they were programmed to follow, with an audit record for each step. Furthermore, bots have the capacity to replay their actions if any stage in a process has to be evaluated. Bots' controlled nature improves transparency and reduces fraud.
- Better Reporting—RPA provides a large amount of data, which enterprises may use to analyse and discover process bottlenecks and inefficiencies. Organisations can use the Bots to provide operational insight, to help businesses streamline their operations.
- Higher Value—Robotic Process Automation lets companies simplify and standardise their processes, occasioning in less data imprecisions. This reduction in faults effects in advanced class data that is aimed at further reliable study.

14.4 General Applications of RPA

General applications in RPA can be customised to fit unique requirements. It can be used in a variety of industries. It can be used for a variety of tasks, including (1) validating application documents, processing credit card transactions and insurance claims, connecting with customers, organising records and emails, and filling up forms and data. (2) Healthcare will retain digital health records, update patient data, insurance claims, and billing during discharge; retail will track orders, update shipping details, manage and audit inventories, and plan logistics; and IT will automate software installation. The most typical RPA uses are depicted in Fig. 14.3.

Vijaya et al. [15] employed RPA towards automating the method of filing insurance claims because it is previously used in motor vehicle insurance. Currently, an agent selling many distinct types of consumer insurance policies needs to list respectively one separately, distributing it by form plus necessitating dissimilar registration processes based on the type of insurance the customer is relating for. Here's a rundown of the RPA methodology: Scope: Determine the target scope depending on the requirements of the context and the resources available. Process Evaluation: Use the scoring matrix below to evaluate the processes. Testing: Run thorough tests on a large data set and keep an eye on the output quality. Model of operation should be pursued: Define suitable governance, sourcing alternatives, and benefit tracking metrics. Build a deployment strategy that includes training and change management.

Kaelble [16] looked at the advantages of automating robotic and desktop procedures, enabling robots more and more boring tasks to execute well without any complaint. It highlights a few of the ideas that can provide to your firm, discusses where it works best, and introduces the idea. Data digitisation was a powerful engine for efficiency and growth, but it also brought with it a new set of routine jobs.

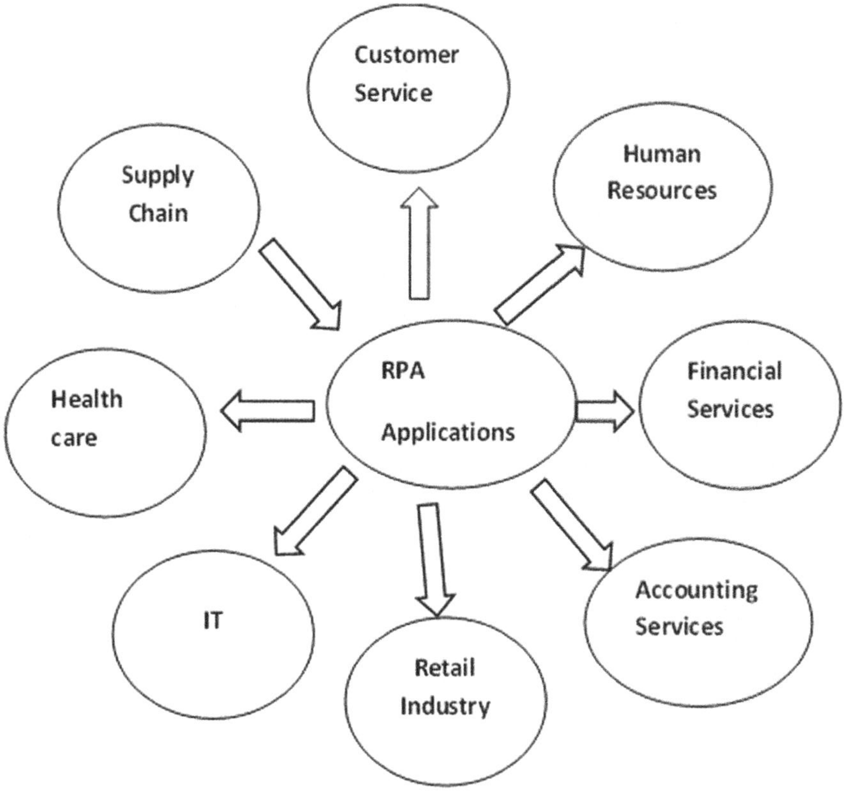

Fig. 14.3 General RPA Applications

Rural areas are home to the majority of wind and hydropower installations. Automation is critical for maximising system performance and increasing industrial efficiency. Every day, smart sensors in numerous locations around a power plant generate massive amounts of data. The development of reports based on data is a typical occurrence. RPA can be used in the energy industry to generate shift duty charts, tariff reports, load schedules, fault reports, energy generation reports, and employee salary slips, and to submit information to the necessary authorities automatically.

14.5 Need for RPA in Healthcare

Clinical studies, health insurance, and medical equipment are all included. In any healthcare system, managing and processing data is very challenging to integrate clinical applications, lab information systems, portals of third-party, insurance and radiology information, and scheduling apps, ERPs, and HR applications from a number of internal and external sources. Because the integration of these technologies is often difficult, healthcare companies must rely on humans to undertake labour-intensive manual procedures in order to process data.

Healthcare organisations experience difficulties in carrying new pharmaceuticals to the marketplace since they must maintain quality while still being efficient and profitable. Regulatory and reporting issues are common in the healthcare industry's innovation processes, which can be solved using process automation technologies. Healthcare organisations can use these technologies to enhance security and bring more successful pharmaceuticals to the marketplace.

RPA procedures are scalable, their deployment is simple and inexpensive, and no programming skills are required. RPA can boost employee morale by automating time-consuming and repetitive technical activities. The author [17] discussed RPA for the healthcare industry, and it is shown in Fig. 14.4. The critical analysis of traditional automation can be improved by RPA.

The Key areas of RPA in healthcare are as shown in Fig. 14.5.

Some of the most important areas in which RPA is altering healthcare are briefly described here.

1. Cycles of Revenue
   One of the most important benefits of RPA in healthcare is that it helps with cost control. RPA's usage in the healthcare industry has the potential to significantly expand back-office activities. RPA is particularly useful in the sales cycle and administrative tasks. Several code changes occur during each billing cycle, and RPA can react and adapt to these changes quickly and simply. RPA will improve billing performance and reduce write-offs by automating time-consuming and common processes like data digitisation and payable accounts, saving healthcare providers much-needed money. In addition, RPA is frequently used in medical

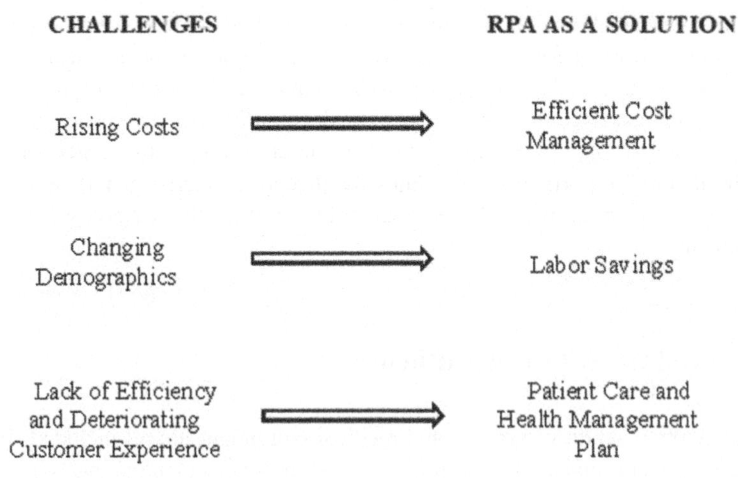

Fig. 14.4 RPA for the healthcare industry

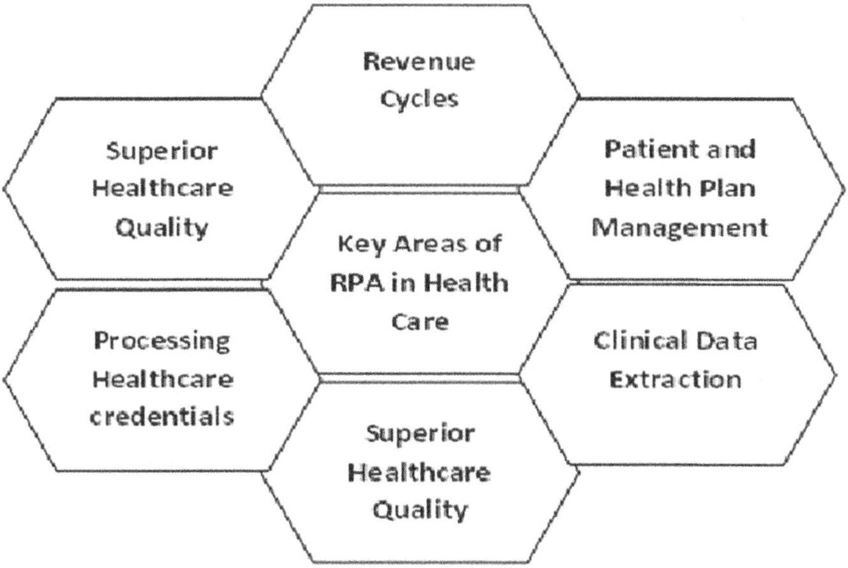

Fig. 14.5 Key areas of RPA in healthcare

reports to provide interpretation and statistics based on a patient's diagnosis and reaction. As a result, the deployment of RPA in the healthcare sector has a significant financial impact on healthcare providers.

2. Patient and Healthcare Plan Administration
   The process involving robots automation in the healthcare business helps with utilisation control, treatment coordination, case administration, and remote patient monitoring within the healthcare system.
3. Extractions of Clinical Data
   Healthcare organisations can use RPA software to extract data from both digital and physical clinical records. Typically, this type of application necessitates the input of login credentials from a human worker in order for it to have access to the network or an EMR device. RPA software may be trained to recognise document metadata like scanned PDF filenames, development dates, and EMR document unique ID numbers. This type of information is usually accessible by looking at the characteristics of a file or folder. RPA software will emulate a human employee after it has seen them do it multiple times. Basic, desktop activities can also be computerised with the RPA solution. RPA collects data from EMRs, medical records, consultation notes, and discharge summaries, among other sources. When a missing detail was discovered in one of these documents, a human employee was required to manually search for it and re-enter it. The above-mentioned issue is likewise resolved by using this RPA solution.

4. Hospital Self-Service Terminals
   The use of RPA in healthcare is automating patient check-in at hospitals, particularly emergency departments. Patients can type their information into self-service kiosks, which can be set up with screens.
5. Healthcare Credentials and Payroll Processing
   Without discussing the transformation, it brings to the processing of healthcare credentials and payrolls any discussion of RPA applications in healthcare would be inadequate. Another use of RPA in healthcare is to automate the processing of credentials and payroll records for healthcare workers, such as timecards and other proof of hours worked.
   Data transfer from a customer-facing manner of data entry, such as an employee mobile app, to the back-end verification process can be automated with RPA software.
6. High-Quality Healthcare
   The availability of higher healthcare quality is one of the primary benefits provided by RPA to the healthcare business. Patient or user happiness is improved by automating certain operations, which saves time, eliminates the risk of human mistakes, and allows staff to focus on more valuable, patient-centric duties. Improved operational efficiency may also increase the applicability of healthcare. As a result, the healthcare system may be able to address the demands of a greater number of individuals.

In the healthcare industry, RPA has the potential to be beneficial [18].

RPA's present use in healthcare will help with the following: Integrating heterogeneous systems to address the challenges that the healthcare sector faces in terms of process complexity, and volume of patient and hospital data from many sources; Employees should be relieved of routine tasks so that they can apply their skills to circumstances that require human interaction; Incorporate rapidity, intellect, efficacy, and eminence interested in healthcare procedures to save money and human resources; Automate procedures for gaining access to information, such as eligibility inquiries. Claim status requests should be automated, and claims should be reviewed to improve the revenue cycle. To achieve increased efficiency, create a digital workforce that will collaborate with employees.

14.6 RPA Tools

The various RPA are explained below.

14.6.1 UiPath

It is a tool which permits the production of Automation features in its structure to design and run programs to make it possible for it to be programmed with a block-based interface and different add-ons to customise industry processes [19]. UiPath Studio, UiPath Robot, and UiPath Orchestrator are the three modules that make up the RPA UiPath platform, with the latter allowing for robot orchestration [20]. This module is an application that helps you design, model, and execute workflows, as well as set up and maintain robot connections, package transfer, and queue management. As a result, log entries are stored and integrated with Microsoft's Information Services Server and SQL Server.

14.6.2 BluePrism

BluePrism is yet another top provider of Tools for automation and problem solvers. It's built on the Java and NET frameworks and uses a drag-and-drop approach to bot creation [21]. Process diagrams are business workflows constructed with the help of fundamental programming ideas. Create, analyse, adjust, and scale business capabilities using these graphical representations of workflows. Process studio is a tool that allows you to design process diagrams using a variety of drag-and-drop activities. Object Studio is used to generate Visual Basic objects that remain used for interfacing by means of further applications. The Application Modeller tool in Object Studio allows you to create application models.

14.6.3 Automation Anywhere

The Automation Anywhere is situated in San Jose, California, and is directed by CEO Mihir Shukla, who is also a Gartner Magic Quadrant leader (Andrade/2020). It is a firm that started with the goal of replacing human scripting applications in process automation. It has a centralised dashboard for developing, configuring, and monitoring bots in great detail using a combination of online services and plug-ins that may be installed on the local computer. It also collaborates closely with BPO firms like Accenture and Deloitte to improve their tools for a variety of industries, including banking and financial services, which accounts for more than half of their revenue, as well as healthcare, automobile manufacturing, and telecommunication behemoths like GM, AT&T, and JP Morgan Chase. It is marketed as a software testing E2E automation solution. As a result, it can handle and customise some, if not all, of the most prevalent software testing difficulties, such as repetitive regression testing, business logic testing, and scalable combinatorial testing. Community, which is free but has limited capabilities, Enterprise on-premises, and Enterprise Cloud are the three variants of Automation Anywhere.

Pros of Automation Anywhere:

- Enhance user experience.
- Support reusability of Library function used for ML.

- Improved security features.
- Analytics of the data can be easily performed.
- Delivery of services can be done through the cloud.
- Improvised strategic planning in the product group, invention, and evaluation.

Cons of Automation Anywhere:

- Experienced issues in upgrading to the latest version, which will now necessitate a formal migration process.
- Earlier version of Automation Anywhere reported a lower level of satisfaction with the support of their deployment experience, and deployed bot automation requires a lot of manual work.
- More expensive.

14.6.4 WinAutomation (26–29)

Email automation, files in various formats such as PDF and Excel, OCR, and other aspects relating to the post employees' work environment are among the features provided by the Win Automation tool, which are incorporated in RPA operations. It was created by Soft Motive, a provider of RPA solutions. In terms of RPA, the tool offers a collection of modules through the "processor bot" module, as well as a relationship with CaptureFast that allows it to extend its RPA capabilities with AI-powered information capture engines, data extraction in documents and systems, and hybrid document categorisation. According to the studied literature, the Cognitive module allows users to combine capabilities with analytical information processing engines like Microsoft, IBM, and Google's Cognitive. The tools, nevertheless, do not appear to provide evidence of AI capabilities availability.

14.7 Research Challenges

Challenges for Process Automation are [22]

- Administrative data entry: Administrative data input is one of the various types of data that can be found in a healthcare setting. Administrative data entry does not often require specialised knowledge, but it is a time-consuming task. RPA can accept data from a variety of sources, some of which may require bots to convert to structured data using natural language processing (NLP), speech recognition, or picture recognition. The information must then be recorded into a database or other repository for usage by the organisation.
- The document digitisation method: RPA can concoct and ingest documents ranging from medical records to insurance claims into a bigger repository for storage or use using intelligent document processing (IDP).

- Manage and schedule patient appointments: Using RPA bots and other processes to engage with customers, appointment scheduling and management can be automated. Patent appointments can also be scheduled, changed, cancelled, or updated as needed or requested.
- Billing and processing: Billing and claims processing are both routine tasks. RPA can automate these operations, with bots handling claims management, including initial inquiries and follow-up.
- Records management: The healthcare business is heavily regulated, with strict processes and reporting requirements for patient records, medical records, and other sensitive data. Regulatory compliance relies on record accuracy, consistency, and security, which RPA can deliver.
- Infection control: RPA can be utilised to assist healthcare personnel in carrying out infection control protocols, such as
  - Managing triage tasks
  - Screening and tracking protocols, including regulatory compliance and CDC protocols
  - Keeping track of inventory and patient flow
  - Keeping track of patient care plans and informing staff when patient data reaches certain criteria.
- Communications: RPA may be utilised in healthcare, as well as any other business, to automate communications such as website responses, first-line customer service calls, front-line administrator enquiries, and email blasts to patients, vendors, and staff.

14.8 Future Research Directions

RPA has been applied in all emerging areas but it also faces a number of hurdles, just like any other new technology [23]. Data input and rekeying jobs will be managed with automated tools and procedures in the near future [24]. This technology will be used to handle all computer-aided operations that are governed by a set of standards. It will be used to increase data accuracy and boost analytics. Because of the set of rules that govern its operation, formatting duties will be managed with the help of Robotic Process Automation. In the following years, there will be more widespread adoption: RPA will be used to manage and integrate a variety of operations, allowing businesses to operate more efficiently. Artificial Intelligence (AI) is a term that refers to the study of artificial intelligence. The next level of this technology will integrate AI features in addition to rule-based technologies.

Smart Process Automation (SPA) Emergence: SPA is seen to be an extension of RPA. The goal of SPA will be to automate the unstructured data tasks that robotics cannot handle on its own. Machine Learning and robotic process automation are projected to be used in SPA. Integration with other tools: RPA will be used in conjunction with other tools and technologies.

14.9 Conclusion

RPA refers to the automation of service processes that mimic human activity. RPA has a wide range of applications in a variety of industries, including healthcare and medicine, and it offers some extremely promising results. RPA is being rapidly adopted by industries such as healthcare, financial services, and government to assist individuals in coping with the financial, health, and practical concerns of living during this epidemic. Thus, this chapter has discussed the utilisation of RPA in the healthcare industries.

References

1. Ivančić Juroš, L., Vugec, D.S., Bosilj-Vuksic, V.: Robotic process automation: systematic literature review. BPM 280–295 (2019)
2. Davenport, T., Kalakota, R.: The potential for artificial intelligence in healthcare. Fut. Healthcare J. **6**(2), 94–8 (2019)
3. Oswal, S., Joshi, P.: Recent trends in automation: robotic process automation. IJRETS: Int. J. Res. Eng. Technol. Sci. **XIII**(VIII), 1–8 (2020)
4. Doğuç, Ö.: Applications of robotic process automation in finance and accounting. BUJSE **14**(1), 51–59 (2021)
5. Dechamma, P.J.R., Shobha, N.S.: A review on robotic process automation. Int. J. Res. Sci. Manag. **3**(5), 237–244 (2020)
6. Pavlov, P.: RPA chief solution architect adastra RPA: how long does it take to automate a business process? https://adastraglobal.co/insights/rpa-how-long-does-it-take-automate-business-process/ (2020)
7. Yahaya, S.A., Lydia, J., Jilantikiri, J., Hassan, K.A., Akande, K.A., Yahaya, A.: Development of an automated healthcare record management system. Adeleke Univ. J. Eng. Technol. **2**(2), 79–90 (2019)
8. Chow, Y.: Preparing for the future of healthcare in light of automation. https://www.automationanywhere.com/company/blog/rpa-thought-leadership/preparing-for-the-future-of-healthcare-in-light-of-automation (2020)
9. Anagnoste, S.: Setting up a robotic process automation center of excellence. Manag. Dyn. Knowl. Econ. **6**(2), 307–332 (2018)
10. Schiraldi, E.: Applying robotic process automation in the pharma industry https://ispe.org/pharmaceutical-engineering/march-april-2021/applying-robotic-process-automation-pharma-industry (2021)
11. Nassir, A.A., Baptiste, M., Mwikarago, I., Habimana, M.R., Ndinkabandi, J., Mrangwa, A., Nyatanyi, T., Muvunyi, C.M., Nsanzimana, S., Leon, M.: Rpa-based method for the detection of SARS-CoV2. MedRxiv 1–23 (2020)
12. Wiggers, K.: How RPA could help distribute and track COVID-19 vaccines. https://venturebeat.com/2020/12/27/how-rpa-could-help-distribute-and-track-covid-19-vaccines/ (2021)
13. Neo, D.: Rpa and COVID-19 - robots helping put employee health first. https://www.uipath.com/blog/how-software-robots-help-companies-keep-employees-safe-from-virus (2021). Accessed 20 March 2021
14. Devarajan, Y.: A study of robotic process automation use cases today for tomorrow's business. Int. J. Comp. Techn. **5**(6), 12–18 (2018)
15. Vijaya, G., Sathish, R., Shivaram, B., Dini Kokila, K., Aishwarya, R.: Claim registration using roboticprocess automation-a survey. Int. J. Sci. Res. Eng. Dev. **1**(2), 347–351 (2019)

16. Kaelble, S.: In: Robotic process automation for dummies. NICE Special Edition https://bpmtraining.net/wp-content/uploads/2020/06/robotic_process_automation_for_dummies.pdf (2018)
17. Pranesh Kathavate: Robotic process automation (RPA) in HEALTHCARE https://mechomotive.com/robotic-process-automation-rpa-in-healthcare/ (2021)
18. Jain, R., Bhatnagar, R.: Robotic process automation in healthcare-a review. Int. Robot. Automat. J. **5**(1), 12–14 (2019)
19. Ribeiro, J., Lima, R., Eckhardt, T., Paiva, S.: Robotic process automation and artificial intelligence in industry 4.0–a literature review CENTERIS–International Conference on ENTERprise information systems/ProjMAN–International Conference on Project MANagement /HCist–International Conference on health and social care information systems and technologies 2020 Procedia Computer Science 181. pp. 51–58 (2021).
20. Tripathi, A.: Learning robotic process automation: Create software robots and automate business processes with the leading RPA tool. Packt Publishing Book Series, UiPath (2018)
21. Khan, S.: Comparative analysis of RPA tools-UIPATH automation anywhere and Blueprism. Int. J. Comp. Sci. Mobile Appl. **8**(11), 1–6 (2020)
22. Williams, P.: Amazing ways that RPA can be used in healthcare https://www.ibm.com/cloud/blog/amazing-ways-that-rpa-can-be-used-in-healthcare (2021)
23. Lahane, K.K., Sutar, A.A.: Automation in Healthcare using RPA. Int. Res. J. Modernizat. Eng. Technol. Sci. **03**(01), 1363–1368 (2021)
24. Gami,.M., Jetly, P., Mehta, N., Patil, S.R.: Robotic process automation–future of business organizations: a review. In: Somaiya, K.J. (ed) 2nd International Conference on Advances in Science & Technology (ICAST-2019) Institute of Engineering & Information Technology. University of Mumbai, Maharashtra, India
25. Andrade, D.: Challenges of automated software testing with robotic process automation RPA–a comparative analysis of UiPath and automation anywhere. Int. J. Intell. Comp. Res. (IJICR) **11**(1), 1066–1072 (2020)
26. WinAutomation: Desktop automation. https://www.winautomation.com/product/all-features/desktop-automation (2020a)
27. WinAutomation: About softomotive. https://www.winautomation.com/about-softomotive/ (2020b)
28. WinAutomation: Installation requirements. https://support.softomotive.com/support/solutions/articles/35000081666-winautomation-installation-requirements (2020c)
29. WinAutomation: Softomotive RPA Review. https://www.rpa-star.com/softomotive-vs-winautomation-rpa-review/ (2020d)

Chapter 15
Cognitive IoT Meets Robotic Process Automation: The Unique Convergence Revolutionizing Digital Transformation in the Industry 4.0 Era

Prasenjit Bhadra, Shilpi Chakraborty, and Subhajit Saha

Abstract The emerging trends of cognitive Internet-of-Things (CIoT) are disrupting industrial process automation by infusing intelligence within the pervasive interactions and process automation of enterprise assets. Robotic Process Automation (RPA) is another fascinating technology trend playing a pivotal role in accelerating operational excellence across industries [1]. RPA solutions are designed to orchestrate service workflows that automate repetitive and rule-driven voluminous tasks. While the CIoT facilitates intelligent cyber-physical integration to enhance ubiquitous operational intelligence, RPA introduces automated workflows within the connected enterprise to maximize agility and resilience. As industrial computing is inclining towards maximizing situational awareness and autonomous operations, the integration of AI-powered IoT and intelligent RPA is paving the path to disrupting innovations in Industry 4.0 era. The paper delves into key technology components and architectural patterns that introduce a new breed of Cognitive enterprise systems enabling intuitive operations and need-based control functions beyond complex decision support and pervasive interlocking of Industrial IoT. We present unique architectural semantics that introduces RPA capabilities within CIoT to transform the actionable insights into context-aware process flows, promote interoperability, and execute prescriptive actions. The objective of the paper is to present the design rationale of next-generation industrial automation, compelling Industrial IoT use cases, and the research directions on autonomous systems achieved through such convergence of CIoT and RPA.

15.1 Introduction

Implementation of smart systems in business-to-business (B2B) and business-to-enterprise (B2E) has triggered radical innovations in integrating Wireless/RFID, mobile technologies, sensors, and actuators with mainstream enterprise systems.

P. Bhadra (✉) · S. Chakraborty · S. Saha
Ranial Systems Inc., 100 Wall Street, 11th Floor, New York, NY 1000, USA
e-mail: prasenjit@ranial.com

S. Bhattacharyya et al. (eds.), *Confluence of Artificial Intelligence and Robotic Process Automation*, Smart Innovation, Systems and Technologies 335,
https://doi.org/10.1007/978-981-19-8296-5_15

Over the last two decades, major developments in semiconductor research, advanced sensor network infrastructure, and AI have transformed our digital experience and augmented hybrid technology platforms aiding automated sensing and intelligent operations. The pervasive interlocking among the machines (Machine-to-Machine or M2M) and seamless interactions between the physical and digital world (Internet of Things or IoT) have unified the sensors and embedded systems with enterprise IT backbone over the wireless networks [2]. Such systems are designed to capture physical events and aggregate and process massive amounts of data to automate industrial operations. The automated sensing of the operating environment and triggering business processes with no or limited human interventions have unlocked the potential of implementing people and asset tracking, electronic surveillance, industrial automation, retail, and logistics operations. However, the purpose-based logic and rules have constrained ubiquitous interactions among the participating cyber-physical objects. The integration of AI with IoT along the maturity cycle of advanced automation has helped organizations to overcome such constraints and facilitated cohesive interactions among the discrete assets and operating technologies. The most common IoT implementations are designed with remote sensing and centralized processing of the sensory feeds. Such implementations are focused on gaining insights into the state of the physical world and taking informed decisions [2]. At an early stage of industrial automation, the purpose-based applications were localized using Programable Logic Controllers (PLC) and Supervisory Control and Data Acquisition (SCADA) systems. As the industries shifted their focus to real-time tracking of assets and processes, the convergence of operating technologies with Information technologies gained significant tractions. A set of edge controllers and mobile edge devices was introduced to implement distributed intelligence at the point of actions as well as enterprise IT systems. Our earlier research on developing wireless/RFID middleware solutions has introduced the 1st generation mobile edge computing framework and edge computing services with Wireless/RFID applications. Such design constructs equipped 1st generation M2M/IoT systems to achieve near-real-time process automation without being constrained by limited network bandwidth and computational capability of distant enterprise systems [3, 4]. However, such automations are designed with predefined rules and/or logics addressing a finite set of operational conditions.

Integration of AI-based decision support systems with IoT applications has been studied extensively [5–7]. There are significant contributions on how to apply AI to the massive amount of sensory data and gain intelligence in healthcare, agriculture, industrial and public sector domains [8–10]. Most of the implementations around AI-centric IoT data analytics have concentrated on the techniques of extracting meaningful insights out of captured measurements and network operating conditions. Such capabilities offer an important breakthrough in terms of crossing the confinement of hardcoded rules and present a broader perspective on operating conditions. However, such analytics developed using the historical data offer limited insights to facilitate preventive actions and/or manage operating conditions. In order to reap the further benefit of AI within IoT application, edge-centric AI models and ad system components are proposed to expedite the process of capturing the operational intelligence and take intuitive actions [11, 12]. The recent developments in Cognitive IoT systems have introduced soft computing-based models within the edge and IoT network to mimic human-like behaviour that enhances the interoperability, interactivity and facilitate need-based actions [13, 14].

The recent developments on the Internet of Robotic Things (IoRT) converge the smart capabilities of any IoT systems with autonomous agents or robots. The convergence of IoRT with cognitive AI that demonstrates autonomous M2M systems achiving decisions autonomy, self-powered machine-to-machine, and machine-to-human collaborations [15–17]. The common discussion on operating principles is lacking the core system-level design semantics to cater a wide range of industrial use cases [18]. With the emergence of cognitive AI and Robotics, the industrial IoT applications are adapting digital twins that can act on the incoming physical events along the process of data aggregation. We argue that the scope of cognitive IoT is not confined the process of gaining operational intelligence rather gain autonomy in enforcing specific actions in real-time or near-real-time using intelligent robotic actions. Moreover, such a seamless process of senescing to execute autonomous operations using cognitive capabilities represents a hybrid software and hardware platform that requires a unified set of technology components self-organizing relevant services within the context of industrial operations. The book chapter focuses on unique IoT solution architecture principles that inherit end-to-end ubiquitous capabilities of sensing, responding, and controlling operations based on cognitive intelligence embedded within edge and cloud application. The proposed integration of AI-driven RPA is responsible for the dynamic binding of autonomous actions performed at various segments of IoT runtime. The reference architecture based on the convergence of Cognitive IoT with intelligent RPA presented in the chapter discusses several aspects of industrial automation imperatives across utilities, manufacturing, and healthcare domains. The complexity of inheriting end-to-end pervasive operations and executing those with human-like efficiency remains the greatest challenge for any autonomous IoT or industrial automation infrastructure. Therefore, articulating a generic set of architectural components and patterns to promote the extensibility and flexibility of an ideal smart industrial IoT system is the most critical challenge in the industry 4.0 era. Within the current scope of discussion, we focus on how to address the diversity and heterogeneity of the physical environment across the industrial value chain, the ability of developing self-learning capabilities and determining operational flow in collaboration with connected systems. We hope the technical foundation and alignment of real-world scenarios will offer new directions to the community and advance further research and development in the related field.

Section 15.2 of the chapter details the background and motivation behind the convergence of emerging trends of CIoT and RPA. Subsequently, we have explained the significance of cognitive IOT and its evolutionary trends over and above the application of AI within IOT application space. The role of RPA in Industrial automation is briefly discussed in Sect. 15.4. The section explains the role of RPA in Industry and the emerging landscape of AI-driven RPA implementation in the recent past. The section includes a few use cases of RPA to illustrate the relevance of integrating cognitive IOT industrial applications. Section 15.5 details out core architectural patterns, unique design frameworks and service integration semantics that allows a seamless integration of Intelligent RPA using CIoT runtime. The relevance of the architectural patterns is explained in Sect. 15.6, through some practical implementation scenarios in health, smart city, smart grid, and heavy industrial application spaces. The chapter is concluded by discussing the theoretic challenges of implementing AI-powered IoT systems in the diverse field of the industrial value chain, and how the need of reliable integration of RPA and Cognitive IoT could elevate the role of intelligent process automation in industry 4.0 era.

15.2 Background and Motivation

The rapid proliferation of emerging technologies has created widespread alterations in enterprise-scale automation. Scholars and industry experts have presented and nurtured unique design frameworks that promote complementary technologies to maximize the return on investment (ROI). The strategic drivers of such amalgamation of technologies unlock potentials of new product and services, address gaps in evolving operational landscape. The benefit of convergence and collaboration of any hybrid digital application is maximized through an efficient structural and functional modelling of respective technology features. The conceptual framework of Industry 4.0 has been promoting both horizontal and vertical integration of operating and information technologies to maximize productivity, organizational performance and competitive advantages. In order to achieve the desired level of operational efficiency, the structural and behavioural synergy among technical capabilities is the most critical step towards shaping pioneering innovation [19]. The scholarly research on the convergence of technologies [20, 21] reveals that the optimal integration of unique and complementary features are critical drivers of business transformation, organizational evolution, and ability to foster innovation. At the core of such amalgamation strategy, the strength and opportunities of complementary technologies need to be analyzed to articulate collaborative value propositions. For example, most common RPA implementations have leveraged structured rule-based process automation to digitize repetitive operations. The emergence AI-powered RPA has widened the scope of pattern-based workflow models that address the need of context-aware operations from within the complex and sophisticated business process. The embedded AI within the CIoT application extends the integrated system with complex decision-support capabilities. The next-generation IoT systems should translate such decisions into operating principles and actionable events managing the pervasive interlocking of industrial infrastructure. The recent research initiatives on Internet-of-Robotic Things or IoRT have presented investigative frameworks, design concepts and implementation of RPA-powered edge computing and autonomous systems [15–17, 22–25]. There is no 'one-size-fits-all' architectural strategy to merge the AI-powered IoT and intelligent RPA to cater to diverse imperatives in Industry 4.0 era. However, the top-down analysis of various industrial applications and use cases reveal a set of unique system behaviours pertaining to industrial IoT implementation and associated ubiquitous operations across numerous industry verticals.

The chapter focuses on the most common implementation semantics and use case patters of CIoT applications where the RPA could be introduced to manage the KPIs (Key Performance Indicator). Based on our experience in CIoT implementation, we believe that the service component modelling, sensory event management and associated AI algorithms can be weaved into reusable frameworks to realize autonomous operations using intelligent RPA and CIoT technologies. The motivation of the paper is to present a set of architectural patterns and relevant AI models that addresses the existing gap of theoretical design constructs of intelligent industrial IoT systems capable of driving autonomous operations. The chapter briefly explains the relevànce of such architectural principles to various use cases in smart city, cleantech, manufacturing, health care and smart grid sectors. We argue that the IIOT equipped with cognitive AI is not self-sufficient terms of enabling intuitive process automation and translating sensing into an action in real-time/near real-time and/or on-demand. The proposed architecture inherits a holistic view of network topologies, sensor integration, and event persistence and dispatch patterns in relation to CIoT applications. The generic design constructs of IoRT and CIoT implementations are emphasizing on cloud managed operational intelligence and automation. In this chapter, the architectural principles are focusing on distributed intelligence that allows multiple layers of automation infrastructure to learn and act independently and in collaboration with carnalized service management hub. Moreover, the existing literatures have identified the challenges of infusing cognitive capabilities within the robotic automation and presented conceptual framework of achieving intuitive operations without explaining the implementation approaches. The design patterns and applications depicted in this chapter emphasize on infrastructure, communication, and intelligent operations as the constituents of extending enterprise scale cognitive operations in line with Industry 4.0 vision. Practically, the use of RPA is broadly limited for rule-based RPA automating routine operations without any intelligence to address exceptions, anomalies, or out-of-the-box operations [26]. Since the intelligent automation can be triggered and managed through any or multiple layers of an industrial IoT implementation, cognitive sensing and controlling measures can be deployed in a decentralized fashion resulting a stateful RPA runtime spanning across multiple nodes. The chapter embraces the practical scenarios of scalable and extensible IIoT application deployments where the cognitive capabilities are optimally combined with robotic processing and controlling measures across the industrial value chain and thereby meet the generalized design standards.

15.3 Significance of Cognitive IoT

The convergence of operating technologies and information technologies constitutes the core foundation of Industry 4.0 revolution. The focus of the industrial automation has shifted from discrete purpose-based infrastructure to process orchestration adaptive to the changing operational environment. The maturity of M2M and IoT has offered seamless integration of physical assets to the enterprise IT while the Integration of cognitive AI for computation merged with IoT runtime promoted intelligent and scalable digital foundation, termed as the Cognitive IoT or CIoT. Cognitive IoT uses smartest computing technique which is new generation of computing systems. The Internet of things (IoT) is basically physical objects which are ordinary appliances by the interface but embedded with sensors, processing power, software, and other technologies that are connected to server to serve additional computation through the internet [5, 6]. Industrial IoT allows various devices to communicate and interact with each other. The System collects and shares large data inside the digital supply chain network to ensure the exchange of information between the system elements [27]. Due to the massive use of the internet and the need of smart solution every day, large volume of data is generated from the edge to the server in distributed system. Hence, the decisions taken by the devices connected to IoT are mostly based on pre-programmed models and they are not fully autonomous systems. Cognitive learning has the ability to get human-like intelligence at the computation. An ideal Cognitive AI model has the ability to do comprehensive analysis from complex data pattern. It also facilitates interpretation of unstructured data, achieve better performance from a small amount of data to make efficient computation. The incremental or adaptive learning capabilities aids CIOT to meet human like cognition by experience over time and gain autonomy in performing ubiquitous operations. The data generated from IoT are mostly unstructured and primarily dependent on centralized server output, which obviously has several issues like data privacy and high latency, and more importantly, any attack at the server will make the whole decentralized IoT device system into failure. For these reasons the popularity of edge computing is growing in recent years. The main problem of edge computing is its resource constraint issues. The advantages of Cognitive AI were discussed earlier in this section. To resolve these issues, Cognitive AI has shown its utility to add efficient computation at the edge. Cognitive IoT or CIoT have enabled the organizations to drive proactive digital transformation. The real time operational intelligence analyze sensor data for making wise decision, enhance production quality to make better customer experience in a complete end to end automation mode. So, in brief, CIoT improves the smart IoT by integrating the human nervous systems into the current system. With the emerging trend of Industry 4.0 adaption, growing number of business processes are integrated through pervasive operations performed within the industrial segments. As the business undergoes cycles of rapid changes, the extensibility and scalability of CIoT play an important role to inherit disruptive service management and managing unaccounted exceptions within the predefined operating conditions.

15.4 Role of RPA in Industrial Automation

Advanced manufacturing technologies (AMTs) like computer aided design (CAD), production and computer guided administration have laid the foundation of modern industrial revolution which we are calling Industry 4.0. Researchers have empirically shown that companies are prone to higher investment on AMT enabled designing than other AMTs [28]. Industries use CAD applications to plan, modify or optimize their project. RPA has to communicate with such applications to complete the designing to manufacturing route. CAD was identified as a leading component of industry 3.0 which enables the manufacturing process to connect with computers. The application was to monitor assets, productions, storage, logistic planning and instruct other plans. But the key imperatives of Industry 4.0 paradigm promote advanced automation where industries are looking for end to end digitalization by reducing human intervention, server connected operating robots [29].

Robots aiding industrial automation were born long time ago. With the growth of robotics technology, they are used in process automation like automatic manufacturing, product automation and quality management systems. At the current post covid economy, every industry is more willing to reduce production cost without compromising the production quality and quantity. So, most of them are moving towards automation, precisely speaking Robotic Process Automation (RPA). The job of RPA is to build software robots to eliminate human effort in automatic production by interacting to server, following the instruction from server, completing the correct keystroke, monitoring the system, and doing other defined action. So, RPA imitates human instructions. RPA has enhanced production quality as it is free from any manual error. Trained RPA bots make the routine production very fast to meet the customer demand [1]. These two utilities make the industry to utilize their manpower in more sophisticated operation. The large data generated by RPA gives a better understanding of the production quality to the management.

In the manufacturing industry, the daily communication between vendors, customers, and the internal labours requires huge manual effort from the customer service. RPA can replace the entire job like opening and reading mail, tracking shipment status from enterprise resource planning (ERP) software, sending update to the customer and move on to the next customer. Similarly in the process of inventory control, RPA performs real-time inventory level monitoring, notifies stock level and reorders in low stock level.

Rule Based RPA has been widely used for less complicated uses which are built just to avoid human error and manual intervention. Very simple tasks which are prone to human error like matching some string of letters or numbers from a text or multiple columns can be executed simply by RPA. RPA is widely used in financial decision making by computing formula-based metrics Internal rate of return (IRR), Return on investment (ROI) etc. or to do very standard rule-based jobs like copying from a spreadsheet to a given customer registry. Precisely, a rule-based RPA is suitable for simple, standard, less intuitive and serving routine tasks. Such RPA services are not fully utilized in automating complex industrial operations yet [23]. Axmann et al. [30] has categorized RPA into three different classes—attended, unattended and hybrid model. The attended model works as an assistance of an industry worker and the worker needs continuous interaction with the robot to start, evaluate and end the process. The unattended model does not need any human intervention except at rule defining or training step. At the hybrid model the worker operates an attended model which operates another unattended model. The unattended model is considered to be the most ideal framework where the end-to-end automation can be managed through cognitive discovery and intuitive decision support capabilities using unstructured, ambiguous IoT datapoints cannot be processed and acted upon in an unattended mode of operation.

Obviously, AI perceives the input data and sends output to the robots for further action for intelligent computation. AI imitate human nervous systems functions. Artificial Neural Network (ANN), Text mining, Natural Language Processing (NLP) are the most common AI techniques in RPA [31].

There should be a kind of partnership relation where robots and humans live together in the same society and the robots will serve the needs of human very efficiently. There are more potential in RPA as most of RPA jobs are used to perform predefined routine actions. More sophisticated and logical jobs still need human intervention. Cognitive IoT can add human like intelligence to overcome this limitation of RPA. To the best of our knowledge, RPA and Cognitive IoT have been treated as two distinct isolated domains and there is not much proposal to bring these two domains together or to integrate them. This paper is aimed to bring these two domains closer together and make the RPA robots smarter and useable in generic scenario. The ultimate goal is to match the robot intelligence same as human capability at least specially for perceptual logic.

15.5 Core Architectural Patterns

Considering the complexity of deployment and expansion of intelligent IIoT applications, both the physical and logical architectures inject intelligence the way sensors, integrated hardware and software are sensing, interpreting, and acting upon the changes in physical environments. Such design principles enable IoT systems to simulate the neuro-motor operations and prioritize decisions to act by weighing the alternatives of execution plans and weaving a set of tasks. An integrated IoT system is devoid of strong partitioning of software and hardware capabilities in terms of realizing intelligent application behaviours. Based on our experience with the first generation IOT systems and RFID/Sensor middleware design [3, 4], we realized that multiple layers of an IoT runtime should host context-aware service components and interaction models to execute time sensitive operations. Considering the positional and computational constraints of participating subsystems, the application services will vary. In our current discussion we will abstain from discussing a set of generic design patterns and component that constitutes an ideal cognitive IoT systems with autonomous operations. We would rather focus on design principles of service components, integration patterns and algorithms that align with operational boundary and computational capacity of participating subsystems. In order to elaborate on the proposed architectural principle, we identify three logical layers across the runtime. Each layer inherits specific frameworks for data persistence, extracting intelligence, and executes autonomous operations.

- Intuitive sensing and controlling in proximity using Edge/ Fog computing infrastructure represents the first layer of the platform runtime that is collocated with the sensors and operating infrastructure. The multichannel data acquisition, embedded persistence and intelligence built on top of Edge/fog computing are extending cohesive frameworks of pervasive operations designed to sense and control multitude of industrial gadgets, actuators, and PLC systems. The most promising developments in edge centric IoT systems are focused on improving responsiveness and promote intuitive operations independent of a centralized cloud backbone. The future state architecture of a ubiquitous edge computing runtime for an intelligent and autonomous industrial IoT system should address 3 critical design imperatives: (1) Accuracy and reliability of actionable insights gained through localized AI models, (2) Uninterrupted incremental learning to improve intelligence, and (3) dynamic service mediation based on the predictive or prescriptive insights generated real-time.
- Near-real-time cognition on stream introduces the centralized event aggregation hub, a critical service integration and delivery capabilities that offer a strong backbone for near real time operational intelligence and intelligent robotic process automation across the enterprise level. The layer acts as a gateway or hub for all the distributed edge/fog nodes being deployed in a high-performance computing environment. The layer is capable of running analytics on stream with massive parallelism and executing multi-dimensional analysis on a diverse set of sensor data as well as correlating those with historical datapoints.
- Offline DSS with deeper intelligence and automated process orchestration capabilities of the proposed architecture are achieved through scalable cloud runtime. The decision support system deployed within the high-performance computing environment manages process intensive training of the models, deploys autonomous operations to reset threshold conditions and introduces automated service orchestration across the implemented system. Our prior research on distributed cognitive IoT systems [32] has presented the core functional modules of a central management hub within the intelligent IoT landscape. By design, the cloud-based runtime doesn't necessarily facilitate all the actions or process automation within the system. However, the visibility of operating conditions and ever-changing system interactions, the DSS can overwrite the models and service integration semantics through continuous learning.

The diagram below presents the core building blocks and key features of the logical layers and delegation of responsibilities (Fig. 15.1).

The intelligent and autonomous interactions among the constituent physical nodes and logical layers leverage a set of reusable frameworks and patterns of capturing

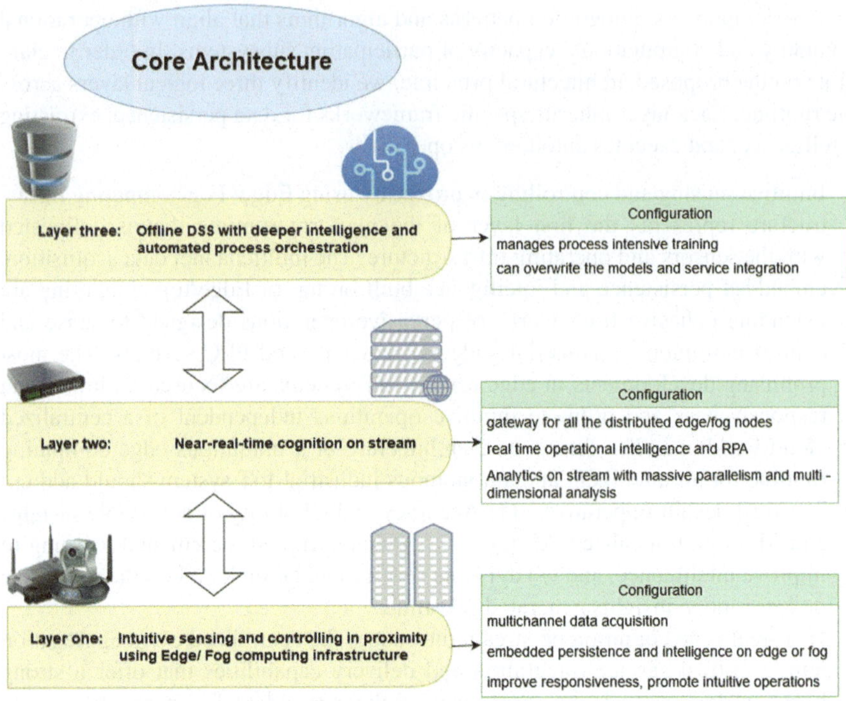

Fig. 15.1 Core architectural layers and services

storing and processing IoT data within the runtime. We present generic persistence models, service mediation and process automation patterns that are relevant to cognitive process automation at various levels.

The ontology of edge-native data aggregation of any cognitive IoT system should follow two broad categories of persistence models: (1) Event based aggregation, and (2) Streaming based aggregation. An event-based model usually encapsulates the context of specific actions. Whereas the streaming-based models stacks up the measurement data points in a time series format. The physical design of such data models varies with the nature of assets and environments managed by the IoT systems. The reusable data structure of an edge native persistence layer usually uses a HashMap, simply storing the property name and measurement as name-value pair and wrapped with the time and source of the event or measurement. Such data model represents a generic pattern that accommodates properties and or extracted features of diverse set of sensors and physical objects that are integrated with across the industrial value chain.

A scalable architecture of an intelligent IoT system inherits an extensible big data analytics foundation to store both pervasive events linked with operational data. The most efficient design of a cognitive IoT database inherits a time series data model on top of a columnar storage pattern. An enterprise scale industry IoT system aggregates data from a diverse set of assets, equipment, and combines those with various operational data. The fixed row/column-based RDBMS data model cannot scale over time. Usually, the time of a sensory event in combination with unique asset or source ID are combined to form a unique key to capture and link both operational data and sensory feeds associated with the respective industrial operations. Often cognitive IoT models leverage time-bound weather data, contextual historical events to gain deeper insights, and situational awareness to act accordingly. Hence, the time series model with high performance querying and linking capabilities is a critical design consideration of an intelligent IoT system. As the system continues to expand its functional and operational features, the training and applications of the cognitive models should extract need-based datapoints at a large quantity, to improve predictive and prescriptive accuracy over time.

Another important design consideration of the cloud persistence layer is the preprocessing and feature extraction of the aggregated data in the cloud. As the cloud runtime receives streams of sensory events and measurements, the data pipeline should introduce a set of mining algorithms to ingest OLAP compliant data sets. Thus, the cognitive models and process automation algorithms can leverage optimized data sets efficiently. Such (Divide-and-Conquer) strategy minimizes the cycle time of preprocessing of the requisite data sets against each lifecycle of testing and executing of AI models.

The ubiquitous interactions among the subsystems would require a robust and configurable service integration to facilitate information interchange and service mediation functions. In order to introduce multi-level cogitation, support real-time operational intelligence, and robotic automation, the logical layer of the runtime introduces three data integration patterns:

15.5.1 Store-Forward Pattern

The edge/fog runtime aggregates the data using common sink or local storage, and dispatches semi-processed or processed data to the cloud at fixed time intervals. Such pattern is critical for edge systems capturing sensory feeds from heterogeneous systems that corresponds to one or more inter-related processing. The pattern is designed to preprocess the raw data to eliminate redundancy, extract essential features, treat missing values, and normalize the data set locally for all types of analytical and autonomous processing. The persistence is utilized as the local Knowledge source gets updated in real-time and, each row has a predefined time-to-live property to ensure that the finite set of history is maintained within the limited storage capacity of the edge/fog controller.

The local persistence is also used for storing updated models that are getting updated locally or centrally. The store-forward integration semantics are primarily used in edge/fog native deployments that can perform intelligent process automation and deliver predictive or prescriptive insights independent from the centralized cloud systems. The persistence and integration semantics of a store-forward

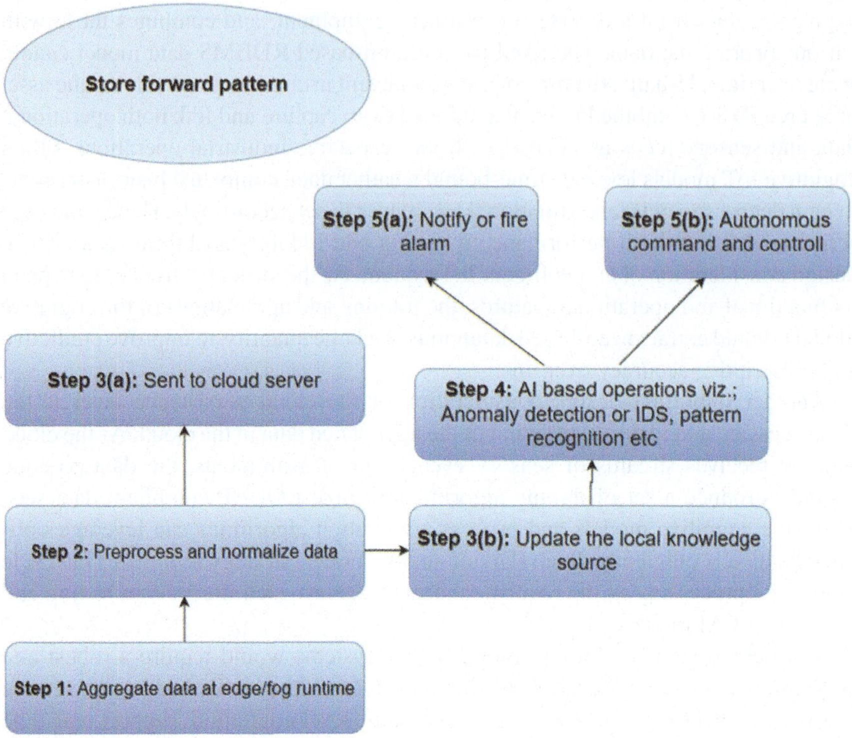

Fig. 15.2 Store forward integration pattern

model introduce the most efficient service delivery and automation framework for complex integration of industrial assets, control systems of utilities and smart cities where ubiquitous operations demand real-time visibility and instantaneous actions. The adaptive edge/fog runtime with self-learning capabilities preferably uses the store-forward integration patterns where cloud systems act as a passive subsystem, execute compute-intensive processes such as data mining, training models, execute complex AI models in batch mode etc. The central management hub or the cloud platforms host system management and monitoring applications. The runtime flow of the store-forward pattern is depicted below (Fig. 15.2).

15.5.2 Streaming Pipeline Pattern

The streaming-based integration is the simplest form of data and event dispatch framework that is heavily dependent on the centralized cloud system for most of the analytical and operational services. Each logical edge unit establishes one or more pipelines to stream the data over the IoT network. The design patterns simplify the overall IoT architecture with no or limited edge capabilities. As depicted in the activity diagram, the deployed services on the streaming channels perform basic validations, applications of light weight models and, filtration of redundant feeds that can be used by the enterprise cloud system for more sophisticated processing and automation.

Usually, the IoT systems are connected through low latency network backbone and don't demand real-time automation from within a streamlined enterprise operations. Most of the conventional IoT systems are designed with such streaming pipeline patterns. Such integration has no or limited persistence capacity. The storage is mostly used to manage failover situations and store lightweight models that can be used on top of streaming sensory feeds. The intelligent process automation in such integration scheme would deal with mission critical failure situation that can be revealed through the lean cognition capabilities deployed within the edge. The diagram below presents the runtime topology of streaming pipeline pattern (Fig. 15.3)

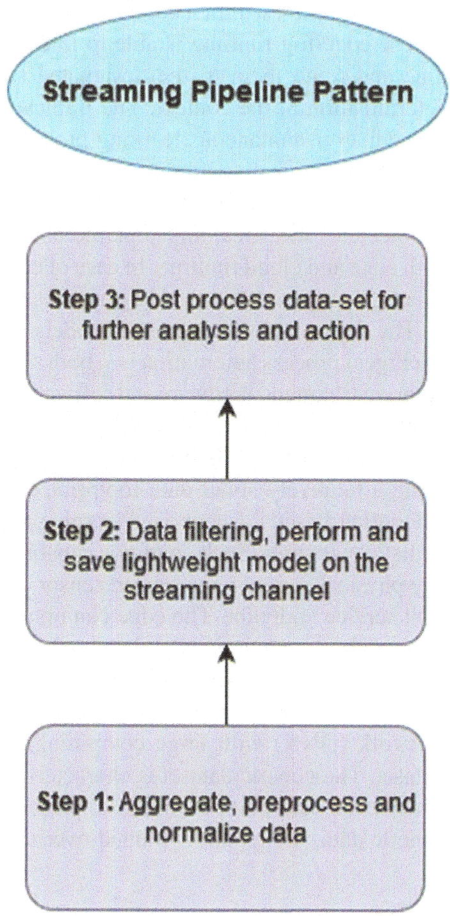

Fig. 15.3 Streaming pipeline pattern

15.5.3 Context Aware Routing and Dispatch Pattern

The primary focus of the design pattern is to introduce intelligent mediation of preprocessed and processed sensory event based on contextual changes in physical conditions. The workflows are similar to streaming pipeline pattern. However the context aware routing algorithms leverage cognitive models to intercept the events, determine the patterns of the sensory feeds or events, and dynamically broadcast those feeds/events to the other peer nodes and/or cloud by relating contextual relevance of the inbound events/feeds. The small footprint of the storage maintains minimum data-points associated with relevant models and a knowledge source to train the deployed models. Such patterns are quite adaptive to changing conditions of the connected systems, however, rely on cloud-based intelligence, learning, and service delivery capabilities. The optimized design balances the load of distributed intelligence and robotic process automation capabilities across vertical line of controls (edge-cloud-edge). The following diagram depicts the workflow of context-aware routing and dispatch runtime within a cognitive IoT platform (Fig. 15.4).

The edge/fog runtime is able to intercept the measurements and events in real-time and route them to the designated service endpoints through a probabilistic determination of the context. The framework has capabilities to update the models through an instantaneous learning process. Such adaptive capabilities can bind the services dynamically without any predefined deterministic rules.

While the store-forward pattern is primarily designed for edge native solution architecture, the streaming pipeline and context-aware routing can be applied to both edge and cloud runtime. In case of could-based integration, the streaming at the cloud gateway correlates and aggregates a diverse set of sensory feeds and events.

The deployment of cognitive models and service orchestrations associated with intelligent process automation i.e., both sensing and controlling depends on the operating environment, hardware and software throughputs, and operating responsibilities of the subsystems. Being constrained with hardware resources, the connected edge nodes should execute lightweight AI models that can offer satisfactory performance using a relatively small data footprint. The ideal automation should leverage soft computing-based unsupervised learning to determine patterns, anomalies, and probabilistic outcomes. Such cognitive capabilities determine the contextual relevance of any physical event or aggregated sensor data and mediate the data/event to the relevant service endpoint. The edge can optionally, deploy sophisticated, low-compute AI models that are trained at the cloud and pushed to the edge.

There are two types of cloud-edge interactive algorithms, (i) Transfer learning and (ii) Federated learning. Transfer learning represents a single Distributed Neural Network (DNN) with large computing resources that are trained over a generic dataset. The generic dataset is characterized as the large volume of time series data where each data-point is similar to preprocessed IoT feeds. After training over that generic data, the model is trained over target dataset just by changing the last few

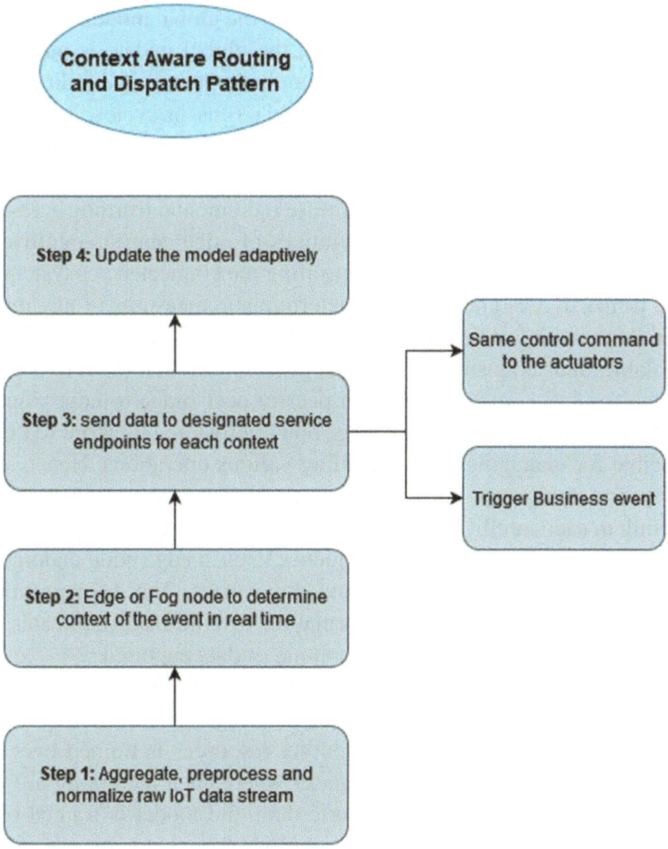

Fig. 15.4 Context-aware routing & dispatch pattern

layers of DNN (we call it as customized training). It allows the training of DNN with very small target dataset size. We propose to use that trained model as a global model within the cloud runtime. Then that model will be sent to the connected edge/for nodes or gateways. Each nodes are designated to process different types of sensory feeds and execute different task depending on physical operating environment. Over the time, edges will execute customized training based on inbound feeds and limited resources of the edge runtime. Finally, each edge operates, predicts and acts using the updated models.

The cloud-managed AI functionalities can inherit Federated learning-based architecture where the layer one captures the client or agent data (possibly heterogeneous) and layer two predicts or decides actions of individual agents. At layer two the models are locally updated, and those local updates (or some parts of those updates) are sent to the centrally located cloud environment. There the global model at the cloud level is tuned from each local models and finally, the global update is pushed to layer two edge gateways. The agents will use the updated global model directly, or after a customizing locally at layer two. This process runs in cycles, and it aids intuitive operations through incremental learning of the robots/agents in the future. The recent developments of federated learning in edge/fog computing have introduced a unique distributed intelligence pattern to infuse incremental learning across multiple layers of the IoT runtime. The key imperatives of edge-centric cognitive automation target low latency in sensing and controlling the connected ecosystems. The AI models are primarily designed to avoid deterministic measures of any machine-to-machine and/or machine-to-human interactions. As presented in our earlier research [33], a scalable and extensible intelligent process automation across the logical layer of edge/fog infrastructure establish peer-to-peer (edge-to-edge), and vertical (edge-to-cloud) interactions. Multiple edge notes across an industrial IoT ecosystem are responsible for managing and controlling various operations. Hence, each edge maintains a service registry, pointer to the knowledge source, and mediation map corresponding to each intelligent processing system. The service registry containing internal and external services and task sequence in each edge node undergo constant changes through incremental learning and discoveries. The edge runtime is also responsible for lightweight data transformation to eliminate duplicates and treat missing values from within the massive volume of data captured.

There are two types of cloud-edge interactive algorithms, namely Transfer learning and Federated learning. The Transfer learning introduces a single DNNDeep Neural Network (DNN) with large computing resources is trained over a generic dataset. The generic dataset is a large dataset where each data-point are similar to target data. After training over that generic data, the model is trained over target dataset just by changing the last few layers of DNN (known as customized training). It allows the training of DNN with very small target dataset size. We propose to use the generic training as a global model at the cloud. Then that model will be sent to different gateways. Each gateway will represent specific intelligent operations at the edge. While each gateway is designated to perform different tasks depending on their operational needs, the respective edge nodes deployed in each physical gateway would interact with each other to perform various stateful workflows. According to the deligation of responsibilities at the respective edge nodes, different sets of customized training will be executed at regular intervals. Finally, each edge module/agent would perform respective predictive and prescriptive operations at its designated gateway to achieve self-organized workflow (Fig. 15.5).

The core architecture mentioned in the Fig. 15.6 is representing Federated learning orchestration, where the layer one will capture the client or agent data (possibly heterogeneous) and layer two predicts or decides actions of individual agents. At layer two the models are locally updated and those local updates (or some of those updates) are sent to the central server or cloud without the respective edge data being pushed to the cloud. The global model hosted on the cloud or the centralized AI engine is incrementally tuned by the edge specific models pushed by the connected edge nodes. Finally, that global update is sent to the respective edge nodes. The process of

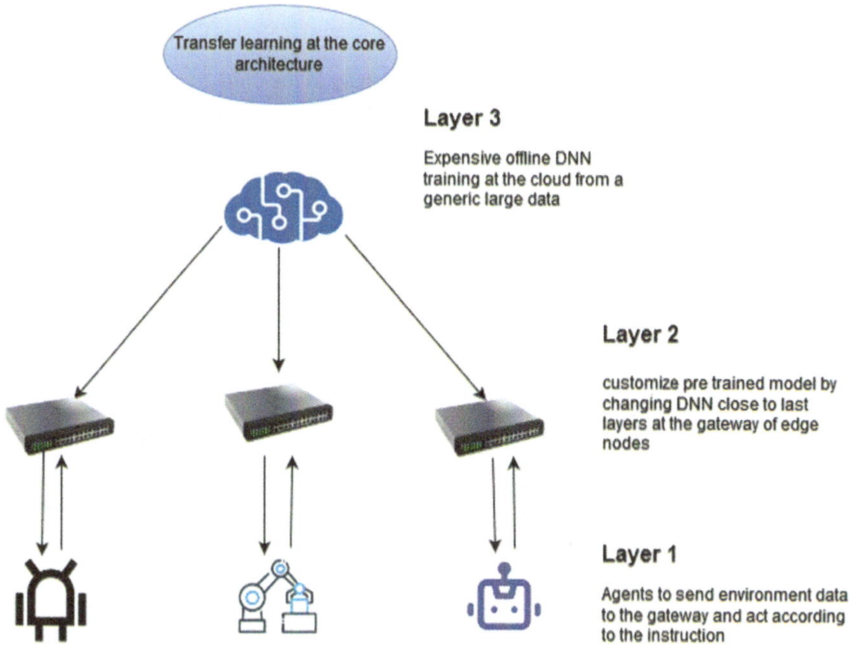

Fig. 15.5 Transfer learning

such discrete learning and aggregated incremental learning tends to improve overall model efficiencies. The agents running at the edges will use that global model directly or after a customization at layer two. This process continuously runs across the distributed AI runtime, and helps intelligent robotic processes to gain situational awareness over time. The diagram below simplified the interactions across the layers to achieve incremental learning and distributed intelligence using federated learning (Fig. 15.6).

The cognitive models deployed on data streams aggregated at the cloud level are usually sophisticated models that can analyze cross-system datapoints. Such models introduce sophisticated data mining techniques to preprocess the data. The stream analytics runtime also introduces regression and classification models to gain deeper insights on system-wide operating conditions. The near real-time insights can trigger alarms/notifications as it predetermines potential faults, issues, and/or concerns. Such high-performance runtime introduces autonomous process-control by pre-determining exceptions or changes in the operating environment. The stream analytics models are trained within the cloud runtime and leverages a wide range of datasets relevant to the application.

The cloud-based decision support system acts as a centralized service management and system monitoring functions. The data and runtime of the IoT cloud executes models in an offline/batch mode. Such models act as a classifier, determinant of boundary conditions and/or segments that are leveraged by all the models deployed

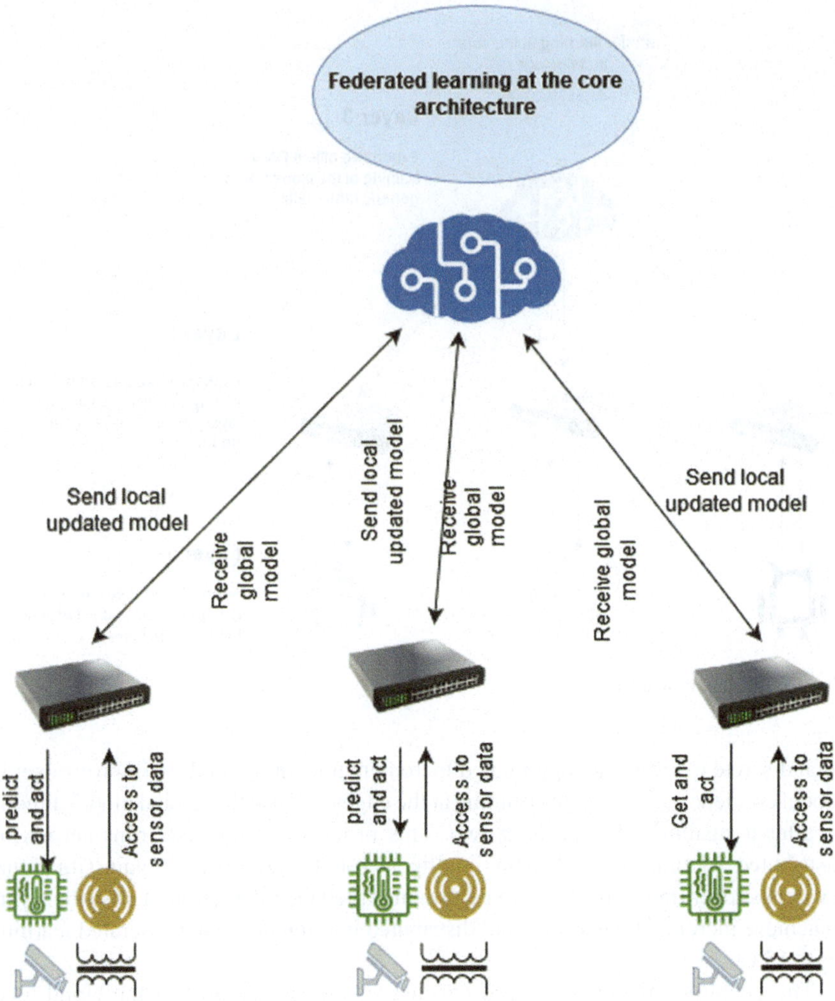

Fig. 15.6 Federated learning

across the IoT system. The cloud is also responsible for training and updating the models at regular intervals. The service integration framework deployed within the cloud frequently changes the event dispatch and task sequence patterns used by various layers of the IoT runtime. Such updated models and service mediation patterns are pushed to the respective systems on a regular basis. Thus, the sophisticated services deployed in the cloud keep the entire runtime up to date with its analytical and autonomic service delivery capabilities.

15.6 Use Cases in Cleantech, Smart Grid, Manufacturing, and Healthcare

In this section, we present a set of use cases that can deliver better outcomes in terms of managing automation and asset management functions. The conventional robotic process automation in the industry is not considered an emerging trend. However, the incubation of AI centric operational intelligence within the industry 4.0 era is disrupting industrial process automation [18]. We have elaborated on a few unique intelligent robotic process automations using prior research conducted in the field.

15.6.1 Process Automation

The journey of process automation started at 70's with the invention of automatic central systems like boilers, chillers, and Air Handling Units (AHU) [34]. The problem of such system was that all sensors and actuators were individually wired to central system which causes extensive cabling or wiring, slow troubleshooting, and unwanted noise. To solve this issue Field bus distributed network was introduced at 90's to distribute multiple inputs and outputs from the field into local output device, then data is transferred through only one cable. This saves insulation, maintenance, and other engineering costs. Till today wireless Fieldbus distributed network is widely used. Digital signals carry more information than analogue per unit time which helps for comprehensive network diagnosis by resolving or preventing downtime situation. Network diagnostic information also helps in sophisticated asset management. To enable further management where large amount of heterogeneous data is arriving from IoT devices, Cognitive IoT is the next generation distributed system.

Robots are the building block for process automation, the Internet of Robotic Things (integration of Robotic technology and IoT) has two main classification based on their functionalities, Smart space, and Robots [35]. Smart space robotics uses cloud computing, wireless sensing for monitoring applications at smart building, smart factories etc. On the other hand, Robots work as the acting agent for indoor activities like control and planning, human-robotic interaction, manipulating etc. Actually, smart space has no active agents and robots work as that agent. Deployment of smart robots and integrating these two robotic classes is not sufficient for sophisticated actions, enablement of smart environment by occupancy sensors, surveillance cameras, magnetic sensors is also essential. In industry multi-robot system, i.e., assigning multiple robotic agents for the same operation under same environment can solve complex operation in more cost-effective way.

YOLO (object detection algorithm) Convolution Neural Network (CNN) was trained with several interfaces used to classify software interfaces in real-time [22]. Also, their developed software takes automated actions by moving the mouse pointer, clicking, editing text, and performing any other necessary action. It is the way to automate any computer task. Graphical objects detected by Graphical User Interference are a key part of many robotic automation. For such jobs popular object detection algorithms YOLO, SSD are very useful. One thing is to be remembered that ideal cognitive robots must have the possibility of further improvement and we need to evaluate their performance, a minimum error made by the robots will cause a huge damage to the production. So, surveillance of those bots is necessary in such situation and smart IoT cameras can be used as the surveillance interference tool. Pre-trained deep learning models at the cloud are efficient to decide the performance efficiency and condition of the robots.

The fifth generation (5G) wireless network with edge computing has made use of IoT more realistic. Framework and implementation of cognitive self-healing has been proposed for current 5G mobile network system. It includes two steps, at first anomalous scores are counted and then case-based reasoning makes the cognitive decision for self-healing [36]. Obviously 5G network is one of the best wireless network with low latency (1 ms) and it offers up to 10 year of battery life for low power IoT devices. We can use such framework in every IoT connection.

Chen et al. [37] proposed and tested an industrial robotic system depending on edge computing. They presented a three-layer architecture by cloud, edge, and physical resource (top to bottom) which was applied on robotic welding of the membrane wall. The cloud works as the system manager which manages the database, keeps track of the available resources and so on. The edge layer works as a data preprocessing unit for the sensors data. The physical resource layer, i.e., the robot controllers studies all the task files received from the cloud and execute movement instructions. They have tested their architecture on a robot for membrane cell wall which is an important part of boiler industry. The membrane wall is a connection of many thin steel rods (cylindrical shape) welded together with many pins. The manual spot-welding procedure is off course not efficient. The automated methodology consists of two stages, namely path planning and pose measurement. At path planning, step pins are vertically welded on the surface plane of the steel tubes in a radial direction. But deformation of the rods can happen due to welding. They use non-compact laser sensor to gather the cross-sectional data of the rods. However, the hybrid edge-cloud model has shown less bandwidth than fully cloud dependent architecture. This is one of the few implementation of edge dependent architecture at heavy industry. The general architecture may be carried out in other heavy industry scenario. The basic jobs of the server like accessing cloud to the users by cloud, systemic service management can be used to design cloud server at other industries. Motion cloud service (MCS) will be used to encapsulate the robotic motion and the process cloud service (PCS) will serve according to the domain of the industry (for example it executes path planning for robotic welding). The edge layer which has basically routers, local PC, gateways will also be used in other industry domain. The physical layer, i.e., the robot controllers are the basic need in other industry domain.

But a complete edge computing in industry faces three main challenges, (a) personalization, i.e., customization of AI models at the edge according to individual device requirements, (b) responsiveness, i.e., adaptation of computing services for new situations and (c) privacy preserving, i.e., preserving the data privacy for the data transmission to the remote cloud. Federated active transfer learning was proposed by Foukalas et al. [38] which is the combination of three algorithms Active learning,

Transfer learning and Federated learning. Active Learning personalizes the AI model according to the task by changing the number of labelled samples for that task. Transfer learning at the edge, i.e., by loading a pre-trained model and then replacing the last layer according to the target completes responsiveness. Finally, for data privacy job, after the local training the local models are sent to the Fog where the FL, i.e., federated learning aggregation is executed. That federated learning is very popular for customized, adaptive edge computing. The privacy preserving approach makes the automation safe from the third party (potential business competitors) access. Therefore, architecture mentioned here is suitable for Federated learning which is mentioned earlier.

Khan et al. [39] proposed a dispersed Federated learning (DFL) framework to minimize FL computation delay and global model accuracy loss due to packet error rate for resource optimization in smart industry. But there are other issues in FL, for an instance if the job is to do a simple classification at N classes and there is a client that has the access to only k class images where k << N, then simple FL would deploy unnecessary large DNN at that node which increases communication and computation costs. Chen et al. [40] proposed such Meta-Federated model and tested on industry dataset with 9369 clients.

Fuzzy logic is motivated by human logic of thinking as on our mind we do not take decision based on quantitative analysis like machine. In our mind we make any quantity in fuzzy form, i.e., whether it is high, low, or medium. Similarly in fuzzy set theory, membership function values in favour of fuzzy sets (such as high, medium, or low) are calculated. PID controller is used in industry to control temperature, pressure, flow, speed etc. For example, PID temperature controller takes temperature input and control the temperature through some heater or fan. Fuzzy logic controller and PID have been proposed for mobile robot navigation [41].

Cooler is an essential component at chemical units, thermal plants, warehouses etc. A Fuzzy logic-based controller was proposed [42] which takes temperature and humidity as inputs and controls the temperature and humidity. The generated fuzzy rules are like if the temperature is "just nice", humidity is in "dry" condition fan speed is "slow". Such rules are transparent to users.

A very simple model based on Fuzzy ANFIS (Adaptive Neuro Fuzzy Inference System) was proposed to navigate the motion of mobile robotic system [43]. The designed robot moves through the optimal path if there is no obstacle. If there is any obstacle then the radial sensor detects that and the proposed algorithm determines the type of obstacle, whether it is static or moving, If the obstacle is static then the robot moves toward the shortest avoiding path else it predicts the probable collision time and position by analyzing that robot's velocity and finds the best path to avoid the obstacle.

Autonomous Underwater Vehicle (AUV) is a robot to monitor underwater wrecks, rocks, obstacles to underwater vehicles and for water environmental survey. Zhu et al. [44] proposed a Biologically Inspired Neural Network (BINN) embedded with Self-Organizing Map (SOM) for task assigning and path planning. The goal is to deploy a team of AUVs to achieve multiple target location and navigate them to avoid obstacles for each AUVs. Such simple Fuzzy Rule based models are convenient to apply in our proposed core architecture. The layer one will obtain the sensor data and layer two will be equipped with predefined fuzzy rules. As the navigation is instructed at gateways close to local edges, it enables real time navigation by the robots. The offline DSS layer will monitor the performance of those robots by keeping the historical sensor data, navigated actions, outcomes and finally it will anticipate the need of further training of those pre trained models. If it requires further training, then the pre trained models are retrained in offline mode.

But if we are interested to make those robots intelligence close to human cognition, reinforcement learning is the best promising option. Some skills learned by human are not taught by other in supervised or even in complete unsupervised manner (like learn to run a bi-cycle, swimming etc.), they learn it by many trials and error and eventually they are trained to that skill after a huge experience of success and failure. Reinforcement learning works exactly like that, where the model performance gets better after many reward and penalty (i.e., success and failure) from the environment. Zuo et al. [45] proposed such reinforcement learning for robotic control. It's better than Conventional PID based robots where rules are predefined and has very limited adaptive skill, also modelling a feedback (capturing contact behaviour of robot and environment) controllers by classic controllers is impossible. The context aware routing and dispatch (Fig. 15.4) pattern can provide the most ideal architecture for multi agent reinforcement learning. Each agent or robot will be designated to either same or different job. For example, in smart industry there are many robots needed dedicated to various job or context. Some robots identify and pick some objects, other robots carry that object and move it in an optimal path. The edge node will determine the job or context of each robots and will send to the designated service point. They do reinforcement learning there and update the strategies of robots for better navigation. In most practical situation the robots need unstructured environment data (image or sound) to determine the current state. CNN extracts feature from such data without any prior knowledge, such feature extraction can be executed at pre trained edge nodes, and the service point will get the extracted feature as current state. Reinforcement learning are data hungry, and it reduces the memory overhead at the service node.

15.6.2 Smart Grid

One such example of cognition is smart micro grid energy management by NSGA-2 (Non-Dominated Sorting Genetic Algorithm-2) optimization [46]. The quantities in smart grid can be fuzzified to make the optimization close to human understanding. In smart grid, multiple sensors are embedded to micro grids which make them an IoT device. All the devices are connected to a server to take the decision. The objective of the decision is to balance load-demand, avoid over and under charging situation and reduce the average peak load (APL). It is obtained by controlling battery charge–discharge rate of Energy Storage System. Such model can be implemented by the mentioned streaming pipeline pattern architecture where each smart micro grid (say each solar panel) is equipped with a sensor, each power station has a designated logical edge unit, which makes simple jobs like fuzzification of inputs, takes decision based on predefined Fuzzy rules. Then each power station will send their individual data like total generation, demand, energy storage, energy loss, forecasting etc. to a centralized cloud server. That cloud server will communicate among all power stations to reduce demand in needy areas, "Islanding" if needed, detect underperforming stations etc.

Due to the increasing number of power loads, controlled and uncontrolled rectifiers (AC to DC converters), electric locomotives, and other switching equipment by complex power electronics technology, the power system faces serious voltage and current harmonics, i.e., their waveforms deviate from respective sine-wave shape. Active power filter (APF) is a popular option to deal such harmonic issue. The nonlinear control problem for APF created by the aging components and variable load is a challenging issue. Hou et al. [47] proposed Type-2 Adaptive Fuzzy Neural Network (AFNN) model that can handle such nonlinear problems.

Microgrids are popular source of energy generation as it has additional self-healing ability. In a distributed energy system "Islanding" helps the micro-grid to protect against widespread grid disturbance and serve the demand at least locally by isolating itself from the distributed system. Grids are proposed to move towards smart grids for self-healing by four steps [48].

Step 1: Building sensor system by installing smart meters at current transformer, Potential transformer and obtaining weather data.
Step 2: Building communication technology in both wired and wireless technology. The wired technology uses mainly fiber optics. In wireless communication cellular, cognitive radio, satellite and Microwave technologies are used very frequently.
Step 3: Here several algorithms from non-AI or AI are used to get predictive output, monitoring and decision making. It serves several jobs like load management, demand response, Volte Var optimization (against voltage sag) and automatic recovery.
Step 4: The Actuator system acts as the reactive part of the architecture. Automatic switches, shunt capacitors, on load tap changers, and hybrid transformers are used for self-healing to control disturbing parameters.

A Cognitive architecture for smart grid was proposed by M R Ahmad et al. [33]. The objective is to improve energy conservation, reduce operational cost and empower the customer from usage analytics. The cognitive architecture is three layered namely Perception layer, Attention-Memory layer, and Decision layer. But there are more possible use-cases of RPA at smart grids. Researchers may design such smart RPA system where robots will study and analyze customer consumption data, deploy smart drones to capture partially discharged Bushing images, show real-time load and demand data, try to avoid power outage situation etc. The core architecture proposed here is useful to create a complete RPA network of a power generation company. Each company will have centralized cloud server (layer three) designated to RPA. Each power station of that company will have separate gateway (layer two) which will send data to centralized cloud for monitoring and offline training. As all the data belongs to that same company, it keeps the data privacy. At layer two the cognitive decisions or navigation instructions are made for the agents. The agents or the robots are the first layer for on ground real time jobs and also sense the environment and send them to layer two to obtain the instruction.

15.6.3 Smart City

Smart city is the ultimate goal of cognitive IoT to enhance human level expertise in automation. It involves smart hospital, smart grid, smart manufacturing, smart traffic and surveillance monitoring and even smart homes. The complete architecture has five levels. At the lowest level, there are smart homes, smart hospital, smart agriculture, smart factory and all the possible places where sensors can be embedded. This level is known as smart city platform. The second layer is called IoT layer which is combined with different sensors from different domains. The third layer is the Data layer which clusters data based on their respective domain. The fourth layer is the computing layers which learn, predict, and take decision for each data by cognitive IoT. The computing is executed by the target server. And finally, the last layer is the service layer which is the interface of the output to all the clients [49].

Cognitive Road traffic Controller was developed to reduce waiting time of the cars at the traffic signal. The researchers proposed an intelligent fuzzy logic-based rule and Local Cognitive Node (LCN) which gives the output of that fuzzy logic controller. In their system they proposed two sensors at each traffic signal, one sensor which is placed at the traffic light, counts the number of cars passing from the signal. The second sensor was installed at some distance from the traffic light, counts the number of cars arriving at that signal. Their researchers simulate several situations, and the cognitive controller reduced the traffic congestion [50].

A Fuzzy Art Map Neural Network was proposed to navigate an autonomous vehicle. The model was proposed for two type of navigation, to recognize target location and avoid the obstacle [51].

An innovative research by Sarder et al. [52] proposed a driver mood repairing system by detecting the mood by cognitive-behavioural mining and recommending a dashboard to keep the drivers mind very active so that he/she avoid any road accident from his/her fault. They proposed to install a camera at the car to detect the activities of the driver; ECG, EEG, EMG, EDA sensors to capture his/her mental situation and other sensors to capture external environment. The activity recognition detects many activities like safe driving, texting, calling, drinking, operating at radio, talking to passengers. Except the first safe driving activity, all the activities are regarded as distracted driving. Data from ECG, EEG, EMG EDA sensors are gathered to classify the mood of the passenger by an emotion recognition engine and environmental factors are extracted to find sequential pattern with driver's mental health. These three results are used to get a priori based sequential behavioural pattern mining. Based on the output it detects the mental state of the driver and Bayesian recommender recommends audio content to improve the negative mental state to positive mental state step by step. Finally, the dashboard notifies the driver, shows mood swing stages towards complete safe driving state.

Self-driving car is gaining tractions in auto-mobile sector and the idea of Cognitive Internet of Vehicle (CIoV) is an important development in the field of RPA powered CIoT implementation. Hasan et al. [53] discussed all the possible challenges and concerns for practical use of CIoT in terms of overcoming cyber physical constraints of analyzing the operating conditions and execute timely operations. CIoT has five distinct layers (Bottom to top) which are (a) sensing and participation, (b) Network communication and data acquisition, (c) Vehicular cloud, (d) Internet cloud and finally (e) Application. Sensing and participation layer is responsible for sensing, communicating and interacting inside the transportation system. This layer has two components smart cars and road infrastructure. The smart car may contain two types of cars, self-driving car and connected car. Most of the Automated vehicles has sensors like cameras (To capture objects like people, traffic signs, other vehicle, road potholes, using 3D cameras distance of the detected object from the autonomous vehicle can also be determined), Lidar (Uses laser to make 360 map of nearby objects and there location), Radar (generates a 3D image of all objects near to the autonomous car and estimates the possible distance and velocity by radio wave) etc. They collect preprocessed data from the outside environment. This data is sent to the server as input to the complex system software which is used by autonomous cars to decide possible courses of actions like, lane changing, slowing down, accelerating, moving left or right, pausing, and overtaking other vehicles. On the other hand, in connected vehicular technology, the vehicles use different wireless communication with both internal and external environments. The second layer in CIoV is responsible for network communications among different transportation components for two types of communications, namely Intra-vehicle and Inter-vehicle communication. Since each smart vehicle is equipped with sensors to monitor driver condition, road and traffic situation, object detection etc., these sensors data are to be communicated together to suggest the correct decision for the driver. At the third layer data preprocessing is executed by edge computing and the real time operations can be performed locally. Edge computing at CIoV is expected to get higher importance by inheriting store forward execution pattern. The fourth layer works like as an integrated cloud that does all the sophisticated decision modelling and simulation correctly and manages incident events. The topmost layer provides high-level services which introduces asynchronous functions to improve operational efficiency of CIoV. But there are many threats to deal with like GPS spoofing (locating false location) and GPS jamming (process of using frequency transmitting device to block radio communications) at autonomous vehicle. Similarly connected vehicle faces reply attack, Sybil attack, and location Privacy attack. Distributed Denial of Service, i.e., DDOS attack to flood the bandwidth or resources of a targeted system is a very common threat at both network communication and vehicular cloud layer. Spyware software and hacking software which install itself and monitor the system are some genuine threats at the application. Most of the IoT devices are not secured and very often exposed to cyber attackers. Proper encryption and building sophisticated Intrusion detection system is needed, a slight attack may be life threatening to the users.

The core architecture proposed here has the potential to build a complete autonomous vehicle network using decentralized multilayer intelligence within the CIoV ecosystem. The Intuitive sensing layer will extract the environment data like weather, speed, condition of the vehicle, degree of safety, trajectory of the vehicle etc. The second layer will navigate the motion of the vehicle seamlessly and the third offline layer will handle more sophisticated jobs. It will train complicated models like complete traffic monitoring, surveillance at important location, track down the problematic vehicle or wanted persons and update those real time models. All the vehicles of a local area will form a clustered network which will be continuously updated based on their location. Each cluster will have a designated cloud node and it orchestrate proper communication among local and nearby vehicles to make the traffic system smooth.

15.6.4 Smart Home

Smart home means an automated home which reduces human afford without his/her instruction to improve his/her quality of living. A very simple smart home automation model based on the decisions made by Fuzzy ANFIS was proposed to control light intensity (used PIR sensor), temperature (used LM35 sensor) and alert at the entry of unauthorized person (used LDR sensor) at the room [54]. Obviously, we can add much more sensors or IoT devices that will sense the dedicated data and Fuzzy ANFIS will make decision.

A more sophisticated Interactive people-centric smart home is proposed by Shuo Feng et al. [55]. The Cognitive Dynamic System (CDS) equipped with IoT censors has two basic parts, perception, and action. Here the cycle begins when the preceptor receives the observation from the environment and in response executor (action) unit acts. A Bayesian model works as the main feature extractor, a Bayesian Filter determines the physical state from the observation, a reinforcement learning decides the correct action to maximize the given reward function value. The authors leveraged IoT devices at the object frequently used by the residents, like sofa where the motion, pressure, temperature etc. can be captured. Controllable devices at blinds, TV, or door act as AI agents. The Bayesian model works as the feature extractor and the Bayesian filter determines the current state of the customer (sleeping, resting, and sitting on the sofa). The reinforcement learning algorithm operates those AI agents to adjust brightness of the room, switch on or off the tv etc. The Context-aware routing and dispatch pattern proposed here is suitable to execute such smart home operations leveraging RPA at the edge and managed through centralized cognitive decision support operation.

15.6.5 Smart Healthcare

AI in healthcare has several applications; healthcare research, disease prediction, cancer detection, automating tasks, chatbot based doctor, AI based robots for assisting surgeons, Detection of bacterial blood disease. Mental health issues like panic disorder, Schizophrenia, bipolar disorder etc. are the serious mental health issues which can be diagnosed by initial stage. But number of psychologists is not sufficient compared to other health specialists especially in the poor countries. Or sometimes the mental health issue is viewed as a social stigma in some non-concern society and the patient does not feel comfortable to express his/her problem to psychologists. To overcome this issue automated mental health diagnosis is required. Speech recognition by NLP can observe any unnatural speech to detect mental health issue [56].

Almost 17.9 million people are died due to cardiovascular diseases (CVDs) in each year. Electrocardiogram (ECG) chart is used by cardiologists to monitor current heart situation. Cognitive Artificial Intelligence (CAI) software was introduced to diagnose heart block and Arrhythmia from ECG report. Based on Knowledge Growing System (KGS) that software makes observation for several time based on eight hypothesis. These hypothesis are normal, Left Axis Deviation (LAD), Right Axis Deviation (RAD), Left Bundle Branch Block (LBBB), Right Bundle Branch Block (RBBB), Arrhythmia, Atrial and Ventricular Tachycardia. These observations are shown graphically by the software and then sent to cardiologists for further treatment [57].

Just like Cardiovascular disease Cancer is another serious deadly disease that have caused 10 million deaths only in 2020. Cognitive computing in cancer treatment promises to accomplish three objectives [58]. First, it can work as a data analyst to extract the data, capture the problem, and visualize the results with users. The next step objective is to link the related data, generate hypothesis and allocate appropriate score to each hypothesis. The final goal is to process and deeply interpret both structured and unstructured data. Obviously, the complete implementation of this project is very complicated to handle this sophisticated diagnosis. IBM Watson is a cognitive technology platform has been applied in Manipal Hospitals of Bangalore for oncology care. Their cognitive clinical decision (CCD) support tool can analyze patient's medical record to identify his/her personalized treatment. It accomplished 90% concordance with the treatment recommendations by CCD of the multidisciplinary tumour board at Manipal Hospitals for over 600 patients suffering from breast cancer.

Not only for cancer treatment, but IBM Watson does several other jobs for medical care. It exploits data from several publicly available and licensed content. Examines data to map relationships within a scientific domain such as gene relationships or relationships of genes to diseases or proteins to compounds, and then visualizes how the interested elements are related to others. Also generates hypotheses which can enhance innovation of drug by examining the relationships between genes, drugs and diseases. It identifies the similarity between observed compounds depending on their structure and context by obtaining statistical correlations. Finally, it brings their outputs and findings to the clients.

Smart watch is one of the most used IoT device by the common people. They are equipped with operating system, memory, screen touch interface; it performs various applications and operates through WIFI connection. That smart watches can work as an edge computing device and has several functionalities. Health monitoring and providing health tips is one of such application. For example, the latest Apple watch series 7 measures our blood oxygen level, takes ECG at any time, and also access mindfulness, sleep even workouts.

Federated learning which allows data privacy preserving learning from many server data can be used for a simple and seamless health monitoring. In public sectors or Government funded hospitals in third world countries has very limited professionals to manually observe the test reports. In such scenario the Government may build a smart server and a separate application dedicated for each medical test or for medical department. Each dedicated application such as the Arthrology application will take an X-ray image from a client and then either use pre-trained model at the application or will be connected to the server to report the damage, treatment advice, and possible expenses, suggest the doctor and suitable hospital, approximated curing time etc. Here each hospital is an individual client and will send medical report at any time. But this approach is not privacy preserving as some medical images may be sensitive to the patient and over-dependency of thousands patients depending on a single server is disastrous at the server maintenance time or if it suffers DDOS attack. To resolve this issue Federated learning can be deployed to each such application. Heterogeneous Federated algorithm allows client specific customized prediction and the continuous learning before decision making, makes the health monitoring more efficient. The professionals will send annotated medical report at each application at some time, the application will send an update of the local model to the server and the server will get updated. Once the server is updated it will send a notification for updating the application. The client or hospital will update the application and by repeating this cycle each server will get better customized output.

15.6.6 Object Detection and Computer Vision

Very deep learning models at the server give very high latency output at the edge which is always problematic. Meta-Cognition-Based object detection method has increased old Yolo v3 model average precision by minimum 2.4% up to maximum 4% [59]. The utility of vision-based application at RPA has discussed earlier but since we are looking to integrate cognition at RPA such Meta-Cognition models are worth to get more attention. Usually most of the deep learning models are data hungry and need huge dataset for training unlike a small boy/girl, even he/she can distinguish cat and cot from a few images. Hence in this scenario human cognition is faster training than conventional deep learning training. But Meta learning uses very small number of images and adaptable to learn at new scenario which is very close to human cognition. In some scenario where the client cannot provide much training sample these meta learning method can be more useful than other. The utility of meta learning at RPA is not fully explored yet. Few shot learning (by meta learning) approach for face recognition where we don't have many images of each faces to train, can be used for robotic surveillance system in defense. Lightweight image classification model like Mobile net embedded with SSD or Yolo algorithm gives very fast output. Such lightweight models can be deployed in each edge units of Streaming pipeline pattern where those edge units preprocess and detect objects from their designated IoT cameras. Each edge unit will send preprocessed data or their output to the cloud for further complicated jobs like surveillance, monitoring and store data for future research and application.

Plastic components are one of the most severe issues to water life and marine ecosystem. Kong et al. [60] proposed a robot to detect floating garbage in real time and collect that garbage. The intelligent water surface cleaner (IWSCR) uses YOLO v3 network for accurate and real time garbage detection. Also, it has six main configuration components namely underwater vehicle, camera cabin, binocular camera, collection box, manipulator, and grasper.

Detecting plastic components at the underwater of ocean is another tough challenge. Hopefully Autonomous underwater vehicles (AUVs) can be embedded to detect plastic garbage from other underwater objects like biological material, fish, plants, biological detritus, and also intentionally placed objects like sensors etc. Such object detection model was evaluated for conventional object detection techniques like SSD, YOLO v2, Tiny YOLO, Faster R-CNN [61].

Xiao et al. [62] proposed adaptive learning for edge computing with the help of cloud computing for image classification. The CNN at the cloud works as the base learner which is trained for large data coming from mobile devices. Some initial convolution layers of a fixed size are then passed to edge server, where additional fully connected layers are added. After that the edge server is trained for a small number of data. While training at edge server the initial convolution layers is frozen and only the fully connected layers are trained. The use of small training sample and less training parameters reduces the computational cost. For adaptive learning the edge server sends the preprocessed and annotated data to cloud for cloud training. From the cloud CNN that fixed initial convolution layers are again sent to edge server to replace its convolution layer. This cycle, i.e., edge server training, cloud server training, edge CNN replacement goes on until the edge server stops getting better with interpretations and gaining actionable insights. Researchers may use similar approach for other domain like speech recognition, text reading etc. These are very useful scenarios in RPA where the robots have to understand streaming videos to extract contextual signals or specific patterns, or interpret human instruction or physical conditions to execute specific operations.

15.7 Conclusion

This chapter has presented the practical challenges and opportunities of industrial IIoT implementation where the cognitive AI-based autonomous operations can be introduced by integrating intelligent RPA components. In order to achieve the industry 4.0 implementation goals, the chapter presents reusable design patterns, frameworks, and service integration models that facilitate robotic process automation on top of cognitive IoT systems. Considering the diversity of applications, infrastructure, and the operational landscape of industrial IoT systems, we cannot define a universal architectural strategy for a cognitive IoT system with autonomous capabilities. However, a set of basic principles and best practices can be followed in designing a responsive IoT system with proactive automation capabilities while designing such systems the stakeholders should thoroughly review the consequence and outcome of such automation based on actionable insights delivered by AI services. We have discussed several AI models and the corresponding execution schemes that can facilitate intelligent process automation. Any intelligent IoT system with cognitive process automation capabilities would not only rely on the low latency and context awareness, but also heavily rely on the accuracy of such models in delivering predictive or prescriptive insights. Therefore, we believe that the basic design principles elaborated in this chapter could be a good start for an IoT system facilitating intelligent process automation however, the nature of the automation and corresponding models should be designed, tested, and deployed based on domain specific application requirements and scope of the automation. The recent developments in IoRT are focusing on the convergence of IoT with Robotic process automation. The logical interactions of sensing capabilities with control operations would require time sensitive and intelligent services that should overcome the network and hardware constraints, and cybersecurity threats. A well-defined separation of concern among the remote edge/fog or robotic units and enterprise systems should endure a seamless interaction among the participating industrial systems. Therefore, the future studies should pay due attention to the rationale on how to organize, deploy and weave the AI modules and service components within the digital footprint of the IIoT systems. Hence comprehensive design strategy and standardization of network and IT architecture are critical to any intelligent IoT system with autonomous operational capabilities. With the novel intent of advancing further research in CIoT and RPA domains, the core design concepts presented in the chapter would establish a foundation of integrated architecture where the convergence of CIoT and RPA transform autonomous industrial operation and replace purpose-based automation infrastructure in the future.

References

1. Rajawat, A.S., Rawat, R., Barhanpurkar, K., Shaw, R.N., Ghosh, A.: Robotic process automation with increasing productivity and improving product quality using artificial intelligence and machine learning. (2021). https://doi.org/10.1016/B978-0-323-85498-6.00007-1

2. Madakam, S., Lake, V., Lake, V., Lake, V.: Internet of Things (IoT): a literature review. J. Comp. Commun. **3**(05), 164 (2015)
3. Bhadra, P.: RFREAD: the generic framework for RFID reader adapter. In: EAIT (International Conference on Emerging Applications of IT), Kolkata, India, pp. 357–360 (2006)
4. Bhadra, P.: An optimized service oriented middleware design for RFID enabled solution. In: Managing Worldwide Operations and Communications with Information Technology, IRMA International Conference, pp. 357–360 (2007)
5. Sassi, M.S.H., Jedidi, F.G., Fourati, L.C.: A new architecture for cognitive internet of things and big data. Proc. Comp. Sci. **159**, 534–543 (2019)
6. Foukalas, F.: Cognitive IoT platform for fog computing industrial applications. Comp. Electr. Eng. **87**, 106770 (2020)
7. Tzafestas, S.G.: Synergy of IoT and AI in modern society: the robotics and automation case. Robot. Automat. Eng. J. **31**(5), 1–15 (2018)
8. Singh, P., Singh, N.: Blockchain With IoT and AI: a review of agriculture and healthcare. Int. J. Appl. Evolut. Comput. (IJAEC) **11**(4), 13–27 (2020)
9. Chakrabarty, S., Engels, D.W.: Secure smart cities framework using IoT and AI. In: 2020 IEEE Global Conference on Artificial Intelligence and Internet of Things (GCAIoT), pp. 1–6. IEEE. (2020)
10. Trakadas, P., Simoens, P., Gkonis, P., Sarakis, L., Angelopoulos, A., Ramallo-González, A.P., Karkazis, P.: An artificial intelligence-based collaboration approach in industrial IoT manufacturing: key concepts, architectural extensions and potential applications. Sensors **20**(19), 5480 (2020)
11. Sun, W., Liu, J., Yue, Y.: AI-enhanced offloading in edge computing: when machine learning meets industrial IoT. IEEE Netw **33**(5), 68–74 (2019)
12. Foukalas, F., Tziouvaras, A.: Edge AI for industrial IoT applications. In: IEEE Industrial Electronics Magazine (2021)
13. Al-Turjman, F.M.: Information-centric sensor networks for cognitive IoT: an overview. Ann. Telecommun. **72**(1), 3–18 (2017)
14. Pramanik, P.K.D., Pal, S., Choudhury, P.: Beyond automation: the cognitive IoT. Artificial intelligence brings sense to the Internet of Things. In: Cognitive Computing for Big Data Systems Over IoT, pp. 1–37. Springer, Cham. (2018)
15. Simoens, P., Dragone, M., Saffiotti, A.: The internet of robotic things: a review of concept, added value and applications. Int. J. Adv. Robot. Syst. **15**(1), 1729881418759424 (2018)
16. Ray, P.P.: Internet of robotic things: concept, technologies, and challenges. IEEE Access **4**, 9489–9500 (2016)
17. Afanasyev, I., Mazzara, M., Chakraborty, S., Zhuchkov, N., Maksatbek, A., Yesildirek, A., Distefano, S.: Towards the internet of robotic things: Analysis, architecture, components and challenges. In: 2019 12th International Conference on Developments in eSystems Engineering (DeSE), pp. 3–8. IEEE (2019)
18. Suresh, A., Nandagopal, M., Raj, P., Neeba, E.A., Lin, J.W.: Industrial IoT Application Architectures and Use Cases. Auerbach Publications
19. Ghobakhloo, M.: Industry 4.0, digitization, and opportunities for sustainability. J. Clean. Prod. **252**, 119869 (2020)
20. Siriram, R.: Convergence of technologies. S. Afr. J. Ind. Eng. **22**(1), 13–28 (2011)
21. Olawuyi, J.O., Friday, M.: Technological convergence. Sci. J. Phys. (2012)
22. Martins, P., Sá, F., Morgado, F., Cunha, C.: Using machine learning for cognitive Robotic Process Automation (RPA). In: 2020 15th Iberian Conference on Information Systems and Technologies (CISTI), pp. 1–6. (2020). https://doi.org/10.23919/CISTI49556.2020.9140440
23. Ivančić, L., Suša Vugec, D., Vuksic, V.: Robotic process automation: systematic literature review (2019). https://doi.org/10.1007/978-3-030-30429-4_19
24. Vermesan, O., Bahr, R., Ottella, M., Serrano, M., Karlsen, T., Wahlstrøm, T., Sand, H.E., Ashwathnarayan, M., Gamba, M.T.: Internet of robotic things intelligent connectivity and platforms. Front. Robot. AI **7** https://doi.org/10.3389/frobt.2020.00104

25. Liu, X.J.: Research toward IoT and robotics in intelligent manufacturing: a survey. Int. J. Mater. Mech. Manuf. **7**:128–132 (2019). https://doi.org/10.18178/ijmmm.2019.7.3.445
26. Sibalija, T., Jovanović, S., Đurić, J.: Robotic process automation: overview and opportunities. (2019).
27. Kovács, G., Benotsmane, R., Dudás, L.: The concept of autonomous systems in industry 4.0. Adv. Logist. Syst. Theory Pract. **12**, 77–87 (2019). https://doi.org/10.32971/als.2019.006
28. Boyer, K.K., Ward, P.T., Leong, G.K.: Approaches to the factory of the future. An empirical taxonomy. J. Operat. Manag. **14**(4), 297–313 (1996), ISSN 0272–6963. https://doi.org/10.1016/S0272-6963(96)00093-9
29. Kansara, H.: The rise of robotic process automation and its application in a business model. Int. J. Eng. Res. Technol. (IJERT) **08**(04) (2019)
30. Axmann, B., Harmoko, H.: Robotic process automation: an overview and comparison to other technology in industry 4.0. (2020). https://doi.org/10.1109/ACIT49673.2020.9208907.
31. Ribeiro, J., Lima, R., Eckhardt, T., Paiva, S.: Robotic process automation and artificial intelligence in industry 4.0—a literature review. Procedia Computer Science, 181:51–58, 2021. ISSN 1877–0509. doi: https://doi.org/10.1016/j.procs.2021.01.104. https://www.sciencedirect.com/science/article/pii/S1877050921001393. CENTERIS 2020–International Conference on Enterprise Information Systems/ProjMAN 2020–International Conference on Project Management/HCist 2020–International Conference on Health and Social Care Information Systems and Technologies 2020. CENTERIS/ProjMAN/HCist 2020
32. Bhadra, P.:US Patent No. 10,362,113. US Patent and Trademark Office, Washington, DC (2019)
33. Pranaya, Y.C., Himarish, M.N., Baig, M.N., Ahmed, M.R.: Cognitive architecture based smart grids for smart cities. In: 2017 3rd International Conference on Power Generation Systems and Renewable Energy Technologies (PGSRET), pp. 44–49 (2017). https://doi.org/10.1109/PGSRET.2017.8251799
34. Ploennigs, J., Ba, A., Barry, M.: Materializing the promises of cognitive IoT: how cognitive buildings are shaping the way. IEEE Int. Things J. 1–1 (2017). https://doi.org/10.1109/JIOT.2017.2755376
35. Afanasyev I., et al.: Towards the internet of robotic things: analysis, architecture, components and challenges. In: 2019 12th International Conference on Developments in eSystems Engineering (DeSE), pp. 3–8 (2019). https://doi.org/10.1109/DeSE.2019.00011
36. Chernogorov, F., Repo, I., Räisänen, V., Nihtilä, T., Kurjenniemi, J.: Cognitive self-healing system for future mobile networks. In: 2015 International Wireless Communications and Mobile Computing Conference (IWCMC), pp. 628–633. (2015). https://doi.org/10.1109/IWCMC.2015.7289156
37. Chen, Y., Feng, Q., Shi, W.: An industrial robot system based on edge computing: an early experience. In: USENIX Workshop on Hot Topics in Edge Computing (HotEdge 18). (2018)
38. Foukalas, F., Tziouvaras, A.: Edge artificial intelligence for industrial internet of things applications: an industrial edge intelligence solution. IEEE Indus. Electr. Magaz. **15**(2), 28–36 (2021). https://doi.org/10.1109/MIE.2020.3026837
39. Khan, L.U., Alsenwi, M., Yaqoob, I., Imran, M., Han, Z., Hong, C.S.: Resource optimized federated learning-enabled cognitive internet of things for smart industries. IEEE Access **8**, 168854–168864 (2020). https://doi.org/10.1109/ACCESS.2020.3023940
40. Chen, F., Dong, Z., Li, Z., He, X.: Federated meta-learning for recommendation. CoRR abs/1802.07876 (2018). https://arxiv.org/abs/1802.07876
41. Babunski, D., Berisha, J., Zaev, E., Bajrami, X.: Application of fuzzy logic and PID controller for mobile robot navigation. In: 2020 9th Mediterranean Conference on Embedded Computing (MECO), pp. 1–4 (2020). https://doi.org/10.1109/MECO49872.2020.9134317
42. Islam, M.S., Sarker, M.S.Z., Rafi, K.A.A., Othman, M.: Development of a fuzzy logic controller algorithm for air-conditioning system. In: 2006 IEEE International Conference on Semiconductor Electronics, pp. 830–834 (2006). https://doi.org/10.1109/SMELEC.2006.380753
43. Kamil, F., Hong, T.S., Khaksar, W., et al.: An ANFIS-based optimized fuzzy-multilayer decision approach for a mobile robotic system in ever-changing environment. Int. J. Control Autom. Syst. **17**, 253–266 (2019). https://doi.org/10.1007/s12555-017-0068-4

44. Zhu, D., Cao, X., Sun, B., Luo, C.: Biologically inspired self-organizing map applied to task assignment and path planning of an AUV system. IEEE Trans. Cogn. Dev. Syst. **10**(2), 304–313 (2018). https://doi.org/10.1109/TCDS.2017.2727678
45. Zuo, B., Chen, J., Wang, L., Wang, Y.: A reinforcement learning based robotic navigation system. In: 2014 IEEE International Conference on Systems, Man, and Cybernetics (SMC), pp. 3452–3457 (2014). https://doi.org/10.1109/SMC.2014.6974463
46. Teo, T.T., Logenthiran, T., Woo, W.L., Abidi, K., John, T., Wade, N.S., Greenwood, D.M., Patsios, C., Taylor, P.: Optimization of Fuzzy Energy-Management System for Grid-Connected Microgrid Using NSGA II. IEEE Trans. Cybernet. 1–12 (2020). https://doi.org/10.1109/TCYB.2020.3031109
47. Hou, S., Chu, Y., Fei, J.: Adaptive type-2 fuzzy neural network inherited terminal sliding mode control for power quality improvement. IEEE Trans. Industr. Inf. **17**(11), 7564–7574 (2021). https://doi.org/10.1109/TII.2021.3049643
48. Elgenedy, M., Masoud, A., Ahmed, S.: Smart grid self-healing: functions. Appl. Dev. (2015). https://doi.org/10.1109/SGRE.2015.7208737
49. Park, J.-H., Salim, M.M., Jo, J.H., Sicato, J.C.S., Rathore, S., Park, J.H.: CIoT-net: a scalable cognitive IoT basedsmart city network architecture. Hum. Centr. Comput. Inform. Sci. **9**(1):1–20 (2019)
50. Arikumar, K.S., Swetha, R., Swathy, D. (2017). Cognitive road traffic controller using fuzzy logic in IoT. Int. J. Comp. Organ. Trends 41, 1–5. https://doi.org/10.14445/22492593/IJCOT-V41P301
51. Chohra, A., Azouaoui, O.: Navigation behaviors based on fuzzy ArtMap neural networks for intelligent autonomous vehicles. Adv. Artif. Neural Syst. **2011**(11), 523094 (2011). https://doi.org/10.1155/2011/523094
52. Shirajum, M., Abedin, S.F., Kim, K.T., Kim, D.H., Rabiul, G., Alam, M., Hong, C.S.: Drive safe: cognitive-behavioral mining for intelligent transportation cyber-physical system. (2020) https://arxiv.org/abs/2008.10
53. Hasan, K.F., Kaur, T., Hasan, M.M., Feng, Y.: Cognitive internet of vehicles: motivation, layered architecture and security issues. In: 2019 International Conference on Sustainable Technologies for Industry 4.0 (STI), pp. 1–6. IEEE (2019)
54. Reena, K., Venkatesh, V.: Intelligent decision support system for home automation–ANFIS based approach. Int. J. Eng. Technol. (UAE) **7**, 421–427 (2018). https://doi.org/10.14419/ijet.v7i2.24.12127
55. Feng, S., Setoodeh, P., Haykin, S.: Smart home: cognitive interactive people-centric Internet of Things. IEEE Commun. Mag. **55**, 34–39 (2017). https://doi.org/10.1109/MCOM.2017.1600682CM
56. Hong, Z., Magdamo, C.G., Sheu, Y.H., Mohite, P., Noori, A., Ye, E.M., Ge, W., Sun, H., Brenner, L., Robbins, G., Mukerji, S., Zafar, S., Benson, N., Moura, L., Hsu, J., Hyman, B., Westover, M.B., Blacker, D., Das, S.: Natural language processing to detect cognitive concerns in electronic health records using deep learning. (2020)
57. Sereati, C.O., Sumari, A.D.W., Adiono, T., Ahmad, A.S.: Cognitive artificial intelligence (CAI) software based on knowledge growing system (KGS) for diagnosing heart block and arrhythmia. In: IEEE 2017 6th International Conference on Electrical Engineering and Informatics (ICEEI)–Langkawi, Malaysia (2017.11.25–2017.11.27, pp. 1–5. https://doi.org/10.1109/ICEEI.2017.8312368
58. Norden, A.D., Dankwa-Mullan, I., Urman, A., Suarez, F., Rhee, K.: Realizing the Promise of Cognitive Computing in Cancer Care: Ushering in a New Era. JCO Clin. Cancer Inform. **2**, 1–6 (2018)
59. Kumar, S.P., Gautam, C., Sundaram, S.: Meta-cognition-based simple and effective approach to object detection. CoRR abs/2012.01201 (2020). https://arxiv.org/abs/2012.01201
60. Kong, S., Tian, M., Qiu, C., Wu, Z., Yu, J.: IWSCR: an intelligent water surface cleaner robot for collecting floating garbage. IEEE Trans. Syst. Man Cybernet. Syst. **51**(10), 6358–6368 (2021). https://doi.org/10.1109/TSMC.2019.2961687

61. Fulton, M., Hong, J., Islam, M.J., Sattar, J.: Robotic detection of marine litter using deep visual detection models. In: 2019 International Conference on Robotics and Automation (ICRA), pp. 5752–5758 (2019). https://doi.org/10.1109/ICRA.2019.8793975
62. Ding, C., Zhou, A., Ma, X., Wang, S.: Cognitive service in mobile edge computing. In: 2020 IEEE International Conference on Web Services (ICWS), pp. 181–188 (2020). https://doi.org/10.1109/ICWS49710.2020.00031
63. Sriram, G.S.: Edge computing vs. cloud computing: an overview of big data challenges and opportunities for large enterprises. https://doi.org/10.0202/Computin.2022197786

Chapter 16
Confluence of Artificial Intelligence and Robotic Process Automation: Concluding Remarks

Arpita Chakraborty, Siddhartha Bhattacharyya, Debashis De, Panagiotis Sarigiannidis, and Jyoti Sekhar Banerjee

Abstract Increasing software intelligence is a hot issue in the research field today. When seen in this light, the fascinating challenge posed by the integration of RPA with Artificial Intelligence (AI) in many domains of application, such as the execution of unstructured versus organized activities, becomes clear. RPA might benefit from using artificial intelligence (AI) or ideas like Machine Learning and Data Mining so that it does not have to depend only on rule-based approaches. There is a perception that RPA was created to help businesses save money on labour expenditures associated with routine work. To ensure that the expense of implementing and maintaining RPA is offset by the savings it yields, it would be useful to analyse how RPA use impacts other aspects of a company's R&D, development, competency, etc. The authors provide some concluding remarks on various RPA applications in this chapter.

A. Chakraborty
Department of ECE, Bengal Institute of Technology, Kolkata, India

S. Bhattacharyya
Algebra University College, Zagreb, Croatia

Rajnagar Mahavidyalaya, Rajnagar, Birbhum, West Bengal, India

D. De
Department of CSE, Maulana Abul Kalam Azad University of Technology, West Bengal, India

P. Sarigiannidis
Department of Informatics and Telecommunication Engineering, University of Western Macedonia, Kozani, Greece

J. S. Banerjee (✉)
Department of Computer Science and Engineering (AI & ML), Bengal Institute of Technology, Kolkata, India
e-mail: tojyoti2001@yahoo.co.in

S. Bhattacharyya et al. (eds.), *Confluence of Artificial Intelligence and Robotic Process Automation*, Smart Innovation, Systems and Technologies 335,
https://doi.org/10.1007/978-981-19-8296-5_16

16.1 Introduction

Financial institutions [1] have increased their investment in process enhancements and technology since the early 1990s. Despite these expenditures, financial institutions still have challenges in meeting the demands of their customers as they evolve, cost-effective procedures and maintaining systematic on outdated systems, and meeting the slew of recently enacted regulatory mandates and security benchmarks. Keeping up with the competition is becoming more difficult as a result of the exponential development in computer power, technological improvements, and newcomers to the market.

RPA [2–5] is a relatively new technology that has just made its public debut with the intention of assisting businesses in their pursuit of automation [6, 7]. The financial services sector has shown a lot of interest in RPA [8–12], and some companies have already begun using it, in the hopes that it would allow them to reduce operating expenses by as much as 75% while simultaneously increasing the efficiency with which they provide their services to customers.

RPA's [13–17] capability of automating business activities inside technical frameworks and present applications limit the influence on current IT architecture, which is one of its primary advantages. Additionally, technology advancements like RPA include a number of steps needed to improve cognitive solutions like machine learning and "artificial intelligence" (AI) [18–30], which, if developed, will further revolutionize the robotization goals in the financial services industry.

RPA [31–33] is a kind of computer programme used on rules-based sub-processes or processes, repeatable, standardized, and structured data that provide consistent results. Nonetheless, there are more than 39 RPA software vendors, and these companies' offerings might vary widely in terms of features and functionality. The "lightweight IT" features of RPA, such as low cost, low time to adopt, and no tying up of IT (information technology) resources, are largely responsible for its meteoric rise in popularity in 2016/17. In addition to interacting with current systems of record in the same manner a person would via the application user interface, RPA is built to be customized by experts in the field rather than IT programmers.

16.2 Different Application Areas of RPA

To understand the impact of RPA, some of its application areas are mentioned here.

16.2.1 Intelligent Auditing

In the world of business and accounting, auditing is very crucial. At some times of the year, auditing needs a lot of time and labour. It takes a lot of effort to remove a company's papers and records that have become jumbled over time, which also wastes time. We may claim that RPA, or robotic process automation is a requirement for business leaders in the sector of auditing and commerce in general because of the aforementioned problems [34, 35]. If an external agency conducts the audit, there is a potential that there will be corruption and loopholes, defeating the audit's overall goal. Robotic process automation (RPA) may assist in auditing in every area, from automatically processing invoices to automatically calculating credit for a specific client. The term "automation" in the field of auditing is not a new concept to managers because auditing has always required the performance of numerous repetitive and autonomous tasks like reconciliations, internal control testing, and detail testing; however, the earlier systems were automated on the basis of scriptable languages such as Python or R, which required experienced developers, programmers, a significant investment of money, and energy as well. Business Managers are eager to use RPA in the Auditing sector as well, however, as it is being investigated in every other area of Business Process Management. This will lower the cost of automation while also making it considerably simpler and more effective.

16.2.2 Smart Education

In many ways, the educational system is one of the most challenging fields to automate [36, 37]. Academics agree that AI and ML will be very important in bringing about this paradigm shift in the field of education that is related to Industry 4.0 [38]. The conventional technique of learning has often been replaced with a high-tech approach to learning, thanks to RPA. There are areas in the educational industry where automation technologies and robots can do tasks more quickly than humans. Beginning with introducing a new course, RPA may also be utilized to introduce numerous delicate subjects that may arise throughout the course of a classroom session. RPA simplifies a professor's workload [39, 40]. By having a bot or system introduce students to the day's material, teachers may focus on more substantive matters than just setting the stage for the lesson. Incorporating RPA into a classroom may help a first-year educator feel at ease. Using a piece of automated data software, a teacher may quickly and easily discover which pupils need extra assistance with their studies, and the programme will then offer the instructor the tools they need to best assist that kid. Students may use RPA to identify problem areas and supplement their own learning with the software's available resources. If a student is feeling overwhelmed by the mental stress of their studies, an RPA system may help. By asking the student a series of questions designed to assess his or her mental health, the system can determine the best ways to alleviate the student's stress. In the event that the situation is serious, both the classroom teacher and the school administration will be alerted so that appropriate measures may be taken. The whole classroom schedule may be optimized for maximum efficiency and effectiveness with the help of RPA software. Transparency in student data is guaranteed by automated systems, benefiting educators, administrators, and parents in monitoring and assessing students' development. Teenagers may get a deeper understanding of complex topics like gender roles, depression, anger management, and more with the use of RPA.

16.2.3 Cybersecurity

In the realm of cybersecurity [41] and online safety, RPA is shown to be effective [42]. RPA systems are adept at determining whether or not a website is harmful. These systems can sift through vast amounts of data quickly and accurately. RPA bots see widespread usage across industries; they are developed to prevent sensitive information from leaking out. If an email has been flagged as harmful in the past, RPA bots may verify the mail's history and prevent other personnel from reading it, as well as raise an alarm in the system and add another layer to the company's cybersecurity system. By reviewing past assaults on the company, RPA algorithms and bots may assist the cyber branch of a business to anticipate future attacks and uncover vulnerabilities. Additionally, RPA makes the security system more effective, rapid, and rewarding while decreasing the likelihood of human mistakes.

16.2.4 Legal Process Handling

AI and RPA work together very well to solve complicated operations. There are many cases that are pending in the legal system; these cases may be rapidly addressed by employing automated technologies. Many attorneys squander valuable time submitting numerous court paperwork; however, this tedious labour may be readily automated utilizing RPA systems or bots. The government may utilize AI and RPA in conjunction to determine if certain types of regulations are being upheld by a specific corporation, and if the system detects any errors, it will quickly notify the appropriate government authorities so they can take the appropriate action. A lawyer may compile papers for their cases with the use of RPA, which can examine a large amount of data [43, 44]. RPA has the authority to conduct sizable financial transactions and file a report if the situation involves major issues like money laundering and funding terrorism. Nowadays, there are a lot of crypto transactions. In these situations, government agencies may utilize RPA and AI to monitor large crypto transactions and file reports as needed. RPA may be utilized in a court of law to examine numerous case files and separate the instances that need swift justice. RPA may increase a government's and a court's efficiency, clarity, stability, and predictability.

16.2.5 Inventory Management

The process of ordering, storing, utilizing, and selling a company's inventory is referred to as inventory management. Any kind of company, no matter how large or little, has the ability to face down its rivals and outperform them in an effort to become the market leader. Small firms, in particular, have a very limited workforce and must complete a multitude of activities in order to establish themselves as a successful company. Among the many duties they must carry out, controlling the inventory is crucial for a firm. When controlling the inventory was done manually, it was a laborious, boring, repetitive, and unpleasant task. Employees at a firm struggled to keep up with the volume of data, which made it difficult for them to manage inventories effectively. The whole process has become considerably easier and more productive with the use of RPA in the area of inventory management [45, 46]. As a business leader, it is very beneficial since it prevents the loss of human resources on such a boring and laborious activity, as well as errors that staff members could make when dealing with a lot of data. To be more precise about the dynamics of the whole process, this is how it happens: After a successful RPA implementation, the manager receives a virtual text-to-speech assistant bot. This bot runs through the transaction details, notifies the manager of the class of items available in the inventory, creates a file of items to be re-ordered, and, upon the supplier's approval, sends a mail to the supplier.

16.3 Advantages of RPA

RPA implementation, however, has a number of potential advantages, including the capacity to interact with existing systems, quicker output, increased effectiveness and quality, and cost savings. Additionally, Vishnu et al. [47] provided instances of key RPA advantages, including better labour management, better business planning, compliance and scalability, and increased operational agility. While RPA can provide the same advantages as core platform transformation initiatives, it can do so with less risk, significantly more quickly, and for a lot less money than conventional IT integration projects.

Figure 16.1 shows the advantages of the RPA with respect to business and customer points of view.

Figure 16.2 shows the benefits of RPA in a number of industries, such as retail banking, capital markets, wealth and asset management, telecom, utilities, insurance, retail, health care, and so on.

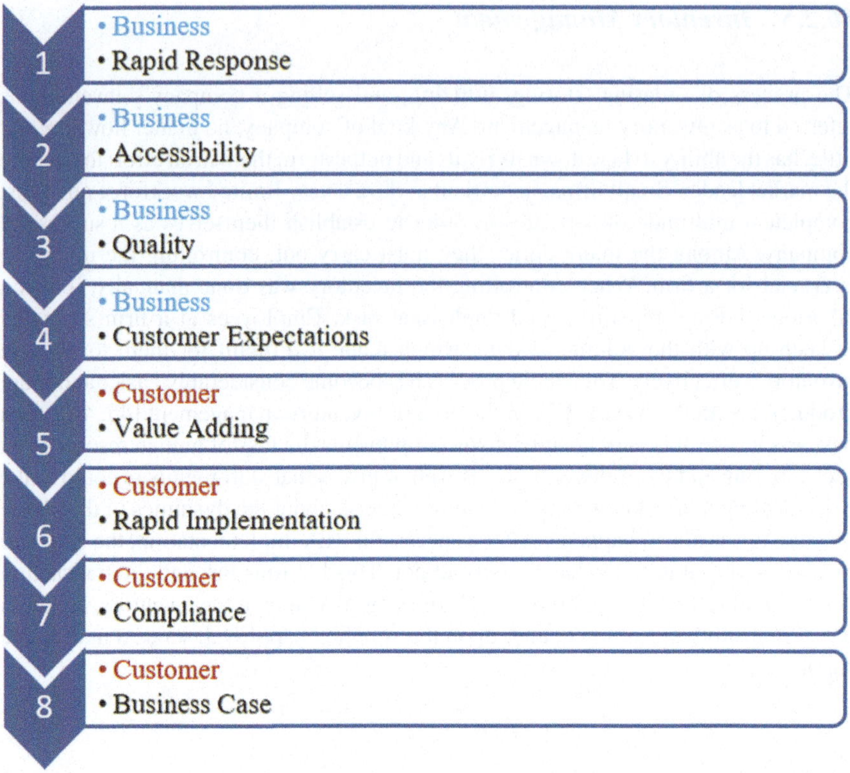

Fig. 16.1 Advantages of the RPA observed by the business and customer

16.4 Why RPA is Falling Short of Expectations

Despite its many advantages, RPA will not eliminate the need for current applications or address all automation issues. However, it is an intriguing choice in a number of contexts that businesses should think about if they want to maintain their relevance and competitiveness and adapt to shifting client expectations [48].

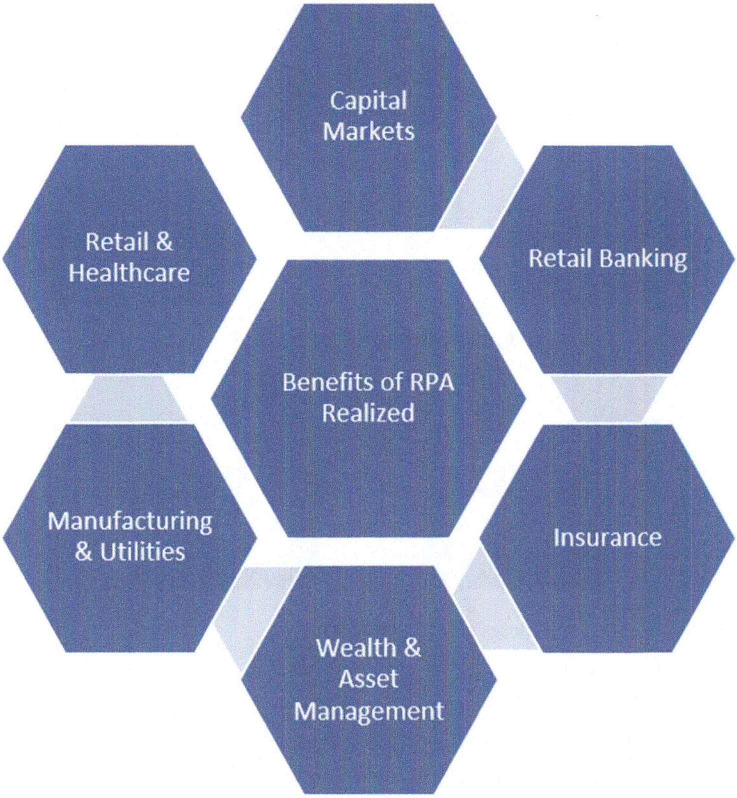

Fig. 16.2 Advantages of RPA across several sectors

Figure 16.3 displays the most frequent causes of RPA project failures.
The following are the top five reasons why RPA projects often fail:

- Improper RPA management at the highest levels
- Choosing Faulty RPA use case selection and unclear metrics
- No Long-Term RPA roadmap or vision
- Delivering RPA benefits to improper recipient
- Deadline for RPA project delivery.

16.5 Conclusion

The need for people to communicate and work collaboratively with robots will increase as organizations keep expanding and start exploring their use of the latest techniques to address the leading issues affecting most businesses, namely customer acquisition and retention, efficiency improvements, and cost reduction. This will

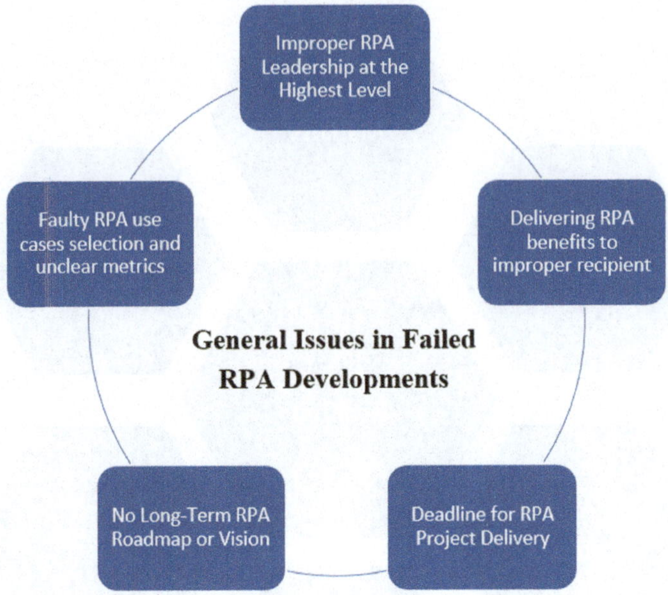

Fig. 16.3 The most frequent causes of RPA project failures

change the necessary skills of the future workforce. It may be difficult to decide which technology to use, where, and how. Organizations need to be aware of how various automation and cognitive tools vary from one another, where these tools have been used, and how they will probably change in the near future. These technologies all include AI or machine learning capabilities. As these technologies are not "plug and play," and as businesses have little understanding of "how" and "where" to utilize them, combining and deploying them into the distinct IT and process environment of a company presents a significant problem. Even though RPA has a great chance of completely replacing the majority of manual and repetitive operations carried out by people, RPA technologies must be compared to other automation and cognitive tools. The best unautomated processes for cognitive technologies or RPA or other automation should be identified and catalogued by organizations using a systematic methodology in order to understand how these tools might help different important business activities.

References

1. Qiu, Y.L., Xiao, G.F.: Research on cost management optimization of financial sharing center based on RPA. Proc. Comp. Sci. **166**, 115–119 (2020)
2. Hildebrandt, T., van Dongen, B.F., Röglinger, M., Mendling, J.: Business process management. Lecture Notes in Computer Science11675 (2019)
3. González Enríquez, J., Jiménez Ramírez, A., Domínguez Mayo, F.J., García García, J.A.:

Robotic process automation: a scientific and industrial systematic mapping study. IEEE Access **8**, 39113–39129 (2020)
4. Ivančić, L., Suša Vugec, D., Bosilj Vukšić, V.: Robotic process automation: systematic literature review. In: International Conference on Business Process Management, pp. 280–295. Springer, Cham (2019)
5. Mendling, J., Decker, G., Hull, R., Reijers, H.A., Weber, I.: How do machine learning, robotic process automation, and blockchains affect the human factor in business process management? Commun. Assoc. Inf. Syst. **43**(1), 19 (2018)
6. Das, D., Pandey, I., Chakraborty, A., Banerjee, J.S.: Analysis of implementation factors of 3D printer: the key enabling technology for making prototypes of the engineering design and manufacturing. Int. J. Comp. Appl. **1**, 8–14 (2017)
7. Das, D., Pandey, I., Banerjee, J.S.: An in-depth study of implementation issues of 3D printer. In: Proceedings MICRO 2016 Conference on Microelectronics, Circuits and Systems (pp. 45–49) (2016)
8. Reddy, K.N., Harichandana, U., Alekhya, T., Rajesh, S.M.: A study of robotic process automation among artificial intelligence. Int. J. Sci. Res. Publ. **9**(2), 392–397 (2019)
9. König, M., Bein, L., Nikaj, A., Weske, M.: Integrating robotic process automation into business process management. In: International Conference on Business Process Management, pp. 132–146. Springer, Cham (2020)
10. Wright, D., Witherick, D., Gordeeva, M.: The robots are ready. Are you? Untapped advantage in your digital workforce. Deloitte 28 (2017)
11. Frank, C.: Introduction to robotic process automation. Proc. Inst. Robot. Process Automat. 35 (2015)
12. Lacity, L., Willcocks, M.: Robotic process automation: the next transformation lever for shared services, School Econ. Political Sci., Outsourcing Unit Working Res. Paper Ser., London, U.K., Tech. Rep. 16/01 (2012)
13. Ansari, W.A., Diya, P., Patil, S., Patil, S.: A review on robotic process automation-the future of business organizations. In: 2nd International Conference on Advances in Science & Technology (ICAST) (2019)
14. Llewellyn Evans, G.: Disruptive technology and the board: the tip of the iceberg. Econ. Bus. Rev. **3**(1) (2017)
15. Gami, M., Jetly, P., Mehta, N., Patil, S.: Robotic process automation–future of business organizations: a review. In: 2nd International Conference on Advances in Science & Technology (ICAST) (2019)
16. Gotthardt, M., Koivulaakso, D., Paksoy, O., Saramo, C., Martikainen, M., Lehner, O.: Current state and challenges in the implementation of smart robotic process automation in accounting and auditing. ACRN J. Finance Risk Perspect. (2020)
17. Aguirre, S., Rodriguez, A.: Automation of a business process using robotic process automation (RPA): a case study. In: Workshop on engineering applications, pp 65–71. Springer, Cham (2017)
18. Banerjee, J., Maiti, S., Chakraborty, S., Dutta, S., Chakraborty, A., Banerjee, J.S.: Impact of machine learning in various network security applications. In: 2019 3rd International Conference on Computing Methodologies and Communication (ICCMC), pp. 276–281. IEEE (2019)
19. Chattopadhyay, J., Kundu, S., Chakraborty, A., Banerjee, J.S.: Facial expression recognition for human computer interaction. In: International Conference on Computational Vision and Bio Inspired Computing, pp. 1181–1192. Springer, Cham (2018)
20. Guhathakurata, S., Kundu, S., Chakraborty, A., Banerjee, J.S.: A novel approach to predict COVID-19 using support vector machine. In: *Data Science for COVID-19* (pp. 351–364). Academic Press, (2021)
21. Guhathakurata, S., Saha, S., Kundu, S., Chakraborty, A., Banerjee, J.S.: A new approach to predict COVID-19 using artificial neural networks. In: *Cyber-Physical Systems* pp. 139–160. Academic Press. (2022)

22. Saha, P., Guhathakurata, S., Saha, S., Chakraborty, A., Banerjee, J.S.: Application of machine learning in app-based cab booking system: a survey on Indian scenario. In Applications of Artificial Intelligence in Engineering (pp. 483–497). Springer, Singapore (2021)
23. Biswas, S., Sharma, L.K., Ranjan, R., Saha, S., Chakraborty, A., Banerjee, J.S.: Smart farming and water saving-based intelligent irrigation system implementation using the internet of things. In: Recent Trends in Computational Intelligence Enabled Research (pp. 339–354). Academic Press. (2021)
24. Mandal, J.K., Misra, S., Banerjee, J.S., Nayak, S. (Eds.).: Applications of Machine intelligence in Engineering. In: Proceedings of 2nd Global Conference on Artificial Intelligence and Applications (GCAIA, 2021), September 8–10, 2021, Jaipur, India. CRC Press. (2022)
25. Guhathakurata, S., Saha, S., Kundu, S., Chakraborty, A., Banerjee, J.S.: South Asian Countries are less fatal concerning COVID-19: a fact-finding procedure integrating machine learning & multiple criteria decision-making (MCDM) technique. J. Instit. Eng. (India) Ser. B **102**(6), 1249–1263 (2021)
26. Guhathakurata, S., Saha, S., Kundu, S., Chakraborty, A., Banerjee, J.S.: South Asian countries are less fatal concerning COVID-19: a hybrid approach using machine learning and M-AHP. In: Computational Intelligence Techniques for combating COVID-19 (pp. 1–26). Springer, Cham. (2021)
27. Chakraborty, A., Singh, B., Sau, A., Sanyal, D., Sarkar, B., Basu, S., Banerjee, J.S.: Intelligent vehicle accident detection and smart rescue system. In: Applications of Machine Intelligence in Engineering (pp. 565–576). CRC Press. (2022)
28. Banerjee, J.S., Mahmud, M. & Brown, D.: Heart Rate Variability-Based Mental Stress Detection: An Explainable Machine Learning Approach. SN COMPUT. SCI **4**, 176 (2023)
29. Chakraborty, A., Banerjee, J.S., Bhadra, R., Dutta, A., Ganguly, S., Das, D., Kundu, S., Mahmud, M., Saha G.: A Framework of Intelligent Mental Health Monitoring in Smart Cities and Societies. IETE J Res (2023)
30. Bhattacharyya, S., Banerjee, J. S., & Köppen, M (eds.).: Human-Centric Smart Computing: Proceedings of ICHCSC 2022, Springer. (2022)
31. Schatsky, D., Muraskin, C., Iyengar, K.: Robotic process automation. A path to the cognitive enterprise. New York: Deloitte Consulting. (2017)
32. Jovanović, S.Z., Đurić, J.S., Šibalija, T.V.: Robotic process automation: overview and opportunities. Int. J. Adv. Qual. **46**(3–4), 34–39 (2018)
33. Met, İ., Kabukçu, D., Uzunoğulları, G., Soyalp, Ü., Dakdevir, T.: Transformation of business model in finance sector with artificial intelligence and robotic process automation. In: Digital Business Strategies In Blockchain Ecosystems (pp. 3–29). Springer, Cham. (2020)
34. Huang, F., Vasarhelyi, M.A.: Applying robotic process automation (RPA) in auditing: a framework. Int. J. Account. Inf. Syst. **35**, 100433 (2019)
35. Eulerich, M., Pawlowski, J., Waddoups, N.J., Wood, D.A.: A framework for using robotic process automation for audit tasks. Contemp. Account. Res. **39**(1), 691–720 (2022)
36. Saha, O., Chakraborty, A., & Banerjee, J. S.: A fuzzy AHP approach to IT-based stream selection for admission in technical institutions in India. In Emerging technologies in data mining and information security (pp. 847–858). Springer, Singapore, (2019)
37. Saha, O., Chakraborty, A., & Banerjee, J. S.: A decision framework of IT-based stream selection using analytical hierarchy process (AHP) for admission in technical institutions. In 2017 4th International Conference on Opto-Electronics and Applied Optics (Optronix) (pp. 1–6). IEEE, (2017)
38. Das, K., Banerjee, J.S.: Green IoT for intelligent cyber-physical systems in industry 4.0: a review of enabling technologies, and solutions. In Applications of Machine Intelligence in Engineering (pp. 463–478). CRC Press. (2022)
39. Turcu, C., Turcu, C.: On robotic process automation and its integration in higher education. In: Conference Proceedings. The Future of Education 2020 (2020)
40. Mosteanu, N.R.: Machine learning and robotic process automation take higher education one step further. Sci. Technol. **25**(1), 92–99 (2022)

41. Banerjee, J. S., Bhattacharyya, S., Obaid, A. J. & Yeh, W. C (eds.).: Intelligent Cyber-Physical Systems Security for Industry 4.0: Applications, Challenges and Management, CRC Press (2022)
42. Adhikari, S.: Integrating deep reinforced learning and robotic process assessment in blockchain digital transformation for autonomous cybersecurity. In: *AIAA Scitech 2021 Forum*, p 0662. (2021)
43. Pagallo, U., Durante, M.: The pros and cons of legal automation and its governance. Eur. J. Risk Regul. **7**(2), 323–334 (2016)
44. Alpers, S., Becker, C., Pieper, M., Wagner, M., Oberweis, A.: Legal challenges of Robotic Process Automation (RPA) in administrative services. In: SEMANTiCS Posters&Demos. (2019)
45. Verma, K., Chahal, A., Kumar, V., Nayak, P., Singh, P.: Automated order management using robotic process automation. In: Proceedings of the International Conference on Innovative Computing & Communications (ICICC). (2020)
46. Sharma, A., Guleria, K.: A framework for hotel inventory control system for online travel agency using robotic process automation. In: 2021 International Conference on Advance Computing and Innovative Technologies in Engineering (ICACITE), pp. 764–768. IEEE (2021)
47. Vishnu, S., Agochiya, V., Palkar, R.: Data-centered dependencies and opportunities for robotics process automation in banking. J. Financ. Transf. **45**(1), 68–76 (2017)
48. Rutaganda, L., Bergstrom, R., Jayashekhar, A., Jayasinghe, D., Ahmed, J.: Avoiding pitfalls and unlocking real business value with RPA. J. Financ. Transf. **46**(11), 104–115 (2017)

Index

S. Bhattacharyya et al. (eds.), *Confluence of Artificial Intelligence and Robotic Process Automation*, Smart Innovation, Systems and Technologies 335,
https://doi.org/10.1007/978-981-19-8296-5

The manufacturer's authorised representative in the EU is Springer Nature Customer Service Centre GmbH, Europaplatz 3, 69115 Heidelberg, Germany. If you have any concerns regarding our products, please contact ProductSafety@springernature.com

Printed and bound by CPI Group (UK) Ltd, Croydon, CR0 4YY
30/09/2026
02225244-0014